Text-Book

on

Spherical Astronomy

Text-Book
on
Spherical Astronomy

By

W. M. SMART, M.A., D.Sc., LL.D.

*Regius Professor of Astronomy in the University
of Glasgow 1937–59, formerly John Couch Adams
Astronomer in the University
of Cambridge*

CAMBRIDGE
AT THE UNIVERSITY PRESS
1971

Published by the Syndics of the Cambridge University Press
Bentley House, 200 Euston Road, London NW1 2DB
American Branch: 32 East 57th Street, New York, N.Y.10022

ISBNs:
0 521 06491 0   cloth edition
0 521 09159 4   paperback

First edition 1931
Second edition 1936
Third edition 1940
Fourth edition 1944
Reprinted 1947 1949 1956 1960
Fifth edition 1962
Reprinted 1965 1971

Printed in Great Britain
at the University Printing House, Cambridge
(Brooke Crutchley, University Printer)

PREFACE

THIS BOOK is based on lectures given annually in the University of Cambridge and on a parallel course of instruction in Practical Astronomy at the Observatory. The recent changes in the almanacs have, in many respects, affected the position of the older text-books as channels of information on current practice, and the present work is intended to fill the gap caused by modern developments. In addition to the time-honoured problems of Spherical Astronomy, the book contains the essential discussion of such important subjects as heliographic co-ordinates, proper motions, determination of position at sea, the use of photography in precise astronomical measurements and the orbits of binary stars, all or most of which have received little attention in works of this kind. In order to make certain subjects as complete as possible, I have not hesitated to cross the traditional frontiers of Spherical Astronomy. This is specially the case as regards the spectroscopic determination of radial velocity which is considered, the physical principles being assumed, in relation to such problems as solar parallax, the solar motion and the orbits of spectroscopic binary stars.

Throughout, only the simplest mathematical tools have been used and considerable attention has been paid to the diagrams illustrating the text. I have devoted the first chapter to the proofs and numerical applications of the formulae of spherical trigonometry which form the mathematical foundation of the subsequent chapters. Although other formulae have been given for reference, I have limited myself to the use of the basic formulae only.

A writer of a text-book on Spherical Astronomy cannot avoid a certain measure of detailed reference to the principal astronomical instruments and, accordingly, general descriptions of instruments have been given in the appropriate places, usually with a simple discussion of the chief errors which must be taken into account in actual observational work.

In numerical applications, the almanac for 1931 has been used.

As regards notation, I have usually followed the recommendations of the International Astronomical Union, but I have made several modifications when the avoidance of confusion or of misconception seemed to me of greater importance than the rigid adherence to the definite system proposed; for example, the angle of stellar parallax has been denoted by Π instead of the usual symbol π which the student is apt to associate on most mathematical occasions with the properties of the circle.

I have also followed the advice of the International Astronomical Union [*Transactions*, vol. III, p. 300 (1928)] as regards the nomenclature of standard mean time, and throughout the book G.C.T. (Greenwich Civil Time) denotes the mean time, for the meridian of Greenwich, reckoned from midnight. (See also footnote to p. 44 *infra*.) This is contrary to the present usage of the British *Nautical Almanac* which, since 1925, has employed the symbol G.M.T. in the sense in which G.C.T. has just been described, notwithstanding the fact that previous to 1925 G.M.T. universally signified the mean time reckoned from mean noon at Greenwich. The latter I have denoted (also in accordance with the recommendations of the I.A.U.) by the letters G.M.A.T. (Greenwich Mean Astronomical Time).

At the end of each chapter there is a collection of exercises many of which have been taken, by permission, from the papers set in the Mathematical Tripos at Cambridge and in the examinations of London University and Cambridge Colleges; several have also been taken from Ball's *Spherical Astronomy*.

I have the pleasant duty of recording my indebtedness to Professor W. E. Anderson, of Miami University, Oxford, Ohio, U.S.A., and to Mr M. J. Dean, B.A., Scholar of Trinity College, Cambridge, who have read the whole of the manuscript with the greatest care and removed many blemishes which might have escaped less vigilant eyes. I am also grateful to Dr F. S. Hogg, formerly of Amherst Observatory, Mass., U.S.A., who has read critically about half of the book, and to Dr L. J. Comrie, Superintendent of the *Nautical Almanac* Office, for helpful suggestions in the chapter on Occultations and Eclipses.

It is also a pleasure to express my thanks to the officials and staff of the University Press for their courtesy and care during the printing of the book.

W. M. S.

OBSERVATORY, CAMBRIDGE
1931 *January* 1

PREFACE TO THE SECOND EDITION

IN this edition several misprints and errors have been corrected and I am very grateful to those readers who were kind enough to notify such errata.

The principal change in the new edition is the addition of three appendices relating to "The Method of Dependences", "Stellar Magnitudes" and "The Coelostat". I hope that these additions will be incorporated in the main text of a subsequent edition.

In the first edition, I used the symbols G.C.T. to signify the mean time, for the meridian of Greenwich, reckoned from midnight; this was in accordance with the recommendations of the *International Astronomical Union* in 1928. In July 1935 the I.A.U., on reconsideration of the subject, have advised the use of the nomenclature "Universal Time" or U.T. It is not yet certain if, or how soon, this advice will be followed in the principal *Nautical Almanacs*; accordingly, I have decided to make no change as regards nomenclature in the present edition.

1936 *March* 25

PREFACE TO THE FIFTH EDITION

(1965 Reprint)

AS regards time-nomenclature, I have now discarded G.C.T. in favour of U.T. for explanatory purposes, postponing the relevance of "Ephemeris Time" (E.T.)—recently adopted in the almanacs—to Appendix F (p. 424). In a text-book dealing with astronomical principles and practices the subtleties existing between E.T. and U.T. lead to unnecessary confusion, and so on this account the use of U.T. may be justified. However, in any astronomical problem, the relevant quantities are to be extracted in terms of E.T. from the appropriate almanac, now known as an *Astronomical Ephemeris*. I have also taken the opportunity to amend the text in a few places.

I am greatly indebted to Mr D. H. Sadler, Superintendent of H.M. Nautical Almanac Office, for expert advice on many points.

W. M. S.

1965 *February* 15

CONTENTS

CHAPTER I

SPHERICAL TRIGONOMETRY

1. *Introduction.*

When we look at the stars on a clear night we have the familiar impression that they are all sparkling points of light, apparently situated on the surface of a vast sphere of which the individual observer is the centre. The eye, of course, fails to give any indication of the distances of the stars from us; however, it allows us to make some estimate of the angles subtended at the observer by any pairs of stars and, with suitable instruments, these angles can be measured with great precision. Spherical Astronomy is concerned essentially with the *directions* in which the stars are viewed, and it is convenient to define these directions in terms of the positions on the surface of a sphere—the *celestial sphere*—in which the straight lines, joining the observer to the stars, intersect this surface. It is in this sense that the usual expression "the position of a star on the celestial sphere" is to be interpreted. The radius of the sphere is entirely arbitrary. The foundation of Spherical Astronomy is the geometry of the sphere.

2. *The spherical triangle.*

Any plane passing through the centre of a sphere cuts the surface in a circle which is called a *great circle*. Any other plane intersecting the sphere but not passing through the centre will also cut the surface in a circle which, in this case, is called a *small circle*. In Fig. 1, EAB is a great circle, for its plane passes through O, the centre of the sphere. Let QOP be the diameter of the sphere perpendicular to the plane of the great circle EAB. Let R be any point in OP and suppose a plane drawn through R parallel to the plane of EAB; the surface of the sphere is then intersected in the small circle FCD. It follows from the construction that OP is also perpendicular to the plane of FCD. The extremities P and Q of the common perpendicular diameter QOP are called the *poles* of the great circle and of the parallel small circle. Now let $PCAQ$ be any great circle passing through the poles P and Q and intersecting the small circle FCD and the great circle EAB in C and A respectively. Similarly, PDB is part of another great circle passing through P and Q. We shall find it convenient to refer to a particular great circle by specifying simply any portion of its circumference. When two great circles intersect at a point they are said to include a *spherical angle* which is defined as follows. Consider the two great circles PA and PB intersecting at P. Draw PS and PT, the tangents to the

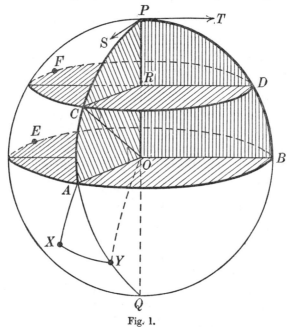

Fig. 1.

circumferences of PA and PB respectively. PT is, by construction, perpendicular to the radius OP of the great circle PB and, being in the plane PBO, is therefore parallel to the radius OB. Similarly PS is parallel to the radius OA. The angle SPT defines the spherical angle at P between the two great circles PA and PB, and it is equal to the angle AOB, AB being the arc intercepted on the great circle, of which P is the pole, between the two great circles PA and PB. It is to be emphasised that a spherical angle is defined only with reference to two intersecting *great circles*.

If we are given any three points on the surface of a sphere, then the sphere can be bisected so that all three points lie in the same hemisphere. If the points are joined by great circle arcs all lying on this hemisphere the figure obtained is called a *spherical triangle*. Thus, in Fig. 1, the three points A, X and Y on the spherical surface are joined by great circle arcs to form the spherical triangle AXY. AX, AY and XY are the *sides* and the spherical angles at A, X and Y are the angles of the spherical triangle. Actually, if R is the radius of the sphere, the length of the spherical arc AY is given by

$$AY = R \times \text{angle } AOY,$$

the angle AOY being expressed in circular measure, i.e. in radians. Now for all great circle arcs on the sphere the radius R is constant and it is convenient to consider its length as unity. The arc AY is then simply the angle which it subtends at the centre of the sphere. If AY is, let us say, one-eighth of the circumference of the complete great circle through A and Y, the side AY is then $\frac{\pi}{4}$ in circular measure and there is no ambiguity if it is expressed as 45°; similarly, for the remaining sides of the triangle. It follows from the definition of a spherical triangle that no side can be equal to or greater than 180°. As another example, PAB is a spherical triangle two of whose sides PA and PB each subtend $\frac{\pi}{2}$ radians or 90° at O; in this instance we say that PA and PB are each equal to $\frac{\pi}{2}$ radians or 90°. But PCD is *not* a spherical triangle, for the arc CD is not a part of a great circle. Accordingly, the formulae which will be derived for spherical triangles will not be applicable to such a figure as PCD.

3. *Length of a small circle arc.*

Consider, in Fig. 1, the small circle arc CD. Its length is given by
$$CD = RC \times \text{angle } CRD.$$
Also, the length of the spherical arc AB is given by
$$AB = OA \times \text{angle } AOB.$$
But since the plane of FCD is parallel to the plane of EAB, then $C\hat{R}D = A\hat{O}B$, for RC, RD are respectively parallel to OA, OB.

Therefore $$CD = \frac{RC}{OA}.AB.$$

But, since $OA = OC$ (radii of the sphere), we have

$$CD = \frac{RC}{OC}.AB.$$

Now RC is perpendicular to OR; $\therefore$ $RC = OC \cos R\hat{C}O$. From the parallelism of RC and OA, $R\hat{C}O = A\hat{O}C$. Hence

$$CD = AB \cos A\hat{O}C.$$

Now AOC is the angle subtended at the centre of the sphere by the great circle arc AC. The formula can then be written as

$$CD = AB \cos AC,$$

or, since $PA = 90°$, $$CD = AB \sin PC \qquad\qquad ......(1).$$

4. *Terrestrial latitude and longitude.*

The concepts introduced so far will now be illustrated with reference to the earth. For many practical problems, the earth can be regarded as a spherical body spinning about a diameter PQ (Fig. 2). P is the *north pole* and Q is the *south pole*. The great circle whose plane is perpendicular to PQ is called the *equator*. Any semi-great circle terminated by P and Q is a

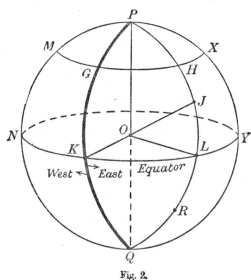

Fig. 2.

meridian. In particular, the meridian which passes through the fundamental instrument (the transit circle) of Greenwich Observatory is, by universal agreement, regarded as the principal or standard meridian; let it be *PGKQ* in Fig. 2, intersecting the equator in *K*. Let *PHLQ* be any other meridian cutting the equator in *L*. The angle *KOL* is defined to be the *longitude* of the meridian *PHQ* and it can be described equally well as the equatorial arc *KL* or the spherical angle *KPL*. Longitudes are measured from 0° to 180° *east* of the Greenwich meridian and from 0° to 180° *west*, following the directions of the arrows near *K* in Fig. 2. Thus, from the figure, the longitude of the meridian *PHQ* is about 100° east (E) and that of the meridian *PMQ* is about 60° west (W). All places on the same meridian have the same longitude and the meridian on which a particular place is situated is specified with reference to the principal meridian *PGQ*. To specify completely the position of a place on the surface of the earth, we require to describe its position on its meridian of longitude. This is done with reference to the equator. Consider a place *J* on the meridian *PHQ*. The meridian through *J* cuts the equator in *L* and the angle *LOJ*, or the great circle arc *LJ*, is called the *latitude* of *J*. If *J* is between the equator and the north pole *P*, as in Fig. 2, the latitude is said to be north (N); a place such as *R*, between the equator and the south pole *Q*, is said to be in south latitude (S). In this way the position of any point on the surface of the earth is referred to the two fundamental great circles, the equator and the meridian of Greenwich.

Let ϕ denote the latitude of *J*; then $L\hat{O}J$ or $LJ = \phi$. Since *OP* is perpendicular to the plane of the equator, $P\hat{O}L = 90°$ and therefore $POJ = 90° - \phi$. The angle *POJ* or the spherical arc *PJ* is the *colatitude* of *J*. We have thus

$$\text{Colat.} = 90° - \text{Lat.}$$

All places which have the same latitude lie on a small circle parallel to the equator, called a *parallel of latitude*. Thus all places with the same latitude as Greenwich lie on the small circle *MGHX*. If θ denotes the latitude of Greenwich, then by formula (1) the length of the small circle arc *HX*, for example, is given in terms of the length of the corresponding equatorial arc *LY* by

$$HX = LY \cos \theta \qquad \ldots\ldots(2).$$

To give greater precision to the meaning of this formula, we consider the units in which distances on the surface of the earth are expressed. The simplest is that defined as the great circle distance between two points subtending an angle of one minute of arc at the centre of the earth—this unit is known as the *nautical mile* and is equivalent to 6080 feet (we neglect the small variations in this value due to the fact that the earth is not quite a sphere). If the difference in longitude between any two places on the same parallel of latitude is known, e.g. LY, then LY can be expressed as so many minutes of arc and this number is the number of nautical miles between the two points L and Y on the equator. The formula (2) then provides the means of calculating the distance between H and X expressed in nautical miles (or minutes of arc) and *measured along the parallel of latitude*.

5. *The fundamental formula of spherical trigonometry.*

Let ABC be a spherical triangle (Fig. 3). Denote the sides BC, CA, AB by a, b and c respectively. Then, by our definition, the

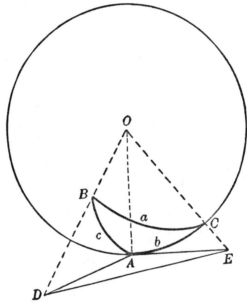

Fig. 3.

side a is measured by the angle BOC subtended at the centre O of the sphere by the great circle arc BC. Similarly, b and c are measured respectively by the angles AOC and AOB. Let AD be the tangent at A to the great circle AB and AE the tangent at A to the great circle AC. Then the radius OA is perpendicular to AD and AE. By construction, AD lies in the plane of the great circle AB; hence, if the radius OB is produced, it will intersect the tangent AD at a point D. Similarly, the radius OC when produced will meet the tangent AE in E. Now the spherical angle BAC is defined to be the angle between the tangents at A to the great circles AB and AC, so that the spherical angle $BAC = D\hat{A}E$. The spherical angle BAC will be denoted simply by A, so that $D\hat{A}E = A$.

Now, in the plane triangle OAD, $O\hat{A}D$ is $90°$ and $A\hat{O}D$, identical with $A\hat{O}B$, is c. We have then

$$AD = OA \tan c; \quad OD = OA \sec c \qquad ......(3),$$

From the plane triangle OAE we have, similarly,

$$AE = OA \tan b; \quad OE = OA \sec b \qquad ......(4).$$

From the plane triangle DAE we have

$$DE^2 = AD^2 + AE^2 - 2AD.AE \cos D\hat{A}E,$$

or $\quad DE^2 = OA^2 [\tan^2 c + \tan^2 b - 2 \tan b \tan c \cos A]$
$$......(5).$$

From the plane triangle DOE,

$$DE^2 = OD^2 + OE^2 - 2OD.OE \cos D\hat{O}E.$$

But $D\hat{O}E = B\hat{O}C = a$;

$$\therefore \quad DE^2 = OA^2 [\sec^2 c + \sec^2 b - 2 \sec b \sec c \cos a]$$
$$......(6).$$

Hence, from (5) and (6),

$\sec^2 c + \sec^2 b - 2 \sec b \sec c \cos a$
$$= \tan^2 c + \tan^2 b - 2 \tan b \tan c \cos A.$$

Now $\quad \sec^2 c = 1 + \tan^2 c; \quad \sec^2 b = 1 + \tan^2 b,$

and after some simplification we obtain

$$\cos a = \cos b \cos c + \sin b \sin c \cos A \qquad ......(A).$$

This is the fundamental formula of spherical trigonometry and it will be referred to in the following pages as the *cosine-formula*

or formula **A**. There are clearly two companion formulae; they
are

$$\cos b = \cos c \cos a + \sin c \sin a \cos B \quad \ldots\ldots(7),$$

$$\cos c = \cos a \cos b + \sin a \sin b \cos C \quad \ldots\ldots(8).$$

From the three formulae—**A**, (7) and (8)—all the other formulae
of spherical trigonometry in use can be derived. The fundamental formula has two direct practical applications:

(1) *If two sides, e.g.* b *and* c, *and the included angle* A *of a
spherical triangle* ABC *are known*, formula A enables the calculation of the third side a to be made.

(2) *If all three sides are known*, the angles of the triangle can
be found successively by means of **A**, (7) and (8).

For suppose the value of A is required; then by **A**

$$\cos A = \operatorname{cosec} b \operatorname{cosec} c \left[\cos a - \cos b \cos c\right] \ldots\ldots(9).$$

Formula (9) can be replaced by one more suitable for logarithmic

calculations as follows. Since $\cos A = 1 - 2\sin^2\dfrac{A}{2}$, we have,

from **A**,

$$\cos a = \cos b \cos c + \sin b \sin c \left(1 - 2\sin^2\frac{A}{2}\right)$$

$$= \cos(b - c) - 2\sin b \sin c \sin^2\frac{A}{2},$$

or $\qquad \cos(b - c) - \cos a = 2\sin b \sin c \sin^2\dfrac{A}{2};$

$$\therefore \ 2\sin\frac{a + (b - c)}{2}\sin\frac{a - (b - c)}{2} = 2\sin b \sin c \sin^2\frac{A}{2}.$$

Let s be defined by $\qquad 2s = a + b + c \qquad \ldots\ldots(10).$

Then $\quad a + b - c = 2(s - c)$ and $a - b + c = 2(s - b).$

Hence $\qquad \sin(s - b)\sin(s - c) = \sin b \sin c \sin^2\dfrac{A}{2};$

$$\therefore \ \sin\frac{A}{2} = \sqrt{\frac{\sin(s - b)\sin(s - c)}{\sin b \sin c}} \quad \ldots\ldots(11).$$

This form is useful in numerical work. There are two similar

equations giving $\sin\dfrac{B}{2}$ and $\sin\dfrac{C}{2}$.

If we write $\cos A = 2\cos^2\dfrac{A}{2} - 1$ in the formula **A** and proceed as before, we shall obtain

$$\cos \frac{A}{2} = \sqrt{\frac{\sin s \sin (s - a)}{\sin b \sin c}} \qquad \ldots\ldots(12)$$

with two similar equations giving $\cos \dfrac{B}{2}$ and $\cos \dfrac{C}{2}$.

From (11) and (12) by division we have

$$\tan \frac{A}{2} = \sqrt{\frac{\sin (s - b) \sin (s - c)}{\sin s \sin (s - a)}} \qquad \ldots\ldots(13).$$

There are two similar equations, giving $\tan \dfrac{B}{2}$ and $\tan \dfrac{C}{2}$.

Any one of (11), (12) and (13) can be used to calculate A, the three sides being known.

6. *The sine-formula.*

We shall now derive what is known as the sine-formula. From the cosine-formula **A**, we have

$$\sin b \sin c \cos A = \cos a - \cos b \cos c.$$

By squaring, we obtain

$$\sin^2 b \sin^2 c \cos^2 A = \cos^2 a - 2 \cos a \cos b \cos c + \cos^2 b \cos^2 c.$$

The left-hand side can be written

$$\sin^2 b \sin^2 c - \sin^2 b \sin^2 c \sin^2 A,$$

or $\qquad 1 - \cos^2 b - \cos^2 c + \cos^2 b \cos^2 c - \sin^2 b \sin^2 c \sin^2 A.$

Hence

$$\sin^2 b \sin^2 c \sin^2 A$$
$$= 1 - \cos^2 a - \cos^2 b - \cos^2 c + 2 \cos a \cos b \cos c.$$

Let a positive quantity X be defined by

$$X^2 \sin^2 a \sin^2 b \sin^2 c$$
$$= 1 - \cos^2 a - \cos^2 b - \cos^2 c + 2 \cos a \cos b \cos c.$$

Then, from the previous equation,

$$\frac{\sin^2 A}{\sin^2 a} = X^2,$$

so that $\qquad\qquad X = \pm \dfrac{\sin A}{\sin a}.$

But in a spherical triangle the sides are each less than 180°, and this applies also to the angles. As $\sin \theta$ is positive for all values of θ between $0°$ and $180°$, the minus sign in the above equation is inadmissible, and we have

$$X = \frac{\sin A}{\sin a}.$$

By treating (7) and (8) in a similar way, we shall obtain

$$X = \frac{\sin B}{\sin b} = \frac{\sin C}{\sin c}.$$

Hence $\qquad$ $\dfrac{\sin A}{\sin a} = \dfrac{\sin B}{\sin b} = \dfrac{\sin C}{\sin c}$ $\qquad$ ......(B).

This result we shall refer to as the *sine-formula* or formula **B**.

Formula **B** gives a relation between any two sides of a triangle and the two angles *opposite* these sides. It has to be used, however, with circumspection in numerical calculations; for, suppose that the two sides a and b and the angle B are given, then by **B**

$$\sin A = \frac{\sin a \sin B}{\sin b},$$

from which the value of $\sin A$ can be calculated. But $\sin (180° - A) = \sin A$, and without further information it is not possible to decide which of the two angles A or $180° - A$ represents the correct solution. The analogous ambiguity in plane trigonometry may be recalled to the reader's attention.

7. *Formula* **C.**

Write equation (7) in the form

$\sin c \sin a \cos B = \cos b - \cos c \cos a$
$\qquad\qquad = \cos b - \cos c (\cos b \cos c + \sin b \sin c \cos A)$
$\qquad\qquad = \sin^2 c \cos b - \sin b \sin c \cos c \cos A.$

Hence, dividing by $\sin c$, we have

$\sin a \cos \underline{B} = \cos \underline{b} \sin c - \sin \underline{b} \cos c \cos A$ ...(C),

a relation involving all three sides and two angles.

We can easily prove in a similar manner, beginning with equation (8), that

$$\sin a \cos \underline{C} = \cos \underline{c} \sin b - \sin \underline{c} \cos b \cos A \quad ...(14).$$

If we regard b and c as the two principal sides then A is the contained angle. As we have seen, the cosine-formula **A** gives $\cos a$ in terms of b, c and the included angle A. Formulae **C** and

(14) are, in some ways, analogous to **A** as they give $\sin a \times$ cosine of one of the two angles B and C, adjacent to the side a, in terms of b, c and A.

The formula **C** can also be proved as follows. Suppose the side c of the triangle ABC to be less than $90°$ (the case when c is between $90°$ and $180°$ is left as an exercise to the student). Produce the great circle arc BA to D so that BD is $90°$ (Fig. 4).

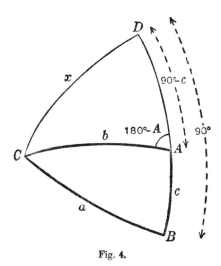

Fig. 4.

Then $AD = 90° - c$ and $C\hat{A}D = 180° - A$. Join C and D by a great circle arc and denote it by x. From the triangle DAC, by **A**,

$$\cos x = \cos(90° - c)\cos b + \sin(90° - c)\sin b\cos(180° - A),$$

or

$$\cos x = \sin c\cos b - \cos c\sin b\cos A \qquad ......(15).$$

From the triangle DBC, by **A**,

$$\cos x = \cos 90°\cos a + \sin 90°\sin a\cos B,$$

or $\qquad \cos x = \sin a\cos B \qquad ......(16),$

and therefore from (15) and (16)

$$\sin a\cos B = \cos b\sin c - \sin b\cos c\cos A,$$

which is formula **C**.

8. *The four-parts formula.*

Another useful formula, known as the four-parts formula, will now be derived. In the spherical tri-
angle ABC (Fig. 5) consider the four consecutive parts B, a, C, b. The angle C is contained by the two sides a and b and is called the "inner angle". The side a is flanked by the two angles B and C and is called the "inner side". Introduce B and C by means of the cosine-formula; then we have

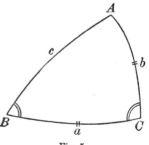

Fig. 5.

$$\cos b = \cos a \cos c + \sin a \sin c \cos B \quad \ldots\ldots (17),$$
$$\cos c = \cos b \cos a + \sin b \sin a \cos C \quad \ldots\ldots (18).$$

Substitute the value of $\cos c$ given by (18) on the right-hand side of (17); then

$$\cos b = \cos a \, (\cos b \cos a + \sin b \sin a \cos C) + \sin a \sin c \cos B;$$
$$\therefore \quad \cos b \sin^2 a = \cos a \sin b \sin a \cos C + \sin a \sin c \cos B.$$

Divide throughout by $\sin a \sin b$; then

$$\cot b \sin a = \cos a \cos C + \frac{\sin c}{\sin b} \cos B.$$

But by the sine-formula **B**,

$$\frac{\sin c}{\sin b} = \frac{\sin C}{\sin B}.$$

Hence $\quad \cos a \cos C = \sin a \cot b - \sin C \cot B \quad \ldots\ldots (D),$

which may be put into words, as an aid to the memory, as follows:

cos (*inner side*).cos (*inner angle*)
    = sin (*inner side*).cot (*other side*)
                − sin (*inner angle*).cot (*other angle*).

9. *Alternative proofs of the formulae* **A,** **B** *and* **C.**

The formulae **B,** **C** and **D** have been derived by algebraic transformations of the fundamental formula. Another proof of each of **A,** **B** and **C** will now be briefly obtained from a simple and instructive geometrical construction. Let ABC (Fig. 6) be a spherical triangle and O the centre of the sphere. Join O to the vertices and take any point P in OC. From P draw PQ perpendicular to OA and PR perpendicular to OB. In the plane OAB, draw QS perpendicular to OA and RS perpendicular to OB. These perpendiculars meet in S. Join PS and OS. If we draw tangents at A to the great circle arcs AB and AC, these tangents, by definition, include the spherical angle A. But QS and QP are by construction parallel to these tangents. Hence $P\hat{Q}S = A$. Similarly $P\hat{R}S = B$. Also $C\hat{O}B = a$, $C\hat{O}A = b$ and $A\hat{O}B = c$.

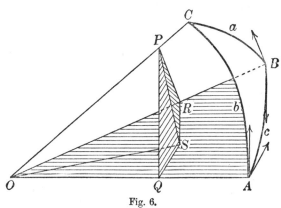

Fig. 6.

The first step is to prove that PS is perpendicular to the plane AOB. By the construction, OQ is perpendicular to both PQ and QS; hence OQ is perpendicular to the plane PQS; therefore OQ is perpendicular to PS which is a line lying in the plane PQS. Similarly, OR is perpendicular to PS. Thus PS is perpendicular to both OQ and OR and is therefore perpendicular to every line in the plane of OQ and OR, that is, PS is perpendicular to the plane OAB and, in particular, to OS, SQ and SR. Thus PQS and PRS are right-angled triangles.

(1) We have, from the right-angled triangles OQP and ORP,
$$PQ = OP \sin b; \quad PR = OP \sin a \qquad \ldots\ldots(19).$$
$$OQ = OP \cos b; \quad OR = OP \cos a \qquad \ldots\ldots(20).$$
Let x denote the angle SOQ; then $R\hat{O}S = c - x$.

Now $\qquad OS = OQ \sec x$ and $OS = OR \sec (c - x)$.

Hence $\qquad\qquad OR \cos x = OQ \cos (c - x)$;

$\therefore$ by (20), $\quad OP \cos a \cos x = OP \cos b \cos (c - x)$;

$\qquad\therefore\ \cos a = \cos b \cos c + \cos b \sin c \tan x.$

But $\qquad \tan x = \dfrac{QS}{OQ} = \dfrac{PQ \cos A}{OQ} = \tan b \cos A,$

and hence $\quad \cos a = \cos b \cos c + \sin b \sin c \cos A,$

which is formula **A**.

(2) Again, from the right-angled triangles PQS and $PRS,$
$$PS = PQ \sin PQS = PQ \sin A,$$
and $\qquad\qquad PS = PR \sin PRS = PR \sin B.$

Hence $\qquad\qquad PQ \sin A = PR \sin B,$

and $\therefore$ by (19),
$$OP \sin b \sin A = OP \sin a \sin B,$$
from which formula **B** follows.

(3) We have, from the right-angled triangles OSQ and $OSR,$
$$QS = OS \sin x \quad \text{and} \quad RS = OS \sin (c - x);$$
$$\therefore \quad RS \sin x = QS (\sin c \cos x - \cos c \sin x),$$
or $\qquad\qquad RS = QS (\sin c \cot x - \cos c).$

Now $\qquad\qquad RS = PR \cos B = OP \sin a \cos B,$

and $\qquad\qquad QS = PQ \cos A = OP \sin b \cos A,$

and $\qquad\qquad QS \cot x = OQ = OP \cos b.$

Hence $\quad \sin a \cos B = \cos b \sin c - \sin b \cos c \cos A,$

which is formula **C**.

10. *Right-angled and quadrantal triangles.*

When one of the spherical angles is 90°, the formulae **A**, **B**, **C** and **D** assume simple forms. This is also the case when one side of a spherical triangle is 90°—the triangle is then said to be *quadrantal*. Rules have been given by Napier according to which the various simple formulae can be written down. The rules, however, impose an additional charge on the memory and it is much simpler to apply one of the main formulae **A** to **D** to the particular right-

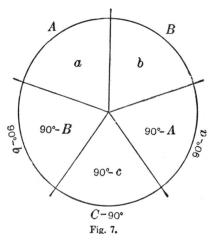

Fig. 7.

angled or quadrantal triangle concerned. The rules are as follows:

(1) Right-angled triangle in which $C = 90°$. Arrange *inside* a circle the five "circular parts" a, b, $90° - A$, $90° - c$, $90° - B$, as in Fig. 7. If any one circular part is chosen as a "middle", the two flanking parts are called "adjacents" and the two others the "opposites". The rules then are:

$$\sin (middle) = \text{product of tangents of } adjacents;$$

$$\sin (middle) = \text{product of cosines of } opposites.$$

(2) Quadrantal triangle in which $c = 90°$. Arrange *outside* the circle (Fig. 7) the five "circular parts" A, B, $90° - a$, $C - 90°$, $90° - b$. The two rules are then the same as for right-angled triangles.

11. *Polar formulae.*

Certain useful formulae can be obtained by means of the polar triangle which is constructed as follows (Fig. 8). Let ABC be a spherical triangle. The great circle of which BC is an arc has two poles, one in each of the hemispheres into which the sphere is divided by the great circle. Let A' be the pole in the hemisphere in which A lies. Similarly B' and C' are the appropriate poles of CA and AB. Produce BC both ways to meet $A'B'$ and $A'C'$ in L and M respectively. Then, since A' is the pole of the great circle $LBCM$, the spherical angle $B'A'C'$ (or simply A') is equal to the arc LM. Again, B' is the pole of AC, that is, the angular distance of B' from any point on AC is $90°$; similarly the angular

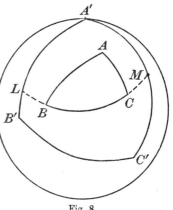

Fig. 8.

distance of A' from any point on BC is $90°$. Hence the angular distance of C from B' and from A' is in each instance $90°$; in other words, C is the pole of $A'B'$. Hence $CL = 90°$, and similarly $BM - 90°$. Now $LM = LB + BM = LB + 90°$. Also $BC = a$; $\therefore LB = 90° - a$. Hence $A' = 180° - a$. Similarly $B' = 180° - b$ and $C' = 180° - c$. We obtain in a similar manner

$$a' = 180° - A; \quad b' = 180° - B; \quad c' = 180° - C.$$

Now apply formula **A** to the triangle $A'B'C'$ and we have, for example,

$$\cos a' = \cos b' \cos c' + \sin b' \sin c' \cos A'.$$

Using the relations just found, we obtain from this equation

$$-\cos A = \cos B \cos C - \sin B \sin C \cos a,$$

which is a formula for the triangle ABC, giving the angle A in terms of the two remaining angles and the included side. The procedure in this instance can be extended to any of the principal formulae which we have already derived, by writing $180° - a$ for A, $180° - b$ for B, etc., in the formulae **A** to **D**.

12. *Numerical example.*

To illustrate the numerical solution of a spherical triangle, we shall consider the following problem. In Fig. 9 let A and B represent two places, in north latitude, on the surface of the earth; their latitudes are respectively 24° 18′ N and 36° 47′ N, and their longitudes 133° 39′ E and 125° 24′ W respectively; it is required to find (i) the length of the great circle arc AB, (ii) the angle PAB, P being the north pole, and (iii) the most northerly point on the great circle AB.

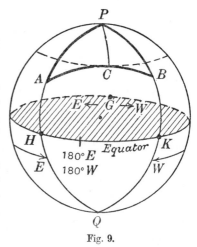

$PAHQ$ is the meridian through A cutting the equator in H. HA measures the latitude of A, i.e. $HA = 24° 18′$. PA is the colatitude of A;
$\therefore PA = 90° - 24° 18′ = 65° 42′.$

Fig. 9.

Similarly $PB = 53° 13′$. Let the Greenwich meridian intersect the equator in G. Then, following the arrows,

$$GH = \text{long. (E) of } A = 133° 39′,$$

and $$GK = \text{long. (W) of } B = 125° 24′.$$

Hence the arc $$HGK = 259° 3′,$$

and therefore HK (the shorter of the two great circle arcs joining H and K) is 100° 57′; that is $A\hat{P}B = 100° 57′$. In the triangle APB we now are given the two sides PA and PB and the contained angle APB.

(i) *Calculation of AB.* By formula **A**, we have

$$\cos AB = \cos PA \cos PB + \sin PA \sin PB \cos APB,$$

which becomes, on inserting the data,

$$\cos AB = \cos 65° 42' \cos 53° 13' - \sin 65° 42' \sin 53° 13' \cos 79° 3'$$
$$\equiv \qquad M \qquad\qquad - N.$$

We shall use five-figure logarithms.

log cos 65° 42'·0	9·61 438	log sin 65° 42'·0	9·95 971
log cos 53° 13'·0	9·77 728	log sin 53° 13'·0	9·90 358
∴ log M =	9·39 166	log cos 79° 3'·0	9·27 864
		∴ log N =	9·14 193
∴ M = 0·24 641;		∴ N = 0·13 865.	

Hence $\qquad\qquad \cos AB \equiv M - N = 0·10\ 776;$

$$\therefore \quad AB = 83° 48'·8 \equiv 5028'·8.$$

Thus the great circle distance between A and B is 83° 48'·8 or 5028·8 nautical miles. To the nearest minute of arc, $AB = 83° 49'$.

(ii) *Calculation of PAB.* By formula **A**,

$$\cos PB = \cos AB \cos PA + \sin AB \sin PA \cos PAB.$$

In this equation, all three sides PB, AB, PA are now known and hence we can derive $P\hat{A}B$. In this instance simple geometrical considerations show that $P\hat{A}B$ is less than 90° and consequently the sine-formula **B** can be used without ambiguity; the appropriate equation is

$$\sin PAB = \frac{\sin APB . \sin PB}{\sin AB},$$

all the quantities on the right-hand side being now known. However, for purposes of illustration, we shall calculate $P\hat{A}B$ by means of formula (11). Denote AB by p, PB by a and PA by b; then

$$2s = p + a + b = 83° 49' + 53° 13' + 65° 42' = 202° 44'.$$

Hence $\quad s = 101° 22'; \quad s - p = 17° 33'; \quad s - b = 35° 40'.$

In this instance, formula (11) is written

$$\sin \frac{A}{2} = \sqrt{\frac{\sin (s - b) \sin (s - p)}{\sin b \sin p}}.$$

$$\log \sin (s - b) \equiv \log \sin \quad 35°\ 40' \quad 9.76\ 572$$
$$\log \sin (s - p) \equiv \log \sin \quad 17°\ 33' \quad 9.47\ 934$$
$$\log \operatorname{cosec} b \quad \equiv \log \operatorname{cosec} 65°\ 42' \quad 0.04\ 029$$
$$\log \operatorname{cosec} p \quad \equiv \log \operatorname{cosec} 83°\ 49' \quad \underline{0.00\ 253}$$

$$\therefore \quad \log \sin^2 \frac{A}{2} = 9.28\ 788$$

$$\therefore \quad \log \sin \frac{A}{2} = 9.64\ 394$$

$$\therefore \quad \frac{A}{2} = 26°\ 8'$$

$$\therefore \quad A = 52°\ 16'.$$

(iii) *Calculation of the most northerly latitude reached by the great circle AB.* Let C be the most northerly point on AB (Fig. 9). Then it is evident that the parallel of latitude through C will touch the great circle at C and that the meridian PC will be perpendicular to the great circle AB at C. Thus $P\hat{C}A$ and $P\hat{C}B$ are each 90°. In the triangle PAC, we now know PA, $P\hat{A}C$ and $P\hat{C}A$ and it is required to find PC. Clearly, formula B can be used; it is

$$\frac{\sin PC}{\sin PAC} = \frac{\sin PA}{\sin PCA},$$

and, since $P\hat{C}A = 90°$, we obtain

$$\sin PC = \sin PA \sin PAC$$

$$\log \sin PA \quad \equiv \log \sin 65°\ 42' \quad 9.95\ 971$$
$$\log \sin PAC \equiv \log \sin 52°\ 16' \quad \underline{9.89\ 810}$$
$$\therefore \quad \log \sin PC = \overline{9.85\ 781}$$
$$\therefore \quad PC = 46°\ 7'.$$

Thus the latitude of C is 43° 53'.

The calculation of the longitude of C is left as an exercise to the reader.

13. *The haversine formula.*

Many calculations are appreciably shortened by the use of "haversines". The *haversine* of an angle θ (written hav θ) is defined by

$$\text{hav } \theta = \tfrac{1}{2}(1 - \cos \theta) = \sin^2 \frac{\theta}{2} \qquad \text{......(21).}$$

Since $\cos \theta = 1 - 2 \sin^2 \dfrac{\theta}{2}$, we have

$$\cos \theta = 1 - 2 \text{ hav } \theta \qquad \text{......(22).}$$

We can now modify formula **A**, which is

$$\cos a = \cos b \cos c + \sin b \sin c \cos A.$$

According to (22) write $(1 - 2\,\mathrm{hav}\,a)$ for $\cos a$, and $(1 - 2\,\mathrm{hav}\,A)$ for $\cos A$. Then

$$1 - 2\,\mathrm{hav}\,a = \cos(b - c) - 2\sin b \sin c\,\mathrm{hav}\,A.$$

Write $1 - 2\,\mathrm{hav}\,(b - c)$ for $\cos(b - c)$. Then we obtain

$$\mathbf{hav}\ \mathbf{a} = \mathbf{hav}\ (\mathbf{b} - \mathbf{c}) + \sin \mathbf{b}\ \sin \mathbf{c}\ \mathbf{hav}\ \mathbf{A} \quad \ldots\ldots(23),$$

which is the form of the fundamental formula expressed in terms of haversines.

From the definition in (21), $\mathrm{hav}\,\theta$ is always positive and $\mathrm{hav}\,(-\,\theta) = \mathrm{hav}\,\theta$.

The haversines and log haversines of angles from $0°$ to $180°$ are found in some collections of mathematical tables among which may be mentioned *Inman's Nautical Tables* (J. D. Potter, 156 Minories, London, E. 1), which, in addition to the usual logarithmic and trigonometrical tables (to five figures), contain several other tables of astronomical value.

The calculation of the side AB (Fig. 9) by means of haversines will now be given in order to show the convenience of the method. We write (23) as follows for the triangle PAB:

$$\mathrm{hav}\,AB = \mathrm{hav}\,(PA - PB) + \sin PA \sin PB\,\mathrm{hav}\,APB$$
$$\equiv \mathrm{hav}\,(PA - PB) + X$$

$$
\begin{array}{lll}
\log \mathrm{hav}\,APB \equiv \log \mathrm{hav} & 100°\ 57' & 9\cdot77\ 450 \\
\log \sin PA \quad \equiv \log \sin & 65°\ 42' & 9\cdot95\ 971 \\
\log \sin PB \quad \equiv \log \sin & 53°\ 13' & 9\cdot90\ 358 \\
\end{array}
$$

$$\therefore\ \log X = 9\cdot63\ 779$$
$$\therefore\ X = 0\cdot43\ 430$$
$$\mathrm{hav}\,(PA - PB) \equiv \mathrm{hav}\,12°\ 29' = 0\cdot01\ 182$$
$$\therefore\ \mathrm{hav}\,AB = 0\cdot44\ 612$$
$$\therefore\ AB = 83°\ 49',$$

which agrees with our result on p. 17.

14. *Another method.*

When two sides and the contained angle of a triangle are given, the following method is sometimes used when it is required to find the third side *and* one of the remaining angles.

To illustrate the method we shall find AB and $P\hat{A}B$ (Fig. 9). Denote AB by p, PB by a, PA by b and $A\hat{P}B$ by P. Then $a = 53° 13'$, $b = 65° 42'$ and $P = 100° 57'$.

By formulae **A**, **C** and **B**, we have

$$\cos p = \cos a \cos b + \sin a \sin b \cos P \quad \ldots\ldots(24),$$
$$\sin p \cos A = \cos a \sin b - \sin a \cos b \cos P \quad \ldots\ldots(25),$$
$$\sin p \sin A = \sin a \sin P \qquad \ldots\ldots(26).$$

Define d (a positive quantity) and D by

$$\cos a = d \cos D \qquad \ldots\ldots(27),$$
$$\sin a \cos P = d \sin D \qquad \ldots\ldots(28).$$

Hence we can write (24)–(26) as follows:

$$\cos p = d \cos (b - D) \qquad \ldots\ldots(29),$$
$$\sin p \cos A = d \sin (b - D) \qquad \ldots\ldots(30),$$
$$\sin p \sin A = \sin a \sin P \qquad \ldots\ldots(31).$$

(i) From (27) and (28), by division,

$$\tan D = \tan a \cos P \qquad \ldots\ldots(32),$$

from which D can be calculated.

(ii) From (30) and (31),

$$\tan A = \frac{\sin a \sin P}{d \sin (b - D)},$$

which, by inserting the value of d given by (28), becomes

$$\tan A = \tan P \sin D \operatorname{cosec} (b - D) \quad \ldots\ldots(33),$$

from which A can be calculated.

(iii) From (29) and (30),

$$\tan p = \tan (b - D) \sec A \qquad \ldots\ldots(34),$$

from which p can be calculated.

The calculations.

(i) $\quad \log \tan a \equiv \log \tan\ 53° 13'\quad 0{\cdot}12\ 631$

$\qquad \log \cos P \equiv \log \cos 100° 57'\quad 9{\cdot}27\ 864\ n$

$\qquad\qquad \therefore\ \log \tan D = \overline{9{\cdot}40\ 495\ n}$

$\cos P$ is *negative* and we attach the letter n to its logarithm to remind us of this fact. It follows that $\tan D$ is *negative*. We have assumed in formulae (27) and (28) that d is a positive quantity. Then, from the given values of a and P, it follows that cos D is positive and sin D is negative; thus D is in the fourth quadrant, and from the value of log tan D which we have found we obtain

$$D = 360° - 14° \ 15'\cdot6 = 345° \ 44'\cdot4.$$

Hence

$$b - D \equiv 65° \ 42' - 345° \ 44'\cdot4 = -280° \ 2'\cdot4 = 79° \ 57'\cdot6.$$

(ii) | log tan P | $\equiv$ log tan $100° \ 57'$ | $0\cdot71 \ 338 \, n$ |
| log sin D | $\equiv$ log sin $345° \ 44'\cdot4$ | $9\cdot39 \ 151 \, n$ |
| log cosec $(b - D)$ | $\equiv$ log cosec $79° \ 57'\cdot6$ | $0\cdot00 \ 670$ |

$$\therefore \ \log \tan A = \overline{0\cdot11 \ 159}$$

and, as A is less than 180°, we have

$$P\hat{A}B \equiv A = 52° \ 16'\cdot9.$$

(iii) | log tan $(b - D)$ | $\equiv$ log tan $79° \ 57'\cdot6$ | $0\cdot75 \ 192$ |
| log sec A | $\equiv$ log sec $52° \ 16'\cdot9$ | $0\cdot21 \ 340$ |

$$\therefore \ \log \tan p = \overline{0\cdot96 \ 532}$$

$$\therefore \ AB \equiv p = 83° \ 49',$$

agreeing with the previous calculations of AB.

15. *The trigonometrical ratios for small angles.*

If θ is a small angle and *expressed in circular measure*, we have the well-known approximate formulae:

$$\sin \theta = \theta \text{ radians}; \quad \cos \theta = 1; \quad \tan \theta = \theta \text{ radians}$$

$$\dots\dots(35).$$

Now

$$1 \text{ radian} = 57° \ 17' \ 45''$$
$$= 3437\tfrac{3}{4}'$$
$$= 206265'',$$

so that

$$1'' = \frac{1}{206265} \text{ radian},$$

and

$$1' = \frac{1}{3438} \text{ radian, approximately.}$$

Hence, by the first equation of (35), when θ is successively $1''$ and $1'$,

$$\sin 1'' = \frac{1}{206265} \qquad \dots\dots(36),$$

and

$$\sin 1' = \frac{1}{3438} \qquad \dots\dots(37).$$

2   s a

If θ'' denotes the *number* of seconds of arc in θ radians, then $\theta = \dfrac{\theta''}{206265}$ and consequently

$$\sin \theta = \frac{\theta''}{206265},$$

which may be written

$$\sin \theta'' = \theta'' \sin 1'' \qquad \qquad \ldots\ldots(38).$$

Similarly, $\qquad \qquad \sin \theta' = \theta' \sin 1' \qquad \qquad \ldots\ldots(39),$

where θ' is expressed in minutes of arc.

In a similar way, we find

$$\tan \theta'' = \theta'' \sin 1''.$$

In spherical astronomy, certain angles are frequently expressed in terms of hours, minutes and seconds of time, according to the following relations:

$$24 \text{ hours} = 360°; \quad 1^h = 15°; \quad 1^m = 15' \text{ and } 1^s = 15''$$
$$\ldots\ldots(40).$$

Thus we obtain the approximate formulae

$$\sin 1^m = \sin 15' = 15 \sin 1' \qquad \ldots\ldots(41),$$
$$\sin 1^s = \sin 15'' = 15 \sin 1'' \qquad \ldots\ldots(42).$$

If H is a small angle, which, when expressed in minutes of time, will be denoted by H^m, then

$$\sin H = H^m \sin 1^m = 15 H^m \sin 1' \qquad \ldots\ldots(43).$$

Similarly, if we express H in terms of seconds of time, we have

$$\sin H = H^s \sin 1^s = 15 H^s \sin 1'' \qquad \ldots\ldots(44).$$

These results will be of use in subsequent chapters.

16. *Delambre's and Napier's analogies.*

For reference, we give the following formulae, originally due to Delambre, and known as Delambre's analogies:

$$\sin \tfrac{1}{2}c \sin \tfrac{1}{2}(A - B) = \cos \tfrac{1}{2}C \sin \tfrac{1}{2}(a - b) \ldots\ldots(45),$$
$$\sin \tfrac{1}{2}c \cos \tfrac{1}{2}(A - B) = \sin \tfrac{1}{2}C \sin \tfrac{1}{2}(a + b) \ldots\ldots(46),$$
$$\cos \tfrac{1}{2}c \sin \tfrac{1}{2}(A + B) = \cos \tfrac{1}{2}C \cos \tfrac{1}{2}(a - b) \ldots\ldots(47),$$
$$\cos \tfrac{1}{2}c \cos \tfrac{1}{2}(A + B) = \sin \tfrac{1}{2}C \cos \tfrac{1}{2}(a + b) \ldots\ldots(48).$$

These formulae are easily derived from the principal formulae already discussed in the previous pages.

Taking these equations in pairs, we obtain Napier's analogies:

$$\tan \tfrac{1}{2}(a + b) = \frac{\cos \tfrac{1}{2}(A - B)}{\cos \tfrac{1}{2}(A + B)} \tan \tfrac{1}{2}c \quad \ldots\ldots(49),$$

$$\tan \tfrac{1}{2}(a - b) = \frac{\sin \tfrac{1}{2}(A - B)}{\sin \tfrac{1}{2}(A + B)} \tan \tfrac{1}{2}c \quad \ldots\ldots(50),$$

$$\tan \tfrac{1}{2}(A + B) = \frac{\cos \tfrac{1}{2}(a - b)}{\cos \tfrac{1}{2}(a + b)} \cot \tfrac{1}{2}C \quad \ldots\ldots(51),$$

$$\tan \tfrac{1}{2}(A - B) = \frac{\sin \tfrac{1}{2}(a - b)}{\sin \tfrac{1}{2}(a + b)} \cot \tfrac{1}{2}C \quad \ldots\ldots(52).$$

EXERCISES

1. In the spherical triangle ABC, $C = 90°$, $a = 119°\ 46'\ 36''$ and $B = 52°\ 25'\ 38''$. Calculate the values of b, c and A.

[Ans. $48°\ 26'\ 49''$, $109°\ 14'\ 0''$ and $113°\ 10'\ 46''$.]

2. In the triangle ABC, $a = 57°\ 22'\ 11''$, $b = 72°\ 12'\ 19''$ and $C = 94°\ 1'\ 49''$. Calculate the values of c, A and B.

[Ans. $83°\ 46'\ 32''$, $57°\ 40'\ 45''$ and $72°\ 49'\ 50''$.]

3. In the triangle ABC, $c = 90°$, $B = 62°\ 20'\ 42$ and $a = 136°\ 19'\ 0''$. Calculate the values of A, C and b.

[Ans. $139°\ 46'\ 13''$, $69°\ 14'\ 45''$ and $71°\ 18'\ 9''$.]

4. Two ships X and Y are steaming along the parallels of latitude 48° N and 15° S respectively, in such a way that at any given moment the two ships are on the same meridian of longitude. If the speed of X is 15 knots,* find the speed of Y.

5. A and B are two places on the earth's surface with the same latitude ϕ; the difference of longitude between A and B is $2l$. Prove that (i) the highest latitude reached by the great circle AB is $\tan^{-1}(\tan \phi \sec l)$, and (ii) the distance measured along the parallel of latitude between A and B exceeds the great circle distance AB by

$$2 \operatorname{cosec} l' \, [l \cos \phi - \sin^{-1}(\sin l \cos \phi)] \text{ nautical miles.}$$

6. The most southerly latitude reached by the great circle joining a place A on the equator to a place B in south latitude ϕ is ϕ_1. Prove that the difference of longitude between A and B is $90° + \cos^{-1}(\tan \phi \cot \phi_1)$.

7. The positions of A and B are respectively: Lat. 39° 20′ S, Long. 110° 10′ E and Lat. 44° 30′ S, Long. 46° 20′ W. Show that, if a ship steams from A to B by the shortest possible route without crossing the parallel of 62° S, the distance steamed is 5847·6 nautical miles.

* The *knot* is the unit of speed in use at sea; it is 1 nautical mile per hour.

8. If the elements a, b, c, A, B, C of a spherical triangle receive increments da, ... dC, show that, if

$$K = \frac{\sin A}{\sin a} = \frac{\sin B}{\sin b} = \frac{\sin C}{\sin c},$$

$$da = \cos C . db + \cos B . dc + K \sin b \sin c . dA,$$

$$db = \cos A . dc + \cos C . da + K \sin c \sin a . dB,$$

$$dc = \cos B . da + \cos A . db + K \sin a \sin b . dC,$$

$$dA = - \cos c . dB - \cos b . dC + \frac{1}{K} \sin B \sin C . da,$$

$$dB = - \cos a . dC - \cos c . dA + \frac{1}{K} \sin C \sin A . db,$$

$$dC = - \cos b . dA - \cos a . dB + \frac{1}{K} \sin A \sin B . dc.$$

CHAPTER II

THE CELESTIAL SPHERE

17. *Introduction.*

In Chapter I we have seen that positions on the surface of the earth are completely specified by reference to two principal great circles, the Greenwich meridian and the equator. The principle of specifying positions on the celestial sphere is fundamentally similar and there are several methods depending on the particular great circles chosen as the principal circles. These methods will now be described.

18. *Altitude and azimuth.*

Let O—the observer on the surface of the earth (supposed spherical)—be the centre of the celestial sphere (Fig. 10). Let Z

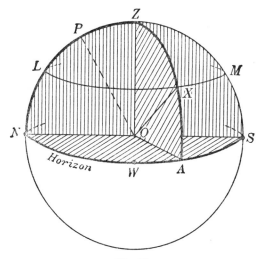

Fig. 10.

(the *zenith*) be the point on the celestial sphere vertically overhead—its direction can be defined by means of a plumb-line. OZ is thus the continuation of the straight line joining the earth's centre to O. The plane through O at right angles to OZ is the plane of the horizon, cutting the celestial sphere in the great circle NAS, called the *celestial horizon* or simply the horizon. Thus, in Fig. 10, the horizon divides the celestial sphere into two hemispheres, of which the upper is the visible hemisphere, the lower being hidden from the observer by the earth. Let X be the position of a star on the celestial sphere at a given moment. Any great circle drawn through Z is called a *vertical circle*; in particular, the vertical circle in Fig. 10 passing through X is ZXA. In the plane of ZXA, the angle AOX or the great circle arc AX is called the *altitude*, which will be denoted by a. Since OZ is perpendicular to the plane of the horizon, the great circle arc ZA is $90°$; hence $ZX = 90° - a$. ZX is called the *zenith distance* (z.d.) of the star X and will be denoted by z. Thus

$$z = 90° - a \qquad \qquad \dots\dots(1).$$

Let LXM be a small circle through X parallel to the horizon; it is called a *parallel of altitude* and is such that all heavenly bodies, whose positions at a given instant lie on this small circle, have the same altitude and also, by (1), the same zenith distance as X. Thus if the altitude or zenith distance of a star is given, the parallel of altitude on which it must lie can be definitely specified. To define its position completely on the celestial sphere, the particular vertical circle on which it lies must also be specified. This is done as follows.

Let OP be parallel to the axis about which the earth spins. If the latitude of the observer is north (as in Fig. 10), the position P is called the *north celestial pole*, or simply the north pole. We are not directly conscious of the earth's rotation, but the effect is shown in the apparent rotation of the celestial sphere. The stars thus appear to travel across the sky and their altitudes and directions are continually changing. In the northern hemisphere there is, however, one star, visible to the naked eye, which appears to change very little. This is Polaris, or the north pole star, whose direction in the sky is very nearly that given by OP. If there happened to be a star exactly situated at P on the celestial sphere, its altitude and direction would be invariable throughout a night. We define the vertical circle through P, that is ZPN (which cuts the horizon in N), as the principal vertical circle and the point N as the *north point of the horizon*.

The point S on the horizon exactly opposite to N is the south point; the west (W) and east (E) points* have directions at right angles to the directions of N and S (E is not shown in Fig. 10). The points N, E, S and W are called the *cardinal points*.

We now specify the position of a star X on the celestial sphere at a given moment by reference to the horizon and the principal vertical circle ZPN. If the star is in the western part of the celestial sphere (as in Fig. 10), the spherical angle PZX (which is formed by the principal vertical circle and the vertical circle through X) or the great circle arc NA is called the *azimuth* (W). If the star is in the eastern part of the celestial sphere, as in Fig. 11, the angle PZX or the arc NB is the azimuth (E). Thus at any instant the position of a heavenly body on the celestial sphere can be described completely by reference to the horizon and the north point of the horizon in terms of altitude and azimuth (E or W) or, alternatively, in terms of zenith distance and azimuth. When the azimuth is 90° E or 90° W, the star is said to

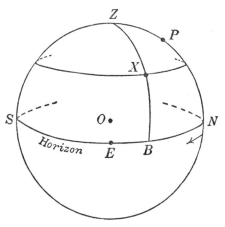

Fig. 11.

be on the *prime vertical*, which is thus the vertical circle through the east point E or the west point W.

Since in Figs. 10 and 11 the angle POZ (or the great circle arc PZ) is equivalent to the angle between the radius of the earth which passes through the observer's position and the earth's axis, then $P\hat{O}Z$ (or PZ) is equal to the colatitude of the observer

or
$$PZ = 90° - \phi \qquad \qquad ......(2),$$

where ϕ is the observer's latitude. Also $PN = 90° - PZ = \phi$; hence the altitude of the pole is equal to the observer's latitude.

* The positions of W and E relatively to N and S are obtained from the consideration that, if the observer faces north, the west point is towards his left hand and the east point towards his right hand.

19. *Declination and hour angle.*

As in the preceding section, suppose that the celestial sphere is drawn for an observer O in latitude ϕ, showing the horizon, the zenith Z and the north pole P (Fig. 12). The great circle RWT whose plane is perpendicular to OP is the *celestial equator* and its plane, clearly, is parallel to that of the earth's equator. The celestial equator and the horizon intersect in two points W and E. Now Z is the pole of the great circle NWS and P is the pole of the great circle RWT; hence W is $90°$ from both Z and P

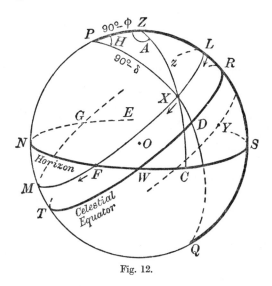

Fig. 12.

and therefore is $90°$ from all points on the great circle through Z and P. In other words, W is the pole of the great circle $NPZSQ$; hence $NW = 90°$ and $WS = 90°$. Similarly $EN = 90°$ and $ES = 90°$. Hence W and E are the two remaining cardinal points, N and S having been previously defined explicitly.

As already stated, the rotation of the earth results in an apparent rotation of the celestial sphere from east to west about OP. It follows that, as the stars are at such great distances from the earth, the angle between the straight line joining the observer at O to any particular star and the straight line OP (parallel to the earth's axis) remains unaltered. If we consider a star X, the earth's rotation makes it appear to describe a small circle LXM, parallel to the celestial equator, in the direction indicated by the arrows in Fig. 12. Let $PXDQ$ be the semi-great circle through X and the poles of the celestial sphere. Then the arc DX is called the *declination* of the star and is north declination if the star is between the celestial equator and the north pole P (as for the star X). The star's declination is south (as for Y) when it is between the celestial equator and the south pole Q. Declination is thus analogous to latitude as defined for points on the earth's surface. Denote the declination of X by δ; then $DX = \delta$ and $PX = 90° - \delta$. PX is called the *north polar distance* (N.P.D.) of the star. It is convenient to treat declination as an algebraic quantity, so that the various formulae to be derived will hold equally for north and south declinations. North declinations carry the positive sign ($+$) and south declinations the negative sign ($-$). Thus the formula for north polar distance, viz. N.P.D. $= 90° - \delta$, is applicable to all stars, whatever their declinations may be.

The declination of a star being known, we can thus specify a small circle, called the *parallel of declination*, on which it must lie. To fix completely its position on the celestial sphere at any moment we require another great circle of reference. This is the semi-great circle $PZRSQ$, called the *observer's meridian*. When the star is at L on the observer's meridian, it is said to *transit* or *culminate*, and it is clear from Fig. 12 that its altitude (that is SL) is then greatest and its zenith distance ZL is least. Thereafter, owing to the earth's rotation, it moves along the small circle LFM crossing the horizon at F where it is said to *set*; its altitude at F is of course $0°$ and its zenith distance is $90°$. During an interval of time depending on its declination, the star is below the horizon, reaching its maximum depression below the horizon at M; eventually it reaches the horizon at G where it is said to *rise*. Its altitude gradually increasing, it returns after an interval, equivalent to that in which the earth makes a complete rotation about its axis, to the observer's meridian at L. At any moment the star's position on the parallel of declination is specified by the angle at P between the observer's meridian and the meridian (PXQ) through the star at this time; this angle is RPX or ZPX or the arc RD on the equator. This angle, denoted by H, is called the *hour angle* and is measured from the observer's meridian *westwards* from 0° (at L) to 360°, when the star again returns to the observer's meridian or, as is more usual, from 0^h to 24^h. We can express this in a slightly different way. When the star is in transit, its meridian coincides with the observer's meridian; thereafter, the star's meridian moves steadily westwards and, when it has made a complete circuit of the celestial sphere, it has described an angle of 360° or 24^h with reference to the observer's meridian. From Fig. 12 it is seen that if the star is *west* of the observer's meridian, that is, if the

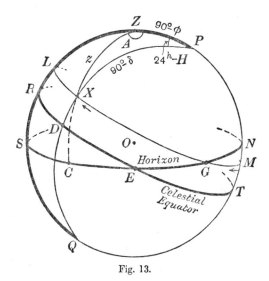

Fig. 13.

azimuth is west, its hour angle is between 0° and 180°, that is, between 0^h and 12^h. Similarly, if the star is *east* of the meridian (azimuth east)—as in Fig. 13—the hour angle is between 12^h and 24^h. We thus have the rules:

If the star's azimuth is west, the hour angle is between 0^h and 12^h (and vice versa); if the star's azimuth is east, the hour angle is between 12^h and 24^h.

20. *Diagram for the southern hemisphere.*

The diagrams described so far in this chapter refer to the celestial sphere for an observer in north latitude. We shall now describe the corresponding diagrams for an observer in the

southern hemisphere. In Fig. 14, we shall place the observer's
zenith as in the previous diagrams. The celestial horizon is then
as indicated. In the southern hemisphere, the south celestial
pole Q is above the horizon. Then, if ϕ denotes the southern
latitude of the observer, $QZ = 90° - \phi$. The principal vertical
circle is now ZQS, intersecting the horizon in the *south* point S.
The north point N can then be placed in the diagram. The
celestial equator and the horizon intersect in the west and east
points W and E (the latter is not shown in Fig. 14) according to

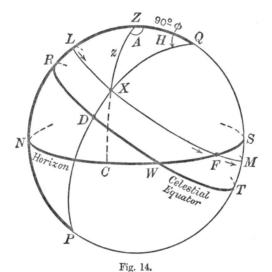

Fig. 14.

the rule in the footnote to page 27. Consider a star X with
south declination. Owing to the earth's rotation it will describe
a small circle LXM, parallel to the celestial equator and lying
between the celestial equator and the south pole Q. At L, the
star will have its greatest altitude—it is *then* on the observer's
meridian, which is the semi-circle $QZRNP$. In consequence of
the earth's rotation the star will move from the observer's
meridian *westwards*, that is, in the direction LXM, as indicated
by the arrows in the diagram. The angle ZQX is the hour angle
measured, as before, from 0^h to 24^h westwards from the observer's
meridian. QZX is the azimuth; in this instance it is west. If δ
is the southern declination of the star, then $DX = \delta$ and

$QX = 90° − δ$. The other parts of the spherical triangle QZX are: $QZ = 90° − φ$, $ZX = z$ (the zenith distance), $QZX = A$ (the azimuth) and $ZQX = H$ (the hour angle). When the star's azimuth is *west*, the hour angle is between 0^h and 12^h. When the star's azimuth is east, the appropriate diagram can be similarly drawn; this is left as an exercise to the student; it will then be found that the hour angle is between 12^h and 24^h. The rules stated at the end of section 19 are seen to hold for southern as well as northern latitudes.

21. *Circumpolar stars.*

Consider the celestial sphere for an observer in northern latitude $φ$ (Fig. 15). The parallels of declination are drawn for two stars X and Y, both of which are always above the horizon and consequently do not set. Such stars are called circumpolar

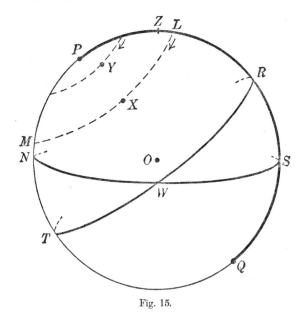

Fig. 15.

stars. It is readily seen from the figure that the condition that a star should not set is: PM must be less than PN; that is, the north polar distance must be less than the latitude, or, in other words, the declination must be greater than the colatitude.

When the star X is on the observer's meridian at L, it is at
upper culmination or in *transit*; when the star reaches M, it is at
lower culmination. The expressions "culmination above pole"
and "culmination below pole" are frequently used. At upper
culmination, the star's zenith distance is ZL or $(PL - PZ)$, that
is, $\phi - \delta$. At lower culmination, the star's zenith distance is ZM
or $(ZP + PM)$, that is, $180° - (\phi + \delta)$. When $\delta = \phi$, the star's
upper culmination occurs in the zenith. When $\delta > \phi$, the upper
culmination occurs between P and Z, as for the star Y; then the
azimuth does not exceed 90°, as can be readily inferred from the
diagram. Southern circumpolar stars can be considered in the
same way.

22. *The standard or geocentric celestial sphere.*

In the previous sections, the declination of a star on the
celestial sphere whose centre is the observer has been defined.
As the stars are at distances almost infinitely great compared
with the dimensions of the earth, the star's declination or polar
distance so defined is inde-
pendent of the observer's
position on the surface of
the earth, as may be readily
seen from Fig. 16. (It is
more convenient for our
present purpose to deal
with the star's north polar
distance than with its declination.) In Fig. 16, P_1CQ_1
is the earth's axis of rotation, C being the centre of
the earth; O is the ob-
server and COZ the direction of the zenith at O; OP
is parallel to CP_1 and the
direction of the star from

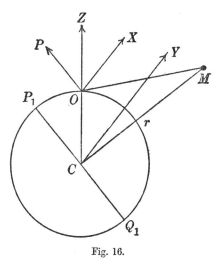

Fig. 16.

O is OX. According to our definition, the north polar distance of
the star for an observer at O is $P\hat{O}X$. If CY is drawn parallel to
OX, then CY is the direction of the star with reference to C, the
earth's centre. Thus $P_1\hat{C}Y = P\hat{O}X$; in other words the north polar

distance of the star (and consequently its declination) is the same
on the celestial sphere centred at O (or any other position on the
earth's surface) as it is on the celestial sphere centred at C. But
when a comparatively near body such as the moon, or sun, or a
planet is observed, the definition of north polar distance (and
therefore of declination) previously given is dependent on the
particular position of the observer on the earth. Thus if M is the
moon (Fig. 16) at the distance r from the centre of the earth, it
is evident that $P\hat{O}M = P_1\hat{C}M + O\hat{M}C$; also $O\hat{M}C$ clearly depends on the position of O, whereas P_1CM is entirely independent

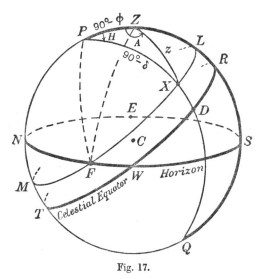

Fig. 17.

of O. $P_1\hat{C}M$ is defined as the north polar distance of M which is
thus the angle between the earth's axis and the straight line
joining the earth's centre to the heavenly body. This definition
is entirely general and is applicable to every heavenly body.
Accordingly, the centre of the standard celestial sphere (or the
geocentric celestial sphere, as it may be called) is taken to be at
C, the earth's centre (Fig. 17). CZ is the direction of the observer's zenith, the diameter QCP is coincident with the earth's
axis, $NWSE$ is the celestial horizon (the great circle whose plane
is perpendicular to CZ), and $RWTE$ is the celestial equator (its
plane is coincident with the plane of the earth's equator). The

arc PX is the north polar distance of the heavenly body according to the definition just given and DX is the declination δ (N.P.D. $= 90° - \delta$). The observer's meridian is $PZRSQ$, the zenith distance of the heavenly body is ZX (denoted by z) and the azimuth A $(P\hat{Z}X)$ and the hour angle H $(Z\hat{P}X)$ are as described previously. The declinations of the principal heavenly bodies (moon, sun, planets and the brightest stars) are tabulated in the *Nautical Almanac* (the British publication), the *American Ephemeris* and in the other national ephemerides.

Hereafter, the celestial sphere will be assumed to be as in Fig. 17, that is, centred at C, the earth's centre.

23. *Solution of the spherical triangle PZX.*

We shall consider two common problems associated with the triangle PZX.

(i) Given the observer's latitude ϕ, the declination δ and hour angle H of the heavenly body, to calculate its zenith distance and azimuth. By formula **A** (the cosine formula), we have, since two sides PZ and PX and the contained angle ZPX are given (Fig. 17),

$$\cos ZX = \cos PZ \cos PX + \sin PZ \sin PX \cos ZPX,$$

or $\quad \cos z = \sin \phi \sin \delta + \cos \phi \cos \delta \cos H \qquad \ldots\ldots(3).$

Thus z can be calculated directly from (3) or by means of the haversine formula (section 13), which in this instance can be written $\quad$ hav $z =$ hav $(\phi - \delta) + \cos \phi \cos \delta$ hav $H \ldots\ldots(4).$

Again, by **A**,

$$\cos PX = \cos PZ \cos ZX + \sin PZ \sin ZX \cos PZX,$$

or $\quad \sin \delta = \sin \phi \cos z + \cos \phi \sin z \cos A \qquad \ldots\ldots(5),$

from which the azimuth A can be calculated. In the haversine form (5) may be written

$$\cos \phi \cos a \text{ hav } A = \text{hav } (90° - \delta) - \text{hav } (\phi - a) \quad \ldots(6),$$

where a is the altitude.

(ii) Given the observer's latitude ϕ, the star's zenith distance and azimuth, to calculate the star's declination and hour angle. We are given ϕ, z and A; hence, by (5), we can calculate the declination. Either equation (3) or (4) is available for calculating the hour angle H. Thus from (4)

$$\cos \phi \cos \delta \text{ hav } H = \text{hav } z - \text{hav } (\phi - \delta) \quad \ldots\ldots(7).$$

Consider now the spherical triangle PZX in Fig. 13. The angle PZX is the azimuth (east). Remembering that the hour angle is measured at the pole from the observer's meridian *westwards*, we see that $Z\hat{P}X = 24^{\text{h}} - H$. The solution of the triangle proceeds as before.

24. *Right ascension and declination.*

In the hour angle and declination method of specifying a star's position on the celestial sphere only one co-ordinate, namely declination, remains constant as the star travels across the sky, whereas the hour angle increases uniformly from 0^{h} to 24^{h}. But

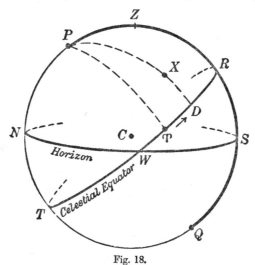

Fig. 18.

the positions of the stars on the celestial sphere may be likened to the positions of fixed points on the surface of the earth and can therefore be specified with reference to the celestial equator and any particular star on the equator. For example, in Fig. 18, let Υ be an equatorial star and X any other star; let the meridian through X cut the celestial equator in D. As the stars pass across the sky we know in particular that the declination of X, that is, DX, remains constant and that the relative configuration of the stars also remains constant. It follows that ΥD is constant; in other words, that the angle between the meridians of Υ and D remains constant. We may regard Υ as a reference point on the celestial equator; with respect to Υ and the celestial equator, we can clearly specify the position of the star X by means of the great circle arc ΥD and the declination DX. The reference point chosen in practice is called the *vernal equinox* or the *first point of Aries*, and it is convenient to regard the position of Υ as specified by a particular star in the sky. Later we shall define Υ more precisely. The arc ΥD or $\Upsilon \hat{P} X$ is called the *right ascension* (R.A.) of the star X (denoted by α) and is measured *eastwards* from Υ from 0^{h} to 24^{h} (in the direction of the arrow near Υ). This direction is opposite to that in which hour angle is measured. From Fig. 18, we see that $R\Upsilon = RD + \Upsilon D$. Now RD (or $R\hat{P}X$) is the hour angle H of X and $R\Upsilon$ is the hour angle of Υ. The hour angle of Υ is called the *sidereal time* (S.T.). We have, accordingly,

$$\text{Sid. time} = \text{H.A. } X + \text{R.A. } X \qquad \ldots \ldots (8),$$

or

$$\text{S.T.} = H + \alpha \qquad \ldots \ldots (9).$$

When Υ is on the observer's meridian, the hour angle of Υ is 0^{h}, that is to say, the sidereal time is 0^{h}. When Υ is next on the observer's meridian, an interval of 24^{h} of *sidereal time* has elapsed. This interval is, of course, the same as that required for the complete rotation of the earth about its axis and it is called a *sidereal day*. The rotating earth is, in fact, the standard time-keeper.

25. *The earth's orbit.*

The earth is a planet revolving around the sun in an elliptical path or orbit, the sun being situated at a focus S of the ellipse (Fig. 19). This is Kepler's first law of planetary motion. The time required for the earth to make a complete revolution of its orbit is a *year*. As the earth progresses in its orbit, the direction of the earth, as viewed from the sun, is continually altering; the angular velocity is, however, not uniform. Since our observations are made from the earth, then relative to the earth the sun appears to describe an elliptical orbit around the earth. In Fig. 20, C is the centre of the earth and the ellipse represents the *apparent* orbit of the sun relative to the earth. The sequence of positions of the sun, namely a, e, f, b, g in this orbit, corresponds to the sequence of positions A, E, F, B, G of the earth in its orbit round the sun (Fig. 19). In the course of the year, the sun thus

appears to make a complete circuit of the heavens against the
background of the stars. The plane of the orbit is called the
plane of the ecliptic, and the great circle in which this plane
intersects the celestial sphere, whose centre is the earth's centre
C, is called the *ecliptic*. In Fig. 21, let *C* be the centre of the
celestial sphere on which the celestial equator *TTR* and the

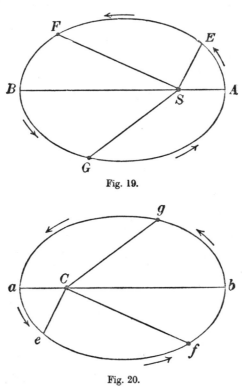

Fig. 19.

Fig. 20.

north pole *P* are drawn. We may imagine that the stars can be
viewed from the centre of the earth, that is, from *C*, and accordingly they will occupy definite positions on the celestial
sphere in Fig. 21. With reference to the stars, the plane of the
ecliptic will have a definite position and, consequently, the
ecliptic will be a particular great circle, which is found from
observations to be inclined at an angle of about $23\frac{1}{2}°$ to the
celestial equator. In Fig. 21, *YTMU* represents the ecliptic and

its inclination to the celestial equator is $M\hat{\Upsilon}R$, which is known as the *obliquity of the ecliptic*. Relative to the earth, the sun appears to move on the celestial sphere along the ecliptic—in the direction $Y\Upsilon M$—and twice yearly, at Υ and at U, its position on the celestial sphere coincides with the intersections of the ecliptic with the celestial equator. Between Υ and M and between M and U the sun is on the north pole side of the equator; its declination is then north. Similarly between U and Y and

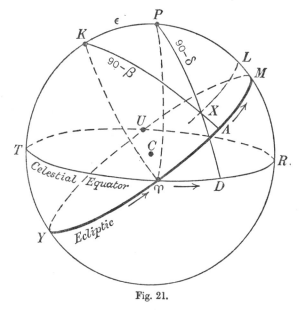

Fig. 21.

between Y and Υ its declination is south. The position Υ, at which the sun's declination changes from south to north, is the *vernal equinox*. It is in this way that the reference point Υ, from which are measured the right ascensions of the stars, is obtained. Thus if X is a star, its right ascension is ΥD or α measured along the equator from Υ eastwards, and its declination δ is DX. From the diagram it is seen that the right ascension and declination of the sun are both changing continually. When the sun is at Υ, its right ascension and declination are both zero (this occurs about March 21—the *vernal equinox*); at M the right ascension is 6^h and declination about $23\frac{1}{2}°$ N (this occurs about June 21—

the *summer solstice*); at U the right ascension is 12^h and declination $0°$ (this occurs about September 21—the *autumnal equinox*) and at Y the right ascension is 18^h and the declination about $23\frac{1}{2}°$ S (this occurs about December 21—the *winter solstice*).

26. *Celestial latitude and longitude.*

The position of a heavenly body can be referred to the ecliptic as fundamental great circle and the vernal equinox Υ as principal reference point. In Fig. 21, let K be the pole of the ecliptic and let KXA be a great circle arc passing through X and meeting the ecliptic in A. The arc ΥA, measured from Υ to A along the ecliptic in the *direction of the sun's annual motion*, i.e. eastwards, is called the *longitude* of the heavenly body X and is measured from $0°$ to $360°$ round the ecliptic. The arc AX is the *latitude* and, like terrestrial latitude, it is measured north or south. If we know the star's right ascension and declination we can obtain its latitude (β) and its longitude (λ) from the triangle KPX; and *vice versa*. Now Υ is the pole of the great circle $KPMR$; hence $K\hat{P}\Upsilon = 90°$, and since $\Upsilon D = \Upsilon\hat{P}X = \alpha$, then $K\hat{P}X = 90° + \alpha$. Also $P\hat{K}\Upsilon = 90°$, and since $\Upsilon A = \Upsilon\hat{K}X = \lambda$, then $P\hat{K}X = 90° - \lambda$. Also $PX = 90° - \delta$ and $KX = 90° - \beta$. Let ϵ denote the obliquity of the ecliptic; it is the angle between the radii CM and CR; thus the arc $RM = \epsilon$. But $KM = 90°$ and $PR = 90°$; hence $KP = \epsilon$. Applying the formulae **A**, **B** and **C**, we have

$$\cos KX = \cos PX \cos KP + \sin PX \sin KP \cos KPX,$$

$$\sin KX \sin PKX = \sin PX \sin KPX,$$

$$\sin KX \cos PKX = \cos PX \sin KP - \sin PX \cos KP \cos KPX,$$

or

$$\sin \beta = \sin \delta \cos \epsilon - \cos \delta \sin \epsilon \sin \alpha \dots\dots(10),$$

$$\cos \beta \cos \lambda = \cos \delta \cos \alpha \qquad \dots\dots(11),$$

$$\cos \beta \sin \lambda = \sin \delta \sin \epsilon + \cos \delta \cos \epsilon \sin \alpha \dots\dots(12).$$

By a similar process, the right ascension α and the declination δ can be expressed in terms of β, λ and ϵ. The formulae are

$$\sin \delta = \sin \beta \cos \epsilon + \cos \beta \sin \epsilon \sin \lambda,$$

$$\cos \delta \cos \alpha = \cos \beta \cos \lambda,$$

$$\cos \delta \sin \alpha = -\sin \beta \sin \epsilon + \cos \beta \cos \epsilon \sin \lambda.$$

27. *Sidereal time.*

Let the earth and the celestial sphere (centred at C) be drawn as in Fig. 22; let g denote the position of Greenwich on the earth's surface and l that of any other place. The angle between the meridians plq and pgq is, of course, the longitude (terrestrial) of l; in this instance l is west of Greenwich. Produce Cg, Cl to meet the celestial sphere in G and L. Then G and L are the zeniths of Greenwich and l respectively. If X is the position of a heavenly body on the celestial sphere at a given moment, $G\hat{P}X$ is the hour angle of X for an observer on the Greenwich meridian and $L\hat{P}X$ is the hour angle for an observer on the

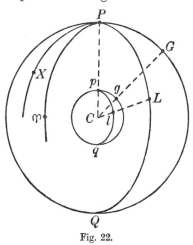

Fig. 22.

meridian of l. But $G\hat{P}X = L\hat{P}X + G\hat{P}L$ and $G\hat{P}L = g\hat{p}l$; hence

H.A. of X at Greenwich = H.A. of X at l + long. (W) of l

$$......(13).$$

In this formula we suppose that the longitude of l is expressed in time-measure ($15° = 1^h$; $15' = 1^m$; $15'' = 1^s$). The formula (13) is a general one and it clearly holds for the vernal equinox Υ. We thus obtain—since sidereal time is the hour angle of Υ—

Sid. time at Greenwich = Sid. time at $l \pm$ long. of l ...(14),

the $+$ sign being taken when l is *west* of Greenwich and the $-$ sign when l is *east* of Greenwich. The sidereal time at l is called the *local sidereal time*.

28. *Mean solar time.*

The sidereal day is an observatory unit of time and is obviously unsuited to the regulation of everyday affairs which in the main are governed according to the position of the sun in the sky. When the sun is on the meridian of a place, it is *apparent noon* there; when the sun is next on the meridian, an *apparent solar day* is said to have elapsed. This interval can be measured, for example, by means of a clock keeping accurate sidereal time and it is found that an apparent solar day is not constant. We have seen that, relative to the earth, the sun appears to describe an elliptic orbit around the earth and the rate at which its direction in the orbit changes is not constant. It follows that the sun appears to describe the ecliptic at a non-uniform rate; in other words, the sun appears to move somewhat irregularly against the background of the stars. Due to this and also to the fact that it is moving in the ecliptic and not along the celestial equator (the fundamental great circle with which the measurement of hour angle or time is associated) its right ascension does not increase uniformly. The average apparent solar day throughout the year is called a *mean solar day* and it is convenient to define the mean solar day as the interval between two successive transits across the observer's meridian of a fictitious body called the *mean sun*. The mean sun is assumed to move in the *celestial equator* at a uniform rate around the earth. This rate is such that the mean sun completes a revolution in the same time as that required by the sun for a complete circuit of the ecliptic. According to this definition, the right ascension of the mean sun (denoted by R.A.M.S.) increases at a uniform rate.

Now if we regard the mean sun as an ordinary celestial body, then at any given moment, we can assume that it has a particular hour angle (H.A.M.S.) at a given place on the earth's surface. At this moment we shall assume that its right ascension is known; hence by (8) or (9),

$$\text{Sid. time} = \text{H.A.M.S.} + \text{R.A.M.S.} \quad \ldots\ldots(15).$$

The time shown by a mean time clock, say, at Greenwich at any moment is simply related to the value of H.A.M.S. there, and if the R.A.M.S. is known, (15) forms the basis of comparison between the sidereal and the mean time clocks. The mean sun is related to the true sun according to certain principles which will be discussed in a later chapter. Meanwhile it will be sufficient to state that the difference at any moment between the right ascension of the mean sun and of the true sun can be calculated; this difference is called the *equation of time** (denoted by E). We thus have

$$E = \text{R.A.M.S.} - \text{R.A.} \odot \quad \ldots\ldots(16),$$

* In older text-books the equation of time is defined by $E = \text{R.A.} \odot - \text{R.A.M.S.}$, but recently the almanacs have adopted the definition in (16).

in which R.A. ⊙ denotes the right ascension of the true sun. E can be positive or negative and it is tabulated in the *Nautical Almanac* (and other ephemerides) for each day of the year. In Fig. 23, let us suppose that at a given instant the right ascension and declination of the sun (⊙) are known. Let Υ be the vernal equinox at this instant so that $R\hat{P}\Upsilon$ or $R\Upsilon$ is the hour angle of Υ, that is, the local sidereal time. If this is known, the position of Υ on the celestial sphere can be definitely specified. The position of the sun can then be indicated on the celestial sphere. $\Upsilon K = $ R.A. ⊙ and K ⊙ is the sun's declination and both of these

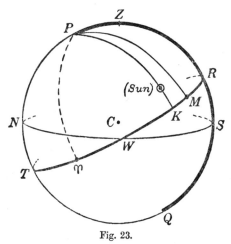

Fig. 23.

are supposed known. Suppose the equation of time is $+$; then by (16), R.A.M.S. is greater than R.A. ⊙, and if E is known the position of the mean sun M at this instant can be indicated in the diagram. $R\hat{P}M$ or RM is the hour angle of M (H.A.M.S.). It is clear from Fig. 23 that, as $RK = RM + MK$, then

$$\text{H.A. ⊙} = \text{H.A.M.S.} + E \qquad \ldots\ldots(17),$$

which is an important relation connecting H.A.M.S. and H.A. ⊙, enabling us to calculate the hour angle of the sun (H.A. ⊙) when the other quantities are known. When the mean sun is on the observer's meridian, it is *local mean noon* there. When the mean sun is on the meridian of Greenwich, it is *Greenwich mean noon*. The hour angle of the mean sun at Greenwich will be denoted in this book by G.M.A.T. (*Greenwich mean astronomical time*). When the mean sun is at T—the H.A.M.S. being then 12^h—it is said to be *mean midnight*. When G.M.A.T. $= 12^h$, it is mean midnight at Greenwich and this is the moment when a new civil day at Greenwich begins. Mean time reckoned from midnight at Greenwich is called *Greenwich Mean Time* (G.M.T.)*, now designated Universal Time (U.T.). It is clear that

$$\text{U.T.} \equiv \text{G.M.T.} = \text{G.M.A.T.} + 12^h \qquad \ldots\ldots(18).$$

Similarly, for any place keeping the mean time appropriate to its meridian, we shall have

$$\text{Local M.T.} = \text{Local M.A.T.} + 12^h \qquad \ldots\ldots(19)$$
$$= \text{H.A.M.S.} \pm 12^h \qquad \ldots\ldots(20).$$

Formula (14) gives the relation between sidereal time at Greenwich and the sidereal time at any place l, and it is clear from Fig. 22 and from (18) and (19) that we shall have a similar relation between mean time at Greenwich and the mean time at the place; it is

$$\text{U.T.} \equiv \text{G.M.T.} = \text{Local M.T.} \pm \text{long. of } l \qquad \ldots\ldots(21),$$

the $+$ sign being taken when the longitude of l is west and the $-$ sign when the longitude is east.

Confusion would be inevitable if every place kept the local mean time appropriate to its meridian, and so in small countries a standard mean time is chosen, corresponding to a particular meridian of longitude (the standard meridian), which is in use uniformly throughout the country. In Great Britain, the standard mean time is G.M.T. In extensive countries such as Russia and the United States of America, two or more standard times are in use in zones of longitude; within each zone, a standard time appropriate to a definite meridian within the zone is kept. The standard time, based on a particular meridian, we shall designate *zone time* (Z.T.). This system is, in effect, kept by ships at sea which are generally less troubled by geographical complications. We have, as in (21),

$$\text{U.T.} \equiv \text{G.M.T.} = \text{Z.T.} \pm \text{long. of standard meridian} \ldots(22).$$

---

* Before 1925, G.M.T. was used in the almanacs to signify Greenwich mean astronomical time (G.M.A.T.). Beginning with 1925, the time used was G.M.T. ($\equiv$ G.C.T.) later superseded, as stated above, by U.T. Recently, for reasons stated in Appendix F (p. 424), U.T. has been replaced in the almanacs by Ephemeris Time (E.T.). The difference between U.T. and E.T. is so minute that we shall use the former generally, unless otherwise stated.

29. *Example.*

We shall use as an illustration the following problem of a common and important type. At a place in longitude 163° 14′ E, it is required to calculate the hour angle of the sun (H.A. ☉) corresponding to an observation made at zone time 8ʰ 46ᵐ 22ˢ on 1931 March 10; the zone time is that of the standard meridian of 165° E (11ʰ E).

The first step is to derive the U.T. at which the observation was made. We have

Zone time   8ʰ 46ᵐ 22ˢ   March 10

Long. of standard meridian − 11ʰ
_____

U.T. = 21ʰ 46ᵐ 22ˢ   *March 9*

We subtract 11ʰ from the zone time in accordance with formula (22). (Clearly, we can write the zone time as 32ʰ 46ᵐ 22ˢ *March 9*.)

We next find the local mean time (that is, the mean time corresponding to the longitude of the place) by means of (21).

U.T.   21ʰ 46ᵐ 22ˢ   March 9

Long. of place (E) + 10ʰ 52ᵐ 56ˢ
_____

Local M.T. = 32ʰ 39ᵐ 18ˢ   March 9

= 8ʰ 39ᵐ 18ˢ   March 10

Formula (20) enables us to write down the H.A.M.S. (the hour angle of the mean sun *at the place*); it is

H.A.M.S. = 20ʰ 39ᵐ 18ˢ.

The next step is to apply the equation of time to H.A.M.S. From the *Nautical Almanac* (1931) it is found, by interpolation, that at U.T. 21ʰ 46ᵐ 22ˢ, *March 9*, E = − 10ᵐ 45ˢ.

Thus, by (17),   H.A. ☉ = 20ʰ 39ᵐ 18ˢ − 10ᵐ 45ˢ,

or   H.A. ☉ = 20ʰ 28ᵐ 33ˢ.

30. *Hour angle of a heavenly body.*

To calculate the hour angle of any heavenly body (X) other than the sun, we proceed as follows. By (8) and (15) we have

Sid. time = H.A. X + R.A. X

and   Sid. time = H.A.M.S. + R.A.M.S.,

whence   H.A. X + R.A. X = H.A.M.S. + R.A.M.S.   ...(23).

In the almanacs, the sidereal time is tabulated at U.T. 0^h for each day throughout the year, that is, when the H.A.M.S. is 12^h. Since, by (15),

$$\text{Sid. time} = \text{H.A.M.S.} + \text{R.A.M.S.},$$

we have

R.A.M.S. at U.T. 0^h for any day = the tabulated sidereal time at U.T. 0^h for that day $- 12^h$.

The R.A.M.S. *increases uniformly at the rate of* $3^m \, 56^s\cdot56$ *per mean solar day or at the rate of* $9^s\cdot857$ *per mean solar hour;* by means of this we can calculate the R.A.M.S. for any given U.T.

Tables are given in the almanacs for facilitating this calculation.

The use of the formula (23) is best illustrated by means of an example. It is required to calculate the hour angle of Betelgeuse (α Orionis) at zone time $18^h \, 35^m \, 46^s$ on 1931 January 26, in a place whose longitude is $64° \, 28' \, 49''$ W. (Zone $+ 4^h$: this means that the standard meridian of the zone is 4^h W or $60°$ W.)

Zone time	$18^h \, 35^m \, 46^s$	January 26
Zone	$+ \; 4^h$	
U.T.	$22^h \, 35^m \, 46^s$	January 26
Longitude of place (W)	$- \; 4^h \, 17^m \, 55^s$	[*Using* (21)]
Local M.T.	$18^h \, 17^m \, 51^s$	
Subtract 12^h	$- \; 12^h$	[*Using* (20)]
H.A.M.S.	$6^h \, 17^m \, 51^s$	
Add R.A.M.S.	$20^h \, 21^m \; \; 0^s$	*From* N.A. *for above*
Sid. time	$26^h \, 38^m \, 51^s$	G.C.T.
Subtract R.A. of Betelgeuse	$5^h \, 51^m \, 27^s$	*From* N.A.
H.A. of Betelgeuse	$20^h \, 47^m \, 24^s$	

31. *Rising and setting.*

Consider Fig. 24. The heavenly body X is said to set at F, the point where it reaches the horizon. Then the zenith distance is $90°$, that is, $ZF = 90°$. Let H be the hour angle of X at setting, so that $ZPF = H$. Also $PF = 90° - \delta$. Let A be the azimuth at setting (PZF) and ϕ the latitude.

From formula **A**,

$$\cos ZF = \cos PZ \cos PF + \sin PZ \sin PF \cos ZPF,$$

or $\cos 90° = \sin \phi \sin \delta + \cos \phi \cos \delta \cos H,$

so that, as $\cos 90° = 0$,

$$\cos H = - \tan \phi \tan \delta \qquad \ldots\ldots(24),$$

from which the hour angle at setting can be calculated.

Also from **A**,

$$\cos PF = \cos PZ \cos ZF + \sin PZ \sin ZF \cos PZF,$$

or $\sin \delta = \qquad 0 \qquad + \cos \phi \cos A,$

whence $\cos A = \sin \delta \sec \phi \qquad \ldots\ldots(25),$

from which the azimuth at setting can be calculated.

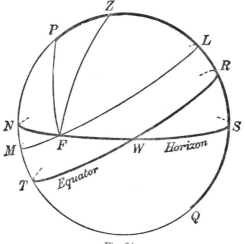

Fig. 24.

In north latitudes, it is seen either from the equations (24) and (25), or from Fig. 24, that if the declination is north the hour angle at setting is between 6^h and 12^h and that the azimuth is less than 90° (that is to say, the body sets between west and north); and that if the declination is south, the hour angle at setting is between 0^h and 6^h and the body sets between south and west. The problem as it concerns the rising of a heavenly body can be treated in a similar way. When the observer's latitude is south, the procedure is similar.

If the heavenly body concerned is a star, the hour angle at setting gives the interval between meridian transit and setting expressed in sidereal time. If the heavenly body is the sun, the interval between meridian transit and setting is expressed in apparent solar time. But during this interval the relative positions of the sun and the mean sun will alter but little (in other words, the change in the equation of time can be usually disregarded unless extreme accuracy is desired), and so the interval can be described, for all practical purposes, in terms of mean time. Thus, if from formula (24) the hour angle H at setting is found to be 7^h 30^m, then the interval between the sun's meridian transit and setting is 7^h 30^m mean solar time. Leaving out of consideration any change in the sun's declination, we infer that this is also the interval between sunrise and meridian passage. Thus the sun is above the horizon for 15^h and below the horizon for 9^h. Actually, of course, the sun's declination is generally slightly different at sunrise from that at sunset owing to its motion along the ecliptic and the effect can be calculated.

Formula (24) shows that if $\phi > 90° - \delta$, cos H is, numerically, greater than unity, so that the equation fails to give a value of H. In this instance, the sun does not set in latitudes and on days such that $\phi > 90° - \delta$, as may also be verified from a diagram. On midsummer day, the sun's north declination is greatest; it is then $23\frac{1}{2}°$ N approximately, so that in latitudes north of $66\frac{1}{2}°$ N, the sun is above the horizon on that day without setting.* At the north pole, since $\phi > 90° - \delta$, provided δ is north, the sun is above the horizon continuously between March 21 and September 21; for the remaining six months it is below the horizon. The parallel of $66\frac{1}{2}°$ N is called the *Arctic Circle* and the corresponding parallel in the southern hemisphere ($66\frac{1}{2}°$ S) is the *Antarctic Circle*.

32. *Rate of change of zenith distance and azimuth.*

Let X in Fig. 25 be the position of a heavenly body on the celestial sphere at a certain instant and Y its position a little later. Assume the declination to be constant so that X and Y lie on the small circle LM (the parallel of declination), of which P is the pole. Draw the great circle arcs PX, PY, ZX, ZY. Let

---

* Hence the expression, the midnight sun.

UX be an arc of a small circle of which Z is the pole; then $ZX = ZU$. Let $Z\hat{P}X = H$ and $Z\hat{P}Y = H + \Delta H$, so that $X\hat{P}Y = \Delta H$. Let $P\hat{Z}X = A$ and $X\hat{Z}Y = \Delta A$; $ZX = z$ and $ZY = z + \Delta z$. Then $UY = \Delta z$. Since XY is supposed to be a small arc, we may assume that UXY is a plane triangle, right-angled at U.

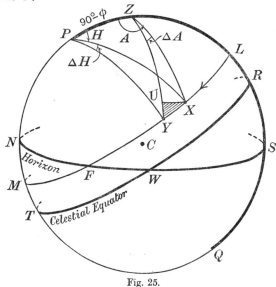

Fig. 25.

As the heavenly body moves, owing to the diurnal motion, from X to Y its zenith distance increases by Δz, its hour angle by ΔH and its azimuth decreases by ΔA.

By formula (1) of section 3 (p. 4),

$$XY = X\hat{P}Y \sin PX = \Delta H \cos \delta,$$

and
$$UX = X\hat{Z}Y \sin ZX = \Delta A \sin z.$$

Let η denote the angle PXZ; η is called the parallactic angle. Then, since Y is very close to X, we may take $P\hat{Y}Z$ to be η. Then

$$UY = XY \cos UYX,$$

and
$$UX = XY \sin UYX.$$

Now $P\hat{Y}Z = \eta$ and $P\hat{Y}X = 90°$; hence

$$UY \equiv \Delta z = \Delta H \cos \delta \sin \eta,$$

and
$$UX \equiv \Delta A \sin z = \Delta H \cos \delta \cos \eta.$$

Now in the spherical triangle PXZ, by formula **B,**

$$\cos \delta \sin \eta = \sin A \cos \phi,$$

and, by formula **C,**

$$\cos \delta \cos \eta = \sin \phi \sin z - \cos \phi \cos z \cos A.$$

Hence $\qquad \Delta z = \Delta H \sin A \cos \phi \qquad\qquad$ ......(26),

and $\qquad \Delta A = \Delta H (\sin \phi - \cos \phi \cot z \cos A) \quad$ ......(27).

In these formulae, ΔH, Δz and ΔA are supposed expressed in circular measure. Let ΔH^s denote the number of seconds of time in ΔH radians; let $\Delta z''$, $\Delta A''$ denote the number of seconds of arc in Δz, ΔA radians respectively. Then, by the principles of section 15, p. 22,

$$\Delta z = \Delta z'' \sin 1''; \quad \Delta A = \Delta A'' \sin 1''; \quad \Delta H = \Delta H^s \sin 1^s,$$

and, since $\sin 1^s = 15 \sin 1''$, we have

$$\Delta z'' = 15 \Delta H^s . \sin A \cos \phi,$$
$$\Delta A'' = 15 \Delta H^s (\sin \phi - \cos \phi \cot z \cos A).$$

If $\Delta H^s = 1$ second, these equations express respectively that the zenith distance increases at the rate of $15 \sin A \cos \phi$ seconds of arc per second of time and that the azimuth is decreasing at the rate of $15 [\sin \phi - \cos \phi \cot z \cos A]$ seconds of arc per second of time.

If the heavenly body is a star, the rates of change of zenith distance and of azimuth are expressed in terms of seconds of arc per second of sidereal time; in the case of the sun, the rates are in terms of seconds of arc per second of apparent solar time or, with sufficient accuracy, of mean solar time.

The results just obtained can be easily derived by calculus methods, as follows. From the triangle PZX, by formula **A,**

$$\cos z = \sin \delta \sin \phi + \cos \delta \cos \phi \cos H,$$

in which δ and ϕ are supposed constant. By differentiation

$$\sin z \frac{dz}{dH} = \cos \delta \cos \phi \sin H.$$

By **B,** $\qquad\qquad \sin z \sin A = \sin H \cos \delta \qquad$ ......(28);

$$\therefore \quad \frac{dz}{dH} = \sin A \cos \phi \qquad\qquad ......(29),$$

which is essentially the same as (26). If z and H are expressed in terms of seconds of arc and seconds of time respectively,

$$\frac{dz}{dH} = 15 \sin A \cos \phi.$$

Differentiate (28)—in which z, A and H are variables—with respect to H. Then

$$\sin z \cos A \frac{dA}{dH} = \cos H \cos \delta - \sin A \cos z \frac{dz}{dH}$$

$$= \cos H \cos \delta - \sin^2 A \cos z \cos \phi,$$

by means of (29).

Also, by C,

$$\cos \delta \cos H = \cos z \cos \phi - \sin z \sin \phi \cos A;$$

$$\therefore \sin z \cos A \frac{dA}{dH} = \cos^2 A \cos z \cos \phi - \sin z \sin \phi \cos A;$$

$$\therefore \frac{dA}{dH} = - (\sin \phi - \cot z \cos A \cos \phi),$$

or, if A and H are expressed in seconds of arc and seconds of time respectively, this last formula becomes

$$\frac{dA}{dH} = - 15 (\sin \phi - \cot z \cos A \cos \phi),$$

which is that already derived.

33. *Twilight.*

After the sun has set, indirect sunlight, reflected and scattered by the upper atmosphere, still continues to illumine the earth, diminishing however as the sun sinks farther below the horizon. When the sun is 18° below the horizon (its zenith distance is then 108°) this indirect illumination—evening twilight—ceases. The interval between sunset and the time when the sun's zenith distance has increased to 108° is called the duration of evening twilight. In a similar way, we define the duration of morning twilight. The duration of evening twilight, for example, can be calculated as follows. In Fig. 26, LFM is the sun's parallel of declination (as no great accuracy is required in this particular calculation, we ignore changes in the sun's declination during the particular day concerned) and JGK is a small circle, parallel to the horizon, every point of which is 108° from Z. This small circle intersects the parallel of declination in G. Then the interval of time required for the sun to travel from F to G, that is $F\hat{P}G$, is the duration of evening twilight. Now $F\hat{P}G = Z\hat{P}G - Z\hat{P}F$; as $Z\hat{P}F$ is the hour angle of sunset it can be calculated by formula (24). Now in the triangle ZPG, we have: $ZG = 108°$, $PZ = 90° - \phi$ and $PG = 90° - \delta$; hence, by **A**,

$$\cos 108° = \sin \phi \sin \delta + \cos \phi \cos \delta \cos Z\hat{P}G,$$

which enables the calculation of $Z\hat{P}G$ to be made. The value of δ, used in this formula, depends of course on the particular day of

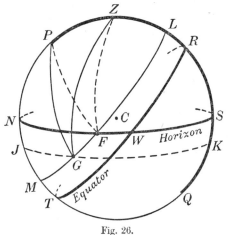

Fig. 26.

the year concerned. Thus the duration of evening twilight is found.

It is clear from Fig. 26 that evening twilight will come to an end if NM is greater than NJ, in other words, if at apparent midnight the sun is more than 18° below the horizon. Now $NT = 90° - \phi$ and $MT = \delta$; $\therefore NM = 90° - \phi - \delta$. Hence evening twilight ends if $90° - \phi - \delta > 18°$, or if $\delta < 72° - \phi$. For example, in latitude 60° N, twilight will end if $\delta < 12°$. When δ is greater than 12°, the sun's zenith distance is less than 108° between sunset and apparent midnight, and also between apparent midnight and sunrise; therefore, in 60° N it is never completely dark on those days of the year when the sun's declination exceeds 12° N. These days are between April 23 and August 22.

EXERCISES

[Symbols used:

ϕ = latitude of observer, z = zenith distance,
A = azimuth of heavenly body, ϵ = obliquity of the ecliptic.]
H = hour angle,

1. If z_1 and z_2 are the zenith distances of a star on the meridian and on the prime vertical respectively, prove that

$$\text{(i)} \quad \cot \delta = \operatorname{cosec} z_1 \sec z_2 - \cot z_1,$$
$$\text{(ii)} \quad \cot \phi = \cot z_1 - \operatorname{cosec} z_1 \cos z_2,$$

where δ is the star's declination. [*Lond.* 1929.]

2. If ψ is the angle which a star's path at rising makes with the horizon, prove that

$$\cos \psi = \sin \phi \sec \delta.$$

3. If h, H are the hour angles of a star, of declination $+ \delta$, on the prime vertical (west) and at setting respectively, for a place in north latitude, show that

$$\cos h \cos H + \tan^2 \delta = 0.$$

Calculate the interval (correct to 0·1 minute of mean solar time) for a place in latitude 36° N between the passage of Aldebaran (declination $+16° 22'$) over the prime vertical (west) and its setting. [*Lond.* 1926.]

4. A boat travelling at 5 knots is steered continually towards a star. Prove that the distance travelled towards the west is approximately $\frac{1}{3} (z_2° - z_1°) \sec \phi$ miles, where $z_1°$ and $z_2°$ are the initial and final zenith distances, in degrees, of the star and ϕ the mean latitude. [*M.T.* 1917.]

5. If the colatitude is C, prove that

$$C = x + \cos^{-1} (\cos z \sec y),$$

where $$\tan x = \cot \delta \cos H,$$
$$\sin y = \cos \delta \sin H,$$

H being the hour angle.

6. Find to the nearest second of mean solar time the interval between the passages across the meridian of two stars whose declinations are 60° N and 60° S, and whose distance apart is $\cos^{-1} (- \frac{5}{8})$. (Assume that 1 year is $365\frac{1}{4}$ days.) [*M.T.* 1923.]

7. If the declination δ of a star is greater than the latitude ϕ, prove that the star's greatest azimuth east or west is

$$\sin^{-1} (\cos \delta \sec \phi).$$

8. At U.T. $21^h 56^m$ on 1927 March 28 a bright star was observed through a break in the clouds as follows: altitude (approximate) 37° 10′; azimuth 136° west. The observer's position was: lat. 50° N, long. 7° 15′ W. Identify the star. (The R.A.M.S. is $0^h 21^m$ approximately.) [*Lond.* 1927.]

3 **SA**

9. The R.A. and declination of Capella at upper transit at Greenwich on 1930 May 30 were $5^h\ 11^m$ and $+45°\ 55'$. Find the altitude and azimuth of the star at the same instant at New York, Columbia University Observatory, Latitude $40°\ 49'$ N, Longitude $4^h\ 56^m$ W.

10. In north latitude $45°$ the greatest azimuth of a circumpolar star is $45°$ (east or west). Prove that the star's declination is $+60°$.

11. If the latitude ϕ and the declination of a star be known, show that the error in the deduced value of the hour angle caused by an erroneous value Δz in the zenith distance is Δz cosec A sec ϕ, where A is the star's azimuth.

12. If the observer increased his latitude by an amount $\Delta\phi$ while the hour angle of a star increased by ΔH, show that the change in altitude is

$$\Delta\phi \cos A - \Delta H \sin A \cos \phi.$$

13. a and $a + \Delta a$ are the altitudes of the sun observed simultaneously at two neighbouring places on the same meridian. If ϕ is the latitude of one of the places and δ is the sun's declination, prove that the difference of latitude between the places is approximately

$$\Delta a \cos a \cos \phi/(\sin \delta - \sin a \sin \phi). \qquad [Ball.]$$

14. Two stars (a, δ) and (a', δ') are observed at the same moment on the same vertical circle. If H is the hour angle of the first star, prove that

$$\cos (\chi + H) = \tan \phi \cos \chi \cot \delta,$$

where χ is given by

$$\tan \tfrac{1}{2} (\alpha - \alpha' - 2\chi) = \frac{\sin (\delta' - \delta)}{\sin (\delta' + \delta)} \cot \tfrac{1}{2} (a' - a).$$

15. If x is the length of the shadow cast on level ground by a vertical pole at apparent noon at an equinox, and if y is the length of shadow cast by the same pole at the summer solstice when the sun is on the prime vertical, show that

$$x = y \tan \psi \tan \phi,$$

where

$$\sin \psi = \sin \epsilon \ \text{cosec}\ \phi. \qquad [Lond.\ 1928.]$$

16. A straight wall of height h runs in the direction θ degrees west of south. Prove that at an equinox the wall casts no shadow when the sun's hour angle H is given by

$$\tan H = \sin \phi \tan \theta,$$

and that at apparent noon the breadth of the shadow is $h \tan \phi \sin \theta$.

17. An observer in latitude $50°$ sees a star set due west behind a low ridge, a mile away, which slopes down to the north at an inclination of $30°$ to the horizontal. Prove that by stepping a yard to his right he will see the star for about 22 seconds longer. $\qquad [M.T.\ 1913.]$

18. Two places are in the same latitude and the polar distance of the great circle through them is equal to the sun's declination. Prove that at these places the length of the night is equal to their difference of longitude.

19. Let a, δ be the co-ordinates of a star with respect to a great circle S, and a', δ' the co-ordinates of the same star with respect to another great circle S'. If i be the inclination of S' to S and if the ascending node of S' on S has co-ordinates $(\theta, 0)$ in the first system and $(\theta', 0)$ in the second, show that

$$\cos \delta' \cos (a' - \theta') = \cos \delta \cos (a - \theta),$$
$$\cos \delta' \sin (a' - \theta') = \sin \delta \sin i + \cos \delta \cos i \sin (a - \theta),$$
$$\sin \delta' = \sin \delta \cos i - \cos \delta \sin i \sin (a - \theta).$$

If $a = 75°$, $\delta = 15°$, $\theta = 215°$, $\theta' = 115°$, $i = 23° \, 30'$, show that from the last equations $a' = 327° \, 12'$, $\delta' = 29° \, 0'$.

20. Show that if a is the altitude of the pole star, H the hour angle and p (in seconds of arc) the polar distance, the latitude is given by

$$\phi = a - p \cos H + \tfrac{1}{2} p^2 \sin^2 H \tan a \sin 1''.$$

21. A heavenly body (declination δ) is at a small angle H from the meridian. Prove that the zenith distance z is given approximately by

$$z = \phi - \delta + a_1 - a_2,$$

where a_1 (expressed in minutes of arc) is given by

$$a_1 = \frac{2 \cos \phi \cos \delta}{\sin (\phi - \delta)} \sin^2 \frac{H}{2} \operatorname{cosec} 1',$$

and

$$a_2 = \tfrac{1}{2} a_1^2 \cot (\phi - \delta) \sin 1'.$$

22. If a is the sun's altitude in the prime vertical at a place in latitude ϕ and L is its longitude, prove that

$$\phi = \sin^{-1} (\sin L \sin \epsilon \operatorname{cosec} a). \qquad [Ball.]$$

23. Prove that, in latitude $45°$, the interval between the moment at which a star's azimuth is $90°$ east and the moment of setting is constant.

24. If δ be a star's declination and A its maximum azimuth, show that in t seconds of time from the moment when the azimuth is A the azimuth has changed by

$$\tfrac{1}{2} \, 15^2 t^2 \sin 1'' \sin^2 \delta \tan A \qquad \text{seconds of arc.}$$

25. If η is the parallactic angle and ϕ and δ are constant, prove that

(i) $\dfrac{d\eta}{dH} = - \cos \phi \cos A \operatorname{cosec} z$;

(ii) $\dfrac{d^2 z}{dH^2} = \dfrac{d\eta}{dH} \cos \delta \cos \eta$;

(iii) $\dfrac{d^2 A}{dH^2} = - \dfrac{\cos \delta}{\sin^2 z} \left(\cos z \cos \eta \, \dfrac{dz}{dH} + \sin z \sin \eta \, \dfrac{d\eta}{dH} \right).$

26. If H is the hour angle of a star at rising, show that

$$\tan^2 \frac{H}{2} = \frac{\cos (\phi - \delta)}{\cos (\phi + \delta)}.$$

27. At a place in north latitude ϕ, two stars A and B (declinations δ and δ_1 respectively) rise at the same moment and A transits when B is setting. Prove that

$$\tan \phi \tan \delta = 1 - 2 \tan^2 \phi \tan^2 \delta_1.$$

28. If two stars (a, δ) and (a_1, δ_1) rise at the same moment at a place in latitude ϕ, show that

$$\cot^2 \phi \sin^2 (a_1 - a) = \tan^2 \delta + \tan^2 \delta_1 - 2 \tan \delta \tan \delta_1 \cos (a_1 - a).$$

[*Ball.*]

29. At a place in latitude ϕ the sun is observed to rise h hours before apparent noon, and the next day it rises m minutes later. Its declination on the first day is δ. Show that the distance in minutes of arc between the two points of rising is

$$15m \cos^2 \delta \operatorname{cosec} \phi. \qquad\qquad [\textit{Coll. Exam.}]$$

30. If evening twilight ends when the sun's centre is $18°$ below the horizon, show that at the equator the duration of evening twilight is given in hours by

$$\frac{12}{\pi} \sin^{-1} (\sin 18° \sec \delta).$$

Use this formula to calculate the duration of evening twilight at the summer solstice. [*Lond.* 1930.]

31. Show that at a place in latitude ϕ the shortest duration of twilight, expressed in hours, is
$$\tfrac{2}{15} \sin^{-1} (\sin 9° \sec \phi),$$

where $\sin^{-1} (\sin 9° \sec \phi)$ is expressed in degrees. [*Ball.*]

32. If twilight begins or ends when the sun is $18°$ below the horizon, show that all places have a day of more than twelve hours, including twilight, so long as the declination of the sun is numerically less than $18°$.

33. If the day is considered to begin and end when the sun is at an angle θ below the horizon, show that the shortest day will not occur at the winter solstice if the latitude is less than ϕ, where

$$\sin \phi = \sin \epsilon \sin \theta,$$

and ϵ is the obliquity of the ecliptic. [*M.T.* 1917.]

34. Assuming that the sun travels uniformly in the ecliptic, completing a revolution in 365 days, show that the number of nights in which there is twilight even at midnight at a place in latitude ϕ is the integer next greater than

$$\tfrac{73}{36} \cos^{-1} \{\cos (\phi + 18°)/\sin \epsilon\},$$

twilight beginning or ending when the sun is $18°$ below the horizon.

[*Coll. Exam.*]

35. If θ denotes the sun's depression below the horizon at the end of evening twilight, and η, η' the parallactic angles at end of twilight and at sunset respectively, prove that the duration (T) of twilight is given by

$$2 \sin^2 \frac{T}{2} \cos^2 \phi = 1 - \cos \theta \cos (\eta' - \eta).$$

36. The right ascension of a star is $5^h 49^m$ and its declination is $+ 7° 23'$, and the obliquity of the ecliptic is $23° 27'$. Show that the longitude and latitude of the star are respectively $87° 10'$, $- 16° 2'$.

37. Two stars (a_1, δ_1) and (a_2, δ_2) have the same longitude; prove that

$$\sin (a_1 - a_2) = \tan \epsilon (\cos a_1 \tan \delta_2 - \cos a_2 \tan \delta_1).$$

38. A star of right ascension a and declination δ has a small latitude β. Prove that the longitude of the sun, when its R.A. is a, differs from the longitude of the star by $\beta \sin \delta \cot a$ approximately.

39. Show that the obliquity of the ecliptic can be determined by making observation of the sun's declination δ at a noon near the summer solstice by means of the formula $\epsilon = \delta + q^2 \sin 2\delta$, where q is one-half the defect from a right angle of the sun's right ascension. [*M.T.* 1924.]

40. The pole of the Milky Way is at R.A. $12^h 48^m$, Dec. $+27°$. About what dates will the sun pass through the Milky Way? (Obliquity of the ecliptic $= 23° 27'$.) [*M.T.* 1925.]

CHAPTER III

REFRACTION

34. *The laws of refraction.*

In astronomical observations, the light from the particular heavenly body observed has to pass through the earth's atmosphere before reaching the observer, and during its passage a ray of light suffers a change in direction, owing to refraction, the amount of which depends on the physical characteristics of the atmosphere and on the altitude (or zenith distance) of the body concerned. It is thus necessary, at the outset, to eliminate from the observations the effects of our terrestrial atmosphere on them. From the study of meteors the deduction is made that the atmosphere extends to a height of at least 100 miles, for even at that height the friction of the air on a rapidly moving meteor is sufficient to render it luminescent. But beyond a height of about 40 miles the air is so tenuous that it has an inappreciable effect on the course of a ray of light.

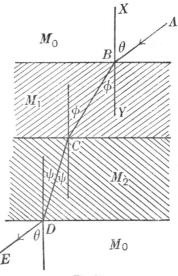

Fig. 27.

We shall first state the laws of refraction.

Consider (Fig. 27) a ray of light AB passing through a transparent medium M_0 (such as air) and falling at B on a slab, with parallel plane faces, of another transparent medium M_1 (such as glass). In the medium M_1 the path of the ray will be along some such line as BC, different in direction from AB; the ray is said to be *refracted* at B. Let YBX be perpendicular to the slab at B. The angle ABX (denoted by θ) is called the angle of incidence and the angle YBC (denoted by ϕ)

the angle of emergence. The laws of refraction are: (i), the incident ray AB, the refracted ray BC and the perpendicular at B to the surface separating the two media M_0 and M_1 lie in the same plane; (ii), the relation between ϕ and θ is

$$\frac{\sin \theta}{\sin \phi} = \mu_1 \qquad \qquad \dots\dots(1),$$

μ_1 being a constant depending on the optical properties of the two media concerned. In this instance, μ_1 is called the *index of refraction* for the two media M_0 and M_1; its value can be determined by laboratory experiment.

Let the ray BC now pass from the medium M_1 into the medium M_2. The angle of incidence is now ϕ and the angle of emergence is ψ. Then

$$\frac{\sin \phi}{\sin \psi} = \mu \qquad \qquad \dots\dots(2),$$

where μ is the refractive index for the two media M_1 and M_2. At D, we shall suppose that the ray emerges into the medium M_0. Its path DE in M_0 is parallel to its original direction AB in the same medium. The path of the ray is reversible, that is to say, a ray in the direction ED will be refracted, at the surface between the media M_0 and M_2, along DC. Hence

$$\frac{\sin \theta}{\sin \psi} = \mu_2 \qquad \qquad \dots\dots(3),$$

where μ_2 is the index of refraction between the media M_0 and M_2. From (1) and (3), we have by division

$$\frac{\sin \phi}{\sin \psi} = \frac{\mu_2}{\mu_1},$$

or $\qquad\qquad\qquad \mu_1 \sin \phi = \mu_2 \sin \psi \qquad \qquad \dots\dots(4).$

It follows from (2) that $\mu = \mu_2/\mu_1$.

Regarding M_0 as a standard medium, we can define μ_1 simply as the refractive index for medium M_1 and μ_2 for medium M_2, and the values of μ_1 and μ_2 may be supposed known. Consider now only the two media M_1 and M_2 in Fig. 27, BC being a ray in M_1 incident at C on the bounding surface between M_1 and M_2 and CD the ray in M_2. ϕ is the angle of incidence, ψ the angle of emergence, and the relation between ϕ and ψ is given by formula (4).

35. *Refraction for small zenith distances.*

As the density of the air diminishes with increasing height above the earth's surface, it is convenient to regard the atmosphere as made up of a large number of thin spherical layers, concentric with the earth's surface regarded as spherical, throughout each of which the density and other physical characteristics are uniform. The simplest case in the investigation of astronomical refraction occurs when the heavenly body observed—for example, a star—is nearly overhead; in this

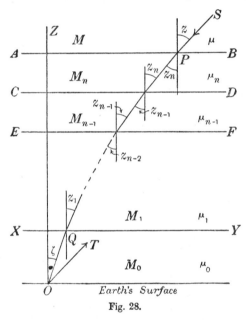

Fig. 28.

instance we can ignore the curvature of the atmospheric strata through which the rays from the star pass and thus we can regard their bounding surfaces simply as a series of parallel planes. Let there be $n + 1$ parallel layers (of which only a few are shown in Fig. 28) and let AB be the upper effective limit of the atmosphere beyond which the air, owing to its extreme tenuity, is ineffective in causing refraction. Each layer has its own optical properties and, in particular, its own index of refraction. Let the medium above AB be denoted by M, the layer between CD and AB by M_n, and so on, with the corresponding refractive indices μ, μ_n, μ_{n-1}, ... μ_0, the last being the refractive index for the lowest layer M_0. We can regard M as the standard medium —effectively, it corresponds to a vacuum—and so we can put $\mu = 1$. Let z denote the angle of incidence, at the surface AB, of a ray from the star that finally reaches the observer at O. Then z is called the true zenith distance of the star; if there were no atmospherical refraction the star would be seen by the observer in the direction OT, which is parallel to PS. Applying formula (4) we have for the successive pairs of layers the following series of equations (the notation is indicated in Fig. 28):

$$\mu \sin z = \mu_n \sin z_n,$$

or, since $\mu = 1$,
$$\sin z = \mu_n \sin z_n,$$
$$\mu_n \sin z_n = \mu_{n-1} \sin z_{n-1};$$
$$\mu_{n-1} \sin z_{n-1} = \mu_{n-2} \sin z_{n-2},$$
$$\vdots \qquad \qquad \vdots$$
$$\mu_1 \sin z_1 = \mu_0 \sin \zeta,$$

in the last of which ζ is the angle between the direction of the zenith OZ and the final element OQ in the path of the ray. ζ is thus the *observed zenith distance* of the star. From these equations we have, clearly,
$$\sin z = \mu_0 \sin \zeta \qquad \qquad \ldots\ldots(5);$$

μ_0 is the index of refraction of the air at the earth's surface. From the first law of refraction it is evident that the path of the ray through the various strata lies in a vertical plane. The values of the refractive index increase from μ in M continuously to μ_0 in M_0, corresponding to the increase in the density of the atmospherical layers from M downwards; accordingly, it follows from (4) that the angles z, z_n, z_{n-1}, ... z_1, ζ form a decreasing sequence and the path of the ray is thus bent in the way indicated in the diagram. In particular ζ is less than z, that is to say, the star is observed *nearer* the zenith than it would be if the atmosphere were non-effective as regards refraction. The angle $z - \zeta$ is called the *angle of refraction*; denote it by R. Then (5) becomes
$$\sin (\zeta + R) = \mu_0 \sin \zeta,$$

or
$$\sin \zeta \cos R + \cos \zeta \sin R = \mu_0 \sin \zeta.$$

Now R is a small angle and we can write $\cos R = 1$ and $\sin R = R$ (R being supposed expressed in circular measure). Thus
$$\sin \zeta + R \cos \zeta = \mu_0 \sin \zeta,$$

or
$$R = (\mu_0 - 1) \tan \zeta \qquad \qquad \ldots\ldots(6).$$

We thus have the result that, *at small zenith distances*, the angle of refraction is proportional to the tangent of the observed zenith distance. Now μ_0 is the index of refraction of the air at the surface of the earth, and its value will be dependent on the density and temperature at any given time. The standard conditions are taken in practice to be: barometric height = 30 inches and temperature = 50° F.; the refraction is then called *mean refraction*. For these conditions, $\mu_0 - 1$ is approximately 0·00029, so that R is approximately 0·00029 tan ζ or, in seconds of arc, $206265 \times 0\cdot00029$ tan ζ. The coefficient of tan ζ is more accurately determined by means of astronomical observations and the value usually adopted is $58''\cdot2$; we then have

$$R = 58''\cdot2 \tan \zeta.$$

The coefficient of tan ζ is called the *constant of mean refraction*; denote it by k. Then
$$R = k \tan \zeta \qquad \ldots\ldots(7).$$

At any barometric pressure P (in inches) and temperature T (in degrees Fahrenheit), the corresponding refraction R' in terms of the mean refraction R is given by

$$\frac{R'}{R} = \frac{17P}{460 + T} \qquad \ldots\ldots(8).$$

For many purposes, the formula (7) is sufficiently accurate for zenith distances not exceeding 45°.

36. *General formula for refraction.*

When the zenith distance of the body observed is considerable, the atmosphere through which the rays pass can no longer be regarded as stratified in plane layers. Assume that the earth is spherical and that the atmosphere is arranged in spherical layers. In Fig. 29, let C be the centre of the earth, O the observer and COZ the direction of his zenith. Let μ', μ be the indices of refraction in two adjacent thin layers M' and M. Let LP be the section of a ray in M' which finally reaches the observer at O. At P it is refracted along PQ. Similarly, it is refracted at the surfaces between successive layers and the final element of its path is TO. If the layers are thin, the path of the ray is curved and the direction in which the observer sees the object is along OT, the tangent to this curve at O. The observed zenith distance is thus $Z\hat{O}T$, denoted by ζ. Draw the radii CP and CE. Let $GPL = \phi'$, $QPF = \psi$, $EQP = \phi$. Then, since the radius CP is perpendicular at P to the bounding surface between the layers M' and M, by the laws of refraction we have

$$\mu' \sin \phi' = \mu \sin \psi \qquad \qquad \ldots \ldots (9).$$

Now from the triangle CQP, in which $CP = r'$ and $CQ = r$ and $CQP = 180° - \phi$, we have

$$r \sin \phi = r' \sin \psi \qquad \qquad \ldots \ldots (10).$$

Eliminate $\sin \psi$ from (9) and (10); then

$$r'\mu' \sin \phi' = r\mu \sin \phi \qquad \qquad \ldots \ldots (11).$$

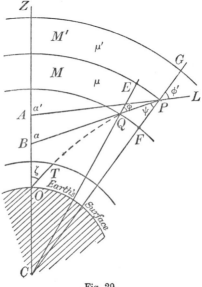

This is a general relation which holds for any two contiguous layers and consequently for any two layers whatever their heights above the earth's surface may be. If r_0, μ_0, ϕ_0 denote the values of r, μ and ϕ for the lowest layer—next the earth's surface—we have, from (11),

$$r\mu \sin \phi = r_0 \mu_0 \sin \phi_0.$$

But $r_0 = a$, the earth's radius, and ϕ_0 is simply the angle ZOT or ζ, the observed zenith distance. Hence

$$r\mu \sin \phi = \mu_0 a \sin \zeta \qquad \ldots \ldots (12).$$

Fig. 29.

Consider now the angle through which the ray is deviated in its passage from one layer through the next. Produce LP and PQ to meet OZ in A and B respectively, and let $Z\hat{A}P$ and $Z\hat{B}P$ be denoted by α' and α. Then the angle through which the ray is refracted at P is $A\hat{P}B$ or $\alpha' - \alpha$; let it be denoted by ΔR, so that

$$\Delta R = \alpha' - \alpha \qquad \qquad \ldots \ldots (13).$$

Let $A\hat{C}P = \theta'$ and $B\hat{C}Q = \theta$, and let

$$\Delta\theta = \theta' - \theta \qquad \qquad \ldots \ldots (14).$$

Then, assuming that the layers are thin, we can write

$$QF = EP = r\Delta\theta.$$

Also if $r' = r + \Delta r$, then $QE = \Delta r$. In the infinitesimal triangle EQP right-angled at E, we have

$$\tan \phi = \frac{EP}{QE} = \frac{r\Delta\theta}{\Delta r} \qquad \ldots\ldots(15).$$

Now, $\alpha' = \theta' + \phi'$ and $\alpha = \theta + \phi$; hence

$$\Delta R \equiv \alpha' - \alpha = (\phi' - \phi) + (\theta' - \theta),$$

or, if $\Delta\phi$ denotes $\phi' - \phi$,

$$\Delta R = \Delta\phi + \Delta\theta \qquad \ldots\ldots(16).$$

Now from (11), writing $\mu' = \mu - \Delta\mu$ (μ decreases as r increases), we obtain

$$(r + \Delta r)\,(\mu - \Delta\mu)\sin(\phi + \Delta\phi) = r\mu \sin \phi,$$

or $\qquad (r + \Delta r)\,(\mu - \Delta\mu)\,(\sin \phi + \Delta\phi \cos \phi) = r\mu \sin \phi,$

since $\Delta\phi$ is a small angle.

Omitting products of the infinitesimal quantities Δr, $\Delta\mu$, $\Delta\phi$ we obtain, after dividing throughout by $r\mu \sin \phi$,

$$\frac{\Delta r}{r} - \frac{\Delta\mu}{\mu} + \Delta\phi \cot \phi = 0 \qquad \ldots\ldots(17).$$

But, by (15), $\qquad \dfrac{\Delta r}{r} = \Delta\theta \cot \phi.$

Hence (17) becomes

$$(\Delta\theta + \Delta\phi)\cot \phi - \frac{\Delta\mu}{\mu} = 0$$

and, by (16),

$$\Delta R = \frac{\Delta\mu}{\mu}\tan \phi \qquad \ldots\ldots(18).$$

Now (12) enables us to express $\tan \phi$ in terms of the variables μ and r and the constants a, μ_0, ζ. When this is done, (18) becomes

$$\Delta R = \frac{\Delta\mu}{\mu} \cdot \frac{a\mu_0 \sin \zeta}{(r^2 \mu^2 - a^2 \mu_0^2 \sin^2 \zeta)^{\frac{1}{2}}} \qquad \ldots\ldots(19).$$

This equation expresses the amount of refraction suffered by a ray in passing from one spherical layer with index of refraction $\mu - \Delta\mu$ to the next lower layer with index of refraction μ. The total refraction R due to the whole atmosphere is given by

$$R = a\mu_0 \sin \zeta \int_1^{\mu_0} \frac{d\mu}{\mu\,(r^2 \mu^2 - a^2 \mu_0^2 \sin^2 \zeta)^{\frac{1}{2}}} \qquad \ldots\ldots(20),$$

the limits of integration being μ_0 at the earth's surface and unity at the highest layer.

Formula (20) is the general expression for refraction. It is to be noted that the integral involves two variables, r and μ. Now the index of refraction of any layer is dependent on the physical characteristics of that layer which, in turn, will depend on the height of the layer above the earth's surface, that is, in effect, on r. Before the integral in (20) can be rigorously calculated the relation between μ and r must be specified. This involves the application of physical laws embracing the pressure, density and temperature of the air. Unfortunately, our knowledge of the physical state of the upper atmosphere is insufficient to indicate the precise dependence of μ on r, and we are thus forced to treat equation (20) by approximate methods.

37. *Development of the general formula for refraction.*

The height of the atmosphere is small in comparison with the radius of the earth, and if we write

$$\frac{r}{a} = 1 + s \qquad \qquad \dots\dots(21)$$

we can regard s as a small quantity; it varies from zero at the earth's surface to about 0·01 at a height of 40 miles, which may be regarded as the limit beyond which the air is ineffective in producing any appreciable refraction.* Using (21) and neglecting terms in s^2, s^3, etc., we can write

$$(r^2\mu^2 - a^2\mu_0^2 \sin^2 \zeta)^{-\frac{1}{2}} = \frac{1}{a} (\mu^2 - \mu_0^2 \sin^2 \zeta + 2s\mu^2)^{-\frac{1}{2}}$$

$$= \frac{1}{a} (\mu^2 - \mu_0^2 \sin^2 \zeta)^{-\frac{1}{2}} \left(1 - \frac{s\mu^2}{\mu^2 - \mu_0^2 \sin^2 \zeta}\right).$$

Hence (20) becomes

$$R = \mu_0 \sin \zeta \int_1^{\mu_0} \frac{d\mu}{\mu (\mu^2 - \mu_0^2 \sin^2 \zeta)^{\frac{1}{2}}} - \mu_0 \sin \zeta \int_1^{\mu_0} \frac{s\mu\, d\mu}{(\mu^2 - \mu_0^2 \sin^2 \zeta)^{\frac{3}{2}}}$$

$$\equiv \qquad\qquad R_1 \qquad\qquad - \qquad\qquad R_2 \qquad \dots\dots(22).$$

The expansion by the binomial theorem is only valid if $2s\mu^2$ is small compared with $(\mu^2 - \mu_0^2 \sin^2 \zeta)$, and since μ_0 and μ are approximately unity it is invalid when the observed heavenly

---

* The earth's mean radius is 3960 miles.

body is on or near the horizon (then ζ is equal to or close to 90° and $\sin \zeta$ is equal to or close to unity). Consider first the term R_1 in (22). It is of the form

$$\alpha \int \frac{d\mu}{\mu \, (\mu^2 - \alpha^2)^{\frac{1}{2}}},$$

which is a well-known integral whose value is $-\sin^{-1}\dfrac{\alpha}{\mu}$. Hence, inserting the limits μ_0 and 1, we obtain

$$R_1 = \sin^{-1} (\mu_0 \sin \zeta) - \zeta.$$

Now μ_0—the index of refraction of air at the earth's surface— is a little greater than unity; denote it by $(1 + x)$, where x is a small quantity. Then

$$R_1 \equiv f(x) = \sin^{-1} [(1 + x) \sin \zeta] - \zeta,$$

and by Maclaurin's theorem, terms in x^2, x^3, etc. being neglected,

$$R_1 \equiv f(x) = f(0) + x \left(\frac{df}{dx}\right)_{x=0}.$$

Now
$$f(0) = \sin^{-1} (\sin \zeta) - \zeta = 0,$$

and
$$\frac{df}{dx} = \frac{\sin \zeta}{\{1 - (1 + x)^2 \sin^2 \zeta\}^{\frac{1}{2}}},$$

so that
$$\left(\frac{df}{dx}\right)_{x=0} = \tan \zeta.$$

Hence to the degree of approximation indicated

$$R_1 = x \tan \zeta,$$

or
$$R_1 = (\mu_0 - 1) \tan \zeta \qquad \ldots\ldots(23),$$

which, on reference to section 35, is seen to be the result obtained for small zenith distances.

Consider now the second integral in (22), namely,

$$R_2 = \mu_0 \sin \zeta \int_1^{\mu_0} \frac{s\mu \, d\mu}{(\mu^2 - \mu_0^2 \sin^2 \zeta)^{\frac{3}{2}}}.$$

In this expression, we regard s as a small quantity varying, as we have seen, from 0 to about 0·01. Now, by Gladstone and Dale's law, we can express μ in terms of the atmospheric density ρ by
$$\mu = 1 + c\rho \qquad \ldots\ldots(24),$$

where c is a constant with a numerical value 0·226. For the air at the earth's surface (density ρ_0)

$$c\rho_0 = 0\cdot00029.$$

From (24), $d\mu = c\,d\rho,$

an equation which expresses the change in refractive index with atmospheric density. We then have

$$R_2 = c\mu_0 \sin \zeta \int_0^{\rho_0} \frac{s\mu\,d\rho}{(\mu^2 - \mu_0^2 \sin^2 \zeta)^{\frac{3}{2}}} \quad \ldots\ldots(25).$$

In this integral, since μ and μ_0 are very nearly unity, and s is small, the value of R_2 will be little affected if we write $\mu = \mu_0 = 1$ in (25). Then

$$R_2 = c \sin \zeta \int_0^{\rho_0} \frac{s\,d\rho}{(1 - \sin^2 \zeta)^{\frac{3}{2}}}$$

$$= c \tan \zeta \sec^2 \zeta \int_0^{\rho_0} s\,d\rho \quad \ldots\ldots(26).$$

Now

$$\int_0^{\rho_0} s\,d\rho = [s\rho]_0^{\rho_0} - \int_{s'}^0 \rho\,ds$$

$$= [s\rho]_0^{\rho_0} + \int_0^{s'} \rho\,ds \quad \ldots\ldots(27),$$

in which s' denotes the value of s at the highest atmospheric level where, of course, $\rho = 0$. Now $s = 0$ when $\rho = \rho_0$, and it follows that $[s\rho]_0^{\rho_0} = 0$.

Now

$$a \int_0^{s'} \rho\,ds = \int_0^{s'} \rho\,d\,(as),$$

and the second integral is the expression for the mass of a column of air, of unit cross section, extending from the earth's surface to the effective limit of the atmosphere; it is therefore independent of the actual law according to which atmospheric density changes with height. The mass of a column of air is, however, related to temperature and barometric pressure, and if we write

$$R_2 = -B \tan \zeta \sec^2 \zeta \quad \ldots\ldots(28),$$

the quantity $-B$ (which is $\dfrac{c}{a} \times$ mass of the column considered) must be regarded as dependent on temperature and pressure. Combining (23) and (28) we obtain the expression for the refraction in the form

$$R = (\mu_0 - 1) \tan \zeta + B \tan \zeta (1 + \tan^2 \zeta),$$

or

$$R = A \tan \zeta + B \tan^3 \zeta \quad \ldots\ldots(29),$$

in which A has been written for $(\mu_0 - 1) + B$.

Instead of calculating the values of A and B from the physical data concerned, it is preferable to assume that the refraction can be expressed by a formula of the type (29), and to derive the values of the coefficients A and B from observations of stars. The numerical expression for the mean refraction (for barometric height 30 inches and temperature 50° F.) is

$$R = 58''\cdot294 \tan \zeta - 0''\cdot0668 \tan^3 \zeta \quad \ldots\ldots(30).$$

The approximations which we introduced in deriving formula (29) are insufficient when the zenith distance exceeds 75° approximately. For observations made near the horizon, special tables of refraction have been prepared (based mainly on observational data) and are in use in observatories where an accurate knowledge of the refraction is essential. Among these may be mentioned the Pulkovo Tables (1870) and the Greenwich Tables (1898).

38. *The determination of the constants* A *and* B *in the formula* (29).

In the next chapter, we shall consider in some detail the instrument by which the zenith distances of heavenly bodies at upper or at lower culmination can be measured; meanwhile we shall take for granted the practical results of the method. Consider a star at upper and lower culmination. At upper culmination it is observed at X_1; the displacement due to refraction is XX_1 (Fig. 30). $ZX_1 = \zeta$ which is obtained from the observation, and ZX is the true zenith distance z, so that

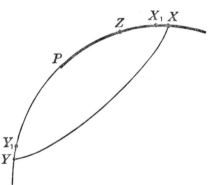

Fig. 30.

$$z = (90° - \delta) - (90° - \phi),$$

or $z = \phi - \delta$. Now $z = \zeta + R$, where R is the refraction corresponding to the observed zenith distance ζ; hence by (29),

$$\phi - \delta = \zeta + A \tan \zeta + B \tan^3 \zeta \quad \ldots\ldots(31).$$

Let ζ' be the observed zenith distance at lower culmination. Then $\zeta' = ZY_1$ and the true zenith distance ZY or z' is given by

$z' = (90° - \phi) + (90° - \delta)$ or $z' = 180° - \phi - \delta$. Also $z' = \zeta' + R'$; hence we have

$$180° - \phi - \delta = \zeta' + A \tan \zeta' + B \tan^3 \zeta' \quad \ldots\ldots(32).$$

If the values of ϕ and δ are known accurately, we have two equations (31) and (32) in which A and B are the only unknowns, for ζ and ζ' have been determined by observation. If we suppose that δ is known, we can eliminate ϕ from (31) and (32) and so obtain

$$180° - 2\delta = \zeta + \zeta' + R + R' \quad \ldots\ldots(33).$$

Observations of another star will lead to an equation similar in form to (33), and from this equation and (33) the values of A and B can be determined. As no observation is free from error, such errors are incorporated on the right-hand side of (33), thus vitiating slightly the deduced values of A and B. To reduce the errors in A and B to a minimum, a large number (n) of stars are observed and the n equations of the type of (33) are solved for A and B by the method of "least squares".

39. *The effect of refraction on the time of sunset.*

When a heavenly body is on the horizon at rising or setting —the zenith distance is then 90°—the numerical value of the refraction is 35' (this is called the *horizontal refraction*). Now the effect of refraction is to make the body appear nearer the zenith than it would be seen if the atmosphere were non-existent or ineffective in deviating rays of light. Hence it follows that at setting, for example, when the observed zenith distance ζ is 90°, the true zenith distance z of the heavenly body is 90° + horizontal refraction, or 90° 35'. If we take the case of the sun, it is clear that the time of visible sunset is *later* than the time of theoretical sunset, which is discussed in section 31 of Chapter II. The interval between theoretical and visible sunset is easily found. Let H be the hour angle when the true zenith distance of the sun's centre is 90° and let $H + \Delta H$ be the hour angle when the sun's centre is seen on the horizon. Then $\zeta = 90°$ and $z = 90°\ 35'$. Then we have, as in section 31,

$$\cos H = -\tan \phi \tan \delta \quad \ldots\ldots(34),$$

and also, by the cosine formula **A**,

$$\cos (90°\ 35') = \sin \phi \sin \delta + \cos \phi \cos \delta \cos (H + \Delta H),$$

which may be written

$$- \sin 35' = \sin \phi \sin \delta + \cos \phi \cos \delta (\cos H \cos \Delta H - \sin H \sin \Delta H).$$

As 35′ and ΔH are small angles, we can reduce the last equation to

$$- 35 \sin 1' = \sin \phi \sin \delta + \cos \phi \cos \delta \cos H$$
$$- 15 \Delta H \sin 1' . \cos \phi \cos \delta \sin H,$$

in which ΔH is now supposed to be expressed in minutes of time. Using (34), we find that

$$\Delta H = \tfrac{35}{15} \sec \phi \sec \delta \operatorname{cosec} H \quad \text{minutes} \quad \dots\dots (35).$$

Let us take a simple example. Suppose the latitude is 60° and $\delta = 0°$ (about March 21 or about September 21). H being the hour angle of theoretical sunset is easily seen to be 6^{h}. Hence from (35), by calculation, $\Delta H = \tfrac{14}{3}$ minutes.

The sun's declination δ which we have introduced into the formulae is the declination of the sun's centre, and the zenith distances concerned are the zenith distances of the sun's centre. To find the hour angle when the sun's upper limb just disappears below the horizon, we notice that the true zenith distance of the sun's centre then is 90° + 35′ + the angle subtended by a radius of the sun. The last quantity is the sun's *semi-diameter*, which is tabulated in the almanacs for each day of the year and, for the purposes under consideration, may be taken to be 16′; thus the true zenith distance of the sun's centre under the circumstances now contemplated is 90° 51′. If ΔH now denotes the interval between the time of theoretical sunset and the time of the disappearance of the sun's upper limb below the horizon, ΔH is given by $\quad \Delta H = \tfrac{51}{15} \sec \phi \sec \delta \operatorname{cosec} H \quad$ minutes.

Hence in latitude 60° when the sun's declination is zero, $\Delta H = 6^{\text{m}}\cdot 8$. There is a similar interval between visible sunrise and theoretical sunrise. The effect of refraction is thus to increase the length of the "day" (by the "day" is here meant the interval during which some part of the sun is above the horizon) by about 13$\frac{1}{2}$ minutes at the latitude and at the dates indicated.

40. *Effect of refraction on the right ascension and declination of a star.*

Consider the position X of a star on the celestial sphere (Fig.31). It is displaced towards the zenith Z by refraction to the position X'. Through X' draw a small circle of which P is the pole to cut PX in Y. Since XX' is small, we can consider $XX'Y$ to be a plane triangle right-angled at Y. Denote the hour angle ZPX and the north polar distance PX of X by H and $90° - \delta$ respectively. X' is the observed position of the star and let ZPX' and PX' be denoted by H' and $90° - \delta'$. It can be supposed that an observation of the star enables the values of H' and δ'

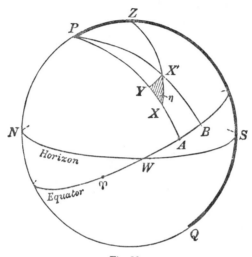

Fig. 31.

to be derived; it is required to find the values of H and δ. From the plane triangle $XX'Y$ in which η denotes the angle $X'XY$, we have

$$X'Y = XX' \sin \eta \qquad \ldots\ldots(36),$$

and

$$XY = XX' \cos \eta \qquad \ldots\ldots(37).$$

Now $\qquad X'Y = X'\hat{P}Y \sin PX' = (H - H') \cos \delta'.$

Also $\qquad\qquad XY = \delta' - \delta.$

Using the simple expression for the refraction in formula (7), we have $XX' = k \tan \zeta$, where $ZX' = \zeta$. Hence (36) and (37) become

$$H - H' = k \tan \zeta \sec \delta' \sin \eta \qquad \ldots\ldots(38),$$

$$\delta - \delta' = -k \tan \zeta \cos \eta \qquad \ldots\ldots(39).$$

In these equations η is the parallactic angle PXZ; as XX' is small, $PX'Z$ will differ little from PXZ and so η may be regarded as defined by $PX'Z$. Thus η can be calculated from the data by means of the cosine formula **A** (we assume $PZ \equiv 90° - \phi$ to be known). Since we have supposed that H' and δ' can be derived from the observation of the star, the right-hand sides of (38) and (39) can be calculated; consequently H and δ can be found. Now if Υ is the vernal equinox the right ascension of the point X is ΥA and of X', ΥB. If α, α' denote the right ascensions of X and of X', $\alpha' - \alpha = AB = H - H'$. Hence

$$\alpha - \alpha' = - k \tan \zeta \sec \delta' \sin \eta \qquad \ldots\ldots(40).$$

(40) and (39) thus enable us to calculate the true right ascension and declination of the observed body.

EXERCISES

1. Find the approximate north latitude where the effect of refraction, at a time when the sun's declination is $10°$ S, is to lengthen the day by 15 minutes. (The horizontal refraction is $35'$.)

2. Assuming that the atmosphere is homogeneous and of height h (*Cassini's hypothesis*), prove that

$$\sin R = \frac{a \sin \zeta}{a + h} (\mu^2 - 2\mu \cos R + 1)^{\frac{1}{2}},$$

where a is the radius of the earth.

3. If the relation between r and μ is $r\mu^{n+1} = \text{constant}$ (*Simpson's hypothesis*), prove that

$$R = \frac{1}{n} \left\{ \zeta - \sin^{-1} \left(\frac{\sin \zeta}{\mu_0^n} \right) \right\},$$

and deduce *Bradley's formula*

$$R = \frac{2}{n} \frac{\mu_0^n - 1}{\mu_0^n + 1} \tan (\zeta - \tfrac{1}{2}nR).$$

4. Assuming that the formula for refraction is $R = k \tan \zeta$, prove that the circular disc of the sun appears, due to refraction, as an ellipse whose semi-major and semi-minor axes are $a(1 - k)$ and $a(1 - k \sec^2 z)$, where k is expressed in circular measure, ζ is the observed zenith distance of the sun's centre, z is the true zenith distance and a is the sun's semi-diameter.

5. The mean of any two perpendicular diameters of the sun is observed to be D. If z is the true zenith distance of the sun's centre, show that the true angular diameter of the sun is

$$D \left\{ 1 + \frac{k}{2}(1 + \sec^2 z) \right\},$$

where k is expressed in circular measure.

6. If the declination of a star is unaffected by refraction at a given moment, prove that the azimuth is then a maximum.

7. X_1 and X_2 are two neighbouring stars, the true angular distance between them being D (in seconds of arc). If Z is the zenith, $Z\hat{X_1}X_2 = \psi$ and the true zenith distance of X_1 is z, prove that the observed angular distance is

$$D - kD\,(1 + \cos^2\psi\,\tan^2 z),$$

in which k is expressed in circular measure.

8. If ϕ is the latitude, H the hour angle and δ the declination of a star, show that refraction diminishes the apparent rate of change of hour angle at the rate of

$$0^{\text{s}}\!\cdot\!51\,\frac{\sin 2\theta}{\sin^2(\delta + \theta)}\,(\tan\delta + \cot\phi\sec H)\ \text{per hour,}$$

where $\tan\theta = \cot\phi\cos H$. (The constant of refraction $= 58''\!\cdot\!2$.)

Show also that the rate of change of refraction in declination is

$$+\,15''\!\cdot\!2\,\cot\phi\,\sin H\,\cos^2\theta\,\operatorname{cosec}^2(\delta + \theta)\ \text{per hour.}$$

CHAPTER IV

THE MERIDIAN CIRCLE

41. *General description.*

In this chapter we shall consider some of the principal features of the fundamental instrument of astronomy—the meridian circle—by which the right ascensions and declinations of the principal heavenly bodies can be determined with great precision. Meridian circle observations also provide the information by which sidereal clocks can be regulated (in practice it is sufficient to derive the error of a sidereal clock), and after a simple step the true mean time at any instant can be deduced so that the error of a clock keeping mean solar time can be easily found. The instrument consists primarily of a refracting telescope which can be rotated about a fixed horizontal axis (the rotation axis) oriented east and west. The telescope itself can thus move only in the plane of the meridian. Fig. 32 shows the main features of the instrument. In the focal plane of the object-glass are two systems of spider threads, or "wires"—as they are generally called—(a) one or sometimes two horizontal wires (a single horizontal wire is shown in Fig. 33), and (b) several wires, at right angles to the horizontal wire, arranged symmetrically about a central wire AB; these wires will be referred to as the vertical wires. In some instruments, attached to the plate carrying the wires are two micrometers, one capable of moving the horizontal wire (or wires) parallel to HK, the other

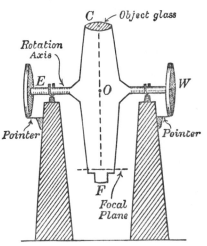

Fig. 32.

capable of moving the system of vertical wires parallel to AB. In other instruments the system of vertical wires is fixed, but there is an additional vertical wire which can be moved by a micrometer across the system of fixed wires.

The plane through the centre C of the object-glass and perpendicular to the rotation axis EW is called the *collimation plane*, and the straight line through C lying on this plane and intersecting the rotation axis is called the *collimation axis* (CO in Fig. 32).

As the telescope rotates about the axis EW, the collimation axis will clearly sweep out the collimation plane. Assume, for a moment, that the instrument is mechanically perfect and that it has been set up accurately, with the central wire AB in the collimation plane; then, as viewed through the eye-piece at F, any star which, at a given instant, is observed on the central wire will, at that moment, be on the meri-

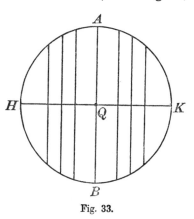

Fig. 33.

dian. If this instant is noted by means of a sidereal clock keeping accurate sidereal time, this sidereal time (otherwise described as the hour angle of the vernal equinox) is clearly equivalent to the star's right ascension. The function of the additional vertical wires in Fig. 33 is to give additional accuracy; for, as the star moves across the field of view owing to the diurnal motion, it coincides with the wires in succession and, if the coincidences with each of the vertical wires are noted, the mean will presumably give a much more accurate determination of the sidereal time of transit—and hence the star's right ascension—than if the observation is restricted to the single coincidence with the central wire.

Attached to the axis of rotation are two finely graduated circles (shown at E and W in Fig. 32), with auxiliary optical arrangements for enabling an accurate measurement of a star's altitude to be made; as the latitude of the meridian circle may be presumed known, the star's declination can then be simply deduced. Such then, in brief, are the main principles according to which the positions of the stars are obtained from meridian circle observations.

42. *Instrumental errors.*

It is impossible, however, to set up an instrument with the precision necessary for the accurate measurement of stellar positions, as just indicated, and it is necessary in actual practice to take into consideration the inherent errors of the instrument. These errors are (i) *azimuth error*—the axis of rotation is not accurately oriented east and west and its angular deviation from the true east and west direction is the azimuth error (sometimes called the deviation error), which will be denoted by a; (ii) *level error*—the axis of rotation is not accurately horizontal and the angular deviation from the horizontal is called the level error, which will be denoted by b; (iii) *collimation error*—the central wire is not quite in the collimation plane, and the angle between the collimation axis and the line joining the middle point of the vertical wire to the centre of the object-glass is called the collimation error (denoted by c). We shall examine in turn the effect of these errors on the observed time of transit of a star.

43. *Azimuth error.*

Assume that this is the only error. Fig. 34 shows the celestial sphere with the horizon and equator and the cardinal points; the centre of the sphere is taken to be the point of intersection of the collimation axis with the rotation axis. If the instrument were perfectly set up, one end of the axis would point to the west point W and the other to the east point E. We shall suppose, with azimuth error presumed, that the west end of the rotation axis points towards the point A of the horizon, WA being the azimuth error; the convention is that the azimuth error is positive when A is between W and S as in Fig. 34. As the telescope is rotated about its axis, the collimation axis will describe a plane perpendicular to the rotation axis and cutting the celestial sphere in a great circle of which A is the pole: this great circle we shall designate the "fictitious meridian". Since A is a point on the horizon, the fictitious meridian will pass through Z. Let X be the position of a star at the moment of its coincidence with the central wire (which is in the collimation plane); then ZX is an arc of the fictitious meridian. Let τ_1 denote the angle ZPX; it is seen from the figure that since X is the position of the star when it is observed on the central wire, the true time of transit will occur later, after an interval τ_1, when the star reaches the true meridian at Y. τ_1 is thus the error in the time of transit of the star.

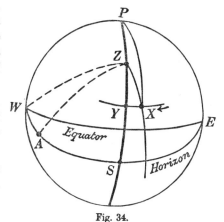

Fig. 34.

Now $W\hat{Z}A \equiv WA = a$, since Z is the pole of $WASE$. Also $W\hat{Z}S = 90°$ and $A\hat{Z}X = 90°$, since A is the pole of the great circle ZX (the fictitious meridian). Hence

$$Y\hat{Z}X = a,$$

and therefore $P\hat{Z}X = 180° - a$. From the triangle PZX in which $PZ = 90° - \phi, PX = 90° - \delta, Z\hat{P}X = \tau_1$ and $P\hat{Z}X = 180° - a$, we have by formula **C** (the sine-formula),

$$\sin PZX \sin ZX = \sin ZPX \sin PX,$$

or $$\sin a \sin ZX = \sin \tau_1 \cos \delta.$$

Also, a and τ_1 are small angles so that, in general, ZX is approximately equal to ZY, that is, to $(\phi - \delta)$; also $\sin a = a$, $\sin \tau_1 = \tau_1$, and consequently,

$$\tau_1 = a \sin (\phi - \delta) \sec \delta \qquad \ldots\ldots(1),$$

in which both τ_1 and a may be supposed expressed in seconds of time. This formula gives the correction to the observed time of transit due to azimuth error alone.

44. *Level error.*

Assume now that the only instrumental error is the level error and that the west end of the axis is tilted upwards by the angle b. In this case the sign of b is, by convention, positive. On the celestial sphere the west end of the axis will be represented by B (Fig. 35) so that $WB = b$; since there is no other error, B will lie in the prime vertical ZW. As the telescope is rotated about its axis, the collimation axis will describe a plane cutting the celestial sphere in a great circle of which B is the pole. In Fig. 35 this great circle is SX (it passes through S, since S is the pole of WZ and consequently $SB = 90°$). SX is now the fictitious meridian. Now $WB \equiv W\hat{S}B = b$; also $W\hat{S}Z = 90°$ and $B\hat{S}X = 90°$; hence $Z\hat{S}X = b$. Consider a star at the observed moment of transit over the central wire; it is then at X, we shall suppose, on the fictitious meridian, and its observed time of transit occurs before it reaches the true meridian

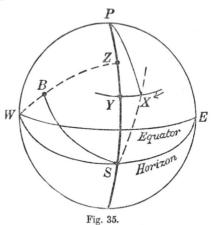

Fig. 35.

at Y. Let $X\hat{P}Y = \tau_2$; then τ_2 is the correction to be applied to the observed time of transit to give the true time of transit at Y. In the triangle PXS, $PX = 90° - \delta$, $PS = PZ + ZS = 180° - \phi$, $S\hat{P}X = \tau_2$ and $P\hat{S}X = b$. By formula C,

$$\sin \tau_2 \sin PX = \sin b \sin SX.$$

Since b is always very small, SX is approximately equal to SY. Also

$$SY = PS - PY = 180° - \phi - (90° - \delta) = 90° - (\phi - \delta).$$

Hence $\sin \tau_2 \cos \delta = \sin b \cos (\phi - \delta).$

Express the small angles τ_2 and b in seconds of time; then

$$\tau_2 = b \cos (\phi - \delta) \sec \delta \qquad \ldots\ldots(2).$$

This formula gives the correction to the observed time of transit due to level error alone.

45. *Collimation error.*

Assume now that the only error is collimation error. In this case the central wire lies outside the collimation plane, and as the telescope is rotated the central wire will describe a plane parallel to the plane of the true meridian; this plane will inter-

sect the celestial sphere in a small circle XU (Fig. 36) which is now the fictitious meridian. The straight line joining the centre of the object-glass to the middle point of the central wire will make a small angle c with the collimation axis and as the latter is, in this case, perpen-
dicular to the rotation axis EW, the angular distance of W from any point on the small circle XU is $90° + c$. (The sign of c is taken to be positive when the fictitious meridian XU cuts the horizon to the east of S, as in Fig. 36.) Consider now a star X at the moment of transit over the central wire; when c is positive, it is seen from the figure that the star is

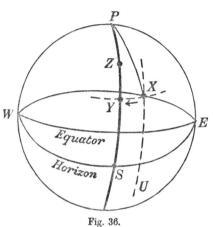

Fig. 36.

on the central wire before it reaches the true meridian at Y. Let $Z\hat{P}X = \tau_3$. In the spherical triangle WPX, $WX = 90° + c$, $WPX = 90° + \tau_3$, $PX = 90° - \delta$ and $WP = 90°$. Then, by formula **A**,

$$\cos WX = \cos WP \cos PX + \sin WP \sin PX \cos WPX,$$

from which $\qquad -\sin c = -\cos \delta \sin \tau_3,$

so that, c and τ_3 being small and both expressed in seconds of time,
$$\tau_3 = c \sec \delta \qquad \qquad \ldots\ldots(3).$$

46. *The total correction to the observed time of transit.*

Let T be the observed time of transit over the central wire according to the sidereal clock—in practice, T would be taken to be the mean of the observed times of transit over, say, seven wires, three on each side of the central wire. Suppose that the clock is in error by ΔT, ΔT being regarded as positive when the clock is slow. Then the true time of transit over the central wire is $T + \Delta T$. The sidereal time of transit over the true meridian is then $T + \Delta T + \tau_1 + \tau_2 + \tau_3$. This is the right ascension (a) of the star. Thus, $\alpha = T + \Delta T + \tau_1 + \tau_2 + \tau_3$, or, from (1), (2) and (3)

$$\alpha = T + \Delta T + \sec \delta \{a \sin (\phi - \delta) + b \cos (\phi - \delta) + c\} \quad \ldots\ldots(4),$$

which is sometimes written in the form

$$\alpha = T + \Delta T + aA + bB + cC \quad \ldots\ldots(5),$$

where $A = \sin (\phi - \delta) \sec \delta$, $B = \cos (\phi - \delta) \sec \delta$, and $C = \sec \delta$. Actually, the error ΔT of the sidereal clock is not constant from day to day, but we shall assume, however, in determining the right ascensions of the stars according to (5), that its value is known for each observation. Also the formula (5) is incomplete in the form just given, as a small term due to aberration (see section 111) has not yet been included; the value of this term is $- 0^s \cdot 021 \cos \phi \sec \delta$.

The formulae given so far refer to the upper culmination of a star. The corresponding formulae for lower culmination can be derived in a similar manner. The formula corresponding to (4) is found to be of similar character and is derivable from (4) by writing $(180° - \delta)$ for δ in that equation.

47. *Bessel's formula.*

The correction τ to the observed time of transit of a star due to all three errors a, b and c, considered together, can be in-

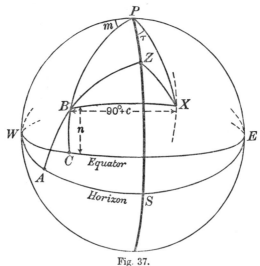

Fig. 37.

vestigated as follows. Suppose that the west end of the rotation axis points to the position B on the celestial sphere. Draw the vertical circle ZBA to meet the horizon in A. Then $WA = a$ (azimuth error) and $AB = b$ (level error). Also let X be the position of a star at the moment of coincidence with the central wire. Then $BX = 90° + c$. Draw the meridian PBC to meet the equator in C and let $WC \equiv W\hat{P}C = m$ and $BC = n$. The quantities m and n are simply related to a and b. In the triangle $PBZ, PZ = 90° - \phi, PB = 90° - n, BZ = 90° - b, P\hat{Z}A = 90° + a$ and $Z\hat{P}B = 90° - m$. By **A**, we have

$$\cos PB = \cos PZ \cos ZB + \sin PZ \sin ZB \cos PZB,$$

or $\qquad \sin n = \sin \phi \sin b - \cos \phi \cos b \sin a,$

which can be written with sufficient accuracy, since n, a and b are small,

$$n = b \sin \phi - a \cos \phi \qquad \qquad \dots\dots(6).$$

Also, by formula **D**,

$$\cos PZ \cos PZB = \sin PZ \cot ZB - \sin PZB \cot ZPB,$$

or $\qquad - \sin \phi \sin a = \cos \phi \tan b - \cos a \tan m,$

from which we have, since m, a and b are small,

$$m = a \sin \phi + b \cos \phi \qquad \qquad \dots\dots(7).$$

Now in triangle $BPX, BX = 90° + c, PB = 90° - n, PX = 90° - \delta$ and $BPX = 90° + \tau - m$, where τ is the angle ZPX (the correction to the observed time of transit at X over the central wire to give the time of transit over the true meridian PZS). By **A**

$$\cos BX = \cos PB \cos PX + \sin PB \sin PX \cos BPX,$$

whence

$$- \sin c = \sin n \sin \delta - \cos n \cos \delta \sin (\tau - m),$$

or, since c, n and $(\tau - m)$ are small angles,

$$- c = n \sin \delta - (\tau - m) \cos \delta,$$

from which we obtain

$$\tau = m + n \tan \delta + c \sec \delta \qquad \qquad \dots\dots(8).$$

This is Bessel's formula and is the form found most convenient in practice. It is easily verified that when the values of m and n, given by (6) and (7) in terms of a and b, are substituted in (8), formula (4) is obtained. Inserting the clock error ΔT and the aberration term we obtain, for the right ascension of the star under observation,

$$a = T + \Delta T + m + n \tan \delta + (c - 0^{s}{\cdot}021 \cos \phi) \sec \delta \dots(9).$$

48. *Wire intervals.*

As already stated, some meridian circles are fitted with a micrometer by means of which the entire system of vertical wires can be moved across the field of view; in other instruments the system of vertical wires is fixed and the micrometer actuates a moving wire, which can be placed successively on any two of the fixed wires. In both types it is necessary to determine

(a) the interval between two consecutive wires in terms of the micrometer scale, and

(b) the value of one revolution of the micrometer in terms of angular measure.

To determine (a) in the first type of instrument, suppose that a distant object is visible in the field of view when the telescope is approximately horizontal. By moving the micrometer so that two wires are successively coincident with the image of the distant object, the interval between the wires is expressed as the difference between the micrometer readings. The observation can then be repeated for any other pair of wires. It is obvious that, in the second type of instrument, the wire intervals can be easily found in terms of the micrometer scale.

To determine (b), the interval of time required by a star to pass from one wire W_1 to another wire W_2 is observed. A star near the pole is selected for this purpose as it moves comparatively slowly across the field of view. As the telescope is rotated about its axis, the wire W_1 will sweep out a plane which will intersect the celestial sphere in a small circle of which the west end B (as in Fig. 37) is the pole. Let the angular distance of B from any point on this small circle be $90° + c_1$; then c_1 is the collimation error of W_1. Then if τ_1 is the interval between the transit of the star X (Fig. 37) over wire W_1 and over the true meridian, we have by (8), taking into account that c_1 and n are small,

$$\tau_1 = m + c_1 \sec \delta + n \tan \delta \qquad \ldots\ldots(10).$$

Now consider wire W_2 and let τ_2 and c_2 be the corresponding quantities. Then we have, as before,

$$\tau_2 = m + c_2 \sec \delta + n \tan \delta \qquad \ldots\ldots(11).$$

If T_1, T_2 are the times by the sidereal clock when the star is on wires W_1 and W_2 respectively, and T_0 is the time by the clock when the star is on the true meridian, we have

$$T_0 = T_1 + \Delta T + \tau_1 \quad \text{and} \quad T_0 = T_2 + \Delta T + \tau_2.$$

Also if $t = T_2 - T_1$ (t being the time in seconds required by the star to pass from W_1 to W_2) then, eliminating T_0 and ΔT,

$$t = \tau_1 - \tau_2,$$

so that, by (10) and (11),

$$t = (c_1 - c_2) \sec \delta,$$

or
$$c_1 - c_2 = t \cos \delta \qquad \qquad \ldots \ldots (12).$$

Now $c_1 - c_2$ is the angular separation of the wires W_1 and W_2 (expressed in seconds of time), and it can consequently be calculated from (12), for t is observed and δ is known. Formula (12) is sufficiently accurate in practice unless the star's declination is very near to $90°$.

As we have explained, the interval between the two wires W_1 and W_2 can be found in terms of the micrometer scale, and thus we are enabled to express one revolution of the micrometer head in terms of seconds of time.

From (12), it is seen that $c_1 - c_2$ (expressed in seconds of time) is the interval required for an equatorial star to pass from wire W_1 to wire W_2.

Example. On 1931 January 3, the interval between the transits of δ Ursae Minoris over two of the wires of the Cambridge Meridian Circle was $3^m 4^s$; to find the wire interval.

From the *Nautical Almanac*, $\delta = + 86° 36' 34''$. Also $t \equiv 3^m 4^s = 184^s$. Thus

$$c_1 - c_2 = 184^s \cos 86° 36' 34''$$
$$= 10^s\cdot 88.$$

49. *Determination of the collimation error.*

The methods of determining the instrumental errors vary according to the size or elaborateness of the meridian circle concerned. It is not our purpose to describe all the different methods in use and for details the student is recommended to study a book dealing specifically with practical astronomy.* There are, however, certain general principles involved in the

* For example, *The Elements of Practical Astronomy*, by W. W. Campbell (Macmillan, 1899).

determination of the various errors and we shall consider these in turn, taking first the collimation error.

As regards large instruments the collimation error is determined by means of two *collimators*, one north of the telescope and the other south. Consider the south collima-

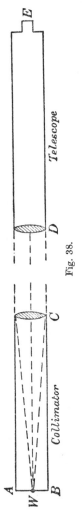

Fig. 38.

tor. It consists of an object-glass mounted vertically, with its axis in or near the meridian and at the same height above the floor as the rotation axis of the telescope. In the focal plane (AB) of the object-glass (C) of the collimator are placed one or two vertical wires and one or two horizontal wires (Fig. 38). (For simplicity, we shall suppose that there is only one vertical wire and one horizontal wire.) These wires can be viewed by means of an eye-piece just outside the focal plane AB. When the meridian circle telescope is pointed horizontally and south, the observer can see in the field of view at E the collimator wire W when it is suitably illuminated. Since the collimator wire is in the focal plane of C, rays passing from W through the collimator will emerge at C as a parallel beam and after entering the object-glass D of the telescope the wire will be seen sharply at E. In this optical arrangement the observation of the collimator wire is equivalent to the observation of an object at an infinite distance. The north collimator is arranged in a similar way, but we shall suppose that the vertical wire in this collimator is movable by means of a micrometer and that the south collimator wire is fixed. When the telescope is rotated to the vertical position, shutters in the framework of the axis can be opened, allowing an unimpeded view from one collimator to the other. Thus the observer, if stationed at the eye-end of the north collimator, can see the wire of the south collimator and by actuating the micrometer can make the north collimator wire coincide with the image of the south collimator wire. When this is done, the combined optical arrangement is such that the south

collimator wire, when viewed in the telescope, appears as if it is
at an infinite distance in a particular direction and the north
collimator wire, when viewed in the telescope, appears as if it
is at an infinite distance in exactly the opposite direction. It is
unlikely that the direction of the south collimator wire is exactly
south; suppose that the direction is θ degrees west of south;
then the north collimator wire will be in the direction θ degrees
east of north. The method of determining the collimation error
c must be such that the unknown quantity θ is eliminated in
the observations.

Suppose that the telescope is first pointed to the south
collimator so that its vertical wire is visible. This wire corresponds to an object whose position on the celestial sphere is
indicated by the point F (Fig. 39), FS being the angle θ. Assume

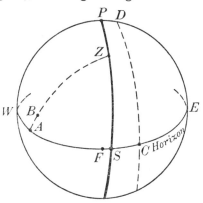

that the collimation error
c is positive; then on the
celestial sphere the central
wire will sweep out the
small circle CD (as the telescope is rotated about its
axis) of which the west axis
B of the instrument is the
pole, and the great circle
arc joining B to C is $90° + c$.
Since the level error is always small, this great circle
arc will be practically coincident with the horizon in

Fig. 39.

the neighbourhood of the south point S; for the present purpose
we can thus ignore level error and assume that the west part of
the axis of the telescope points to A, A being on the horizon
such that WA is the azimuth error a. Then $AC = 90° + c$. Now
$AS = 90° - a$ and $AC = 90° + c$; hence $SC = a + c$ and, since
$FS = \theta$,
$$FC = \theta + a + c \qquad \qquad \text{......(13)}.$$

We shall consider a meridian circle fitted with a fixed system of
vertical wires and a movable wire connected with a micrometer
whose scale has been determined, for example, by the method of
section 48. Place the movable wire over the central wire (corre-

4   SA

sponding to C in Fig. 39) and let the reading be M_1; then place it over the image of the south collimator wire F and let the reading be M_2. Then, in terms of the micrometer scale, $M_2 - M_1 = CF$, and writing $D = M_2 - M_1$ we have, from (13),

$$D = \theta + a + c \qquad \ldots \ldots (14).$$

It is necessary to make some convention with regard to the sign of D; the sign will be assumed to be positive when, as in Fig. 39, C is east of F; as viewed in the telescope (the optical combination is inverting) the central wire C will thus appear to be west of the south collimator wire F.

Now suppose the telescope pointed to the north collimator. Neglecting the level error, the east end of the axis will point to A_1 (Fig. 40) such that $EA_1 = a$. The small circle on the celestial sphere, of which A_1 is the pole, associated with the central wire will intersect the horizon, between the north and east points, at C_1. Since A_1 in Fig. 40 and A in Fig. 39 are diametrically opposite points and since $AC = 90° + c$, then we have $A_1 C_1 = 90° - c$. Also $A_1 N = 90° - a$; hence $C_1 N = c - a$. Again, the vertical wire of the north collimator is in the direction

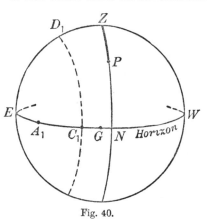

Fig. 40.

θ degrees east of north; on the celestial sphere it will define the position G such that $GN = \theta$. Hence

$$GC_1 = c - a - \theta \qquad \ldots \ldots (15).$$

Now the micrometer reading when the movable wire coincides with the central wire is M_1; let the reading be M_3 when the movable wire coincides with the north collimator wire G. Write $D' = M_3 - M_1$. Then according to Fig. 40, since C_1 is east of G so that in the telescope the central wire appears west of the collimator wire, the sign of D' is positive by our convention. By (15) we have

$$D' = c - a - \theta \qquad \ldots \ldots (16).$$

From (14) and (16), $\qquad c = \tfrac{1}{2}(D + D') \qquad \ldots \ldots (17).$

This is the formula for the collimation error c. D and D' are at first expressed in terms of the micrometer scale, and, from the previously ascertained value of one revolution of the micrometer in terms of seconds of time, we are thus enabled to express c in terms of the latter unit.

It is to be noted that for instruments in which the system of vertical wires is movable the effect of collimation error and the aberration correction (to which reference has been made in section 46) on the time of transit of a star can be removed. To each position of the central wire corresponds a definite value of c, and, if the wires are moved to such a position that

$$c - 0^{\text{s}}\cdot 021 \cos \phi = 0 \qquad \ldots\ldots(18)$$

(in which ϕ is the latitude), the only corrections to be applied to the time of transit over the central wire to give the time of transit over the true meridian are the corrections due to the azimuth error a and the level error b.

50. *Determination of the level error.*

As regards the smaller instruments, this error can generally be derived with sufficient accuracy by means of a striding level, placed on the axis of rotation. For large instruments, a more precise method is employed, which we shall now describe. The telescope is pointed vertically downwards towards a bowl of mercury which is generally placed below the floor level to avoid atmospherical and other disturbances. The surface of the mercury acts as a reflecting horizontal plane, the normal to which defines the direction of the zenith. A special eye-piece— the Bohnenberger eye-piece—enables a beam of light to pass into the telescope; after traversing the object-glass, the beam will be

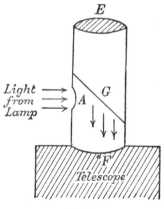

Fig. 41.

reflected at the mercury surface and will then re-enter the object-glass. The central wire and its image formed as a result of reflection at the mercury surface will both be seen in the field of

4-2

view. The special eye-piece (Fig. 41) consists of a tube in the side of which is an aperture A through which the horizontal rays from a lamp, placed some distance away, enter the tube. In the tube is placed opposite A a piece of clear glass G inclined at $45°$ to the axis of the tube. The glass surface acts as a reflector and so part of the beam entering at A is reflected by G into the telescope. The glass being unsilvered, the observer with his eye at E can see the vertical wires and their images due to the reflection at the mercury surface. In Fig. 42, let EW be the rotation axis of the telescope inclined at the angle b to the horizontal, O the centre of the object-glass, C the central wire and D its image, and AB the mercury surface. The ray CO from C to the centre of the object-glass will be undeviated by the latter (supposed thin) and will strike the mercury surface at B. Since C is in the focal plane of the object-glass any other ray CG will emerge from the object-glass along GA parallel to OB. Thus a cone of rays from C will emerge as a parallel beam. This

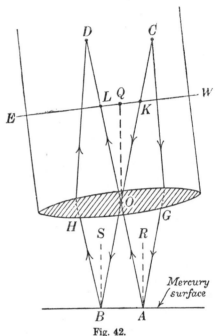

Fig. 42.

beam will be reflected at the mercury surface into another parallel beam which will be brought by the object-glass to a focus at D. Consider the ray COB; at B it is reflected along BH in such a way that $O\hat{B}S = S\hat{B}H$, BS being perpendicular to the mercury surface at B; after passing through the object-glass at H, it passes through D. Produce DO to meet the mercury surface at A. Let CG be the ray from C which is incident on the mercury surface at A. Then this ray is reflected along AO and will pass through D; thus GA is parallel to OB and AO is parallel to BH.

It follows that $C\hat{O}D = G\hat{A}O = O\hat{B}H$ and each is twice $O\hat{B}S$. Let OQ be parallel to BS and AR. Then, since OQ is perpendicular to the mercury surface, $K\hat{Q}O = 90° + b$. Also $W\hat{K}O = 90° + c$, where c is the collimation error. Now

$$K\hat{Q}O = W\hat{K}O - Q\hat{O}K,$$

that is $\qquad\qquad b = c - \tfrac{1}{2}C\hat{O}D.$

The angle COD can be measured. Place the movable wire over the central wire C and then over its image at D; the difference in readings is, say, R, which can be converted into seconds from the known value of the micrometer scale. We thus have

$$b = c - \tfrac{1}{2}R \qquad\qquad ......(19),$$

from which b can be derived, the value of c being supposed known. When the system of vertical wires is movable, the value of $\tfrac{1}{2}R$ is obtained simply by moving the wires until the central wire coincides with its image. The convention as to the sign of R is as follows: when the central wire is seen in the telescope to the west of its image, R is defined to be positive.

51. Determination of the azimuth error.

The azimuth error a is determined from observations of stars, the values of b and c having previously been derived. Let (α_1, δ_1) and (α_2, δ_2) be the equatorial co-ordinates of two stars, T_1 and T_2 the observed times of transit by the sidereal clock. By (5), we have

$$\alpha_1 = T_1 + \Delta T + aA_1 + bB_1 + cC_1,$$
$$\alpha_2 = T_2 + \Delta T + aA_2 + bB_2 + cC_2.$$

In these equations, ΔT is the unknown error of the clock, and if, by a suitable choice of stars, the observations are made within a few minutes of each other, the value of ΔT can be assumed to be the same in the two equations; also $A_1 = \sin(\phi - \delta_1)\sec\delta_1$, etc. By subtraction, ΔT is eliminated and we have

$$a = \frac{(\alpha_2 - \alpha_1) - (T_2 - T_1) - b(B_2 - B_1) - c(C_2 - C_1)}{A_2 - A_1} \qquad ......(20).$$

In this equation,

$$A_2 - A_1 = \sin(\phi - \delta_2)\sec\delta_2 - \sin(\phi - \delta_1)\sec\delta_1.$$

The right-hand side of (20) can be calculated from the known values of b, c, α_1, α_2, δ_1, δ_2 and the observed times T_1, T_2. It remains to indicate how best the choice of stars can be made to ensure an accurate determination of a. It is clear from (20) that a can be most accurately determined if the denominator $(A_2 - A_1)$ is large. This condition is fulfilled if (i) a star of high declination (δ_2) and a star near the equator (declination δ_1) are selected, for then A_2 is large, since sec δ_2 is large, or (ii) a star of high declination (δ_2) observed at upper culmination and a star of high declination (δ_1), differing from the first by about 12 hours in right ascension, observed at lower culmination. In (ii), the values of A_1 and A_2 will both be large but of opposite sign, so that $(A_2 - A_1)$ will be numerically large.

52. *The chronograph.*

In modern instruments the times of transit of a star over the system of vertical wires are recorded electrically by means of a chronograph. A clockwork mechanism causes a paper tape to be drawn out at a uniform rate (or a cylinder covered by a sheet of paper to revolve at a uniform rate). The sidereal clock by which the times of transit are to be recorded is connected electrically with an electro-magnet belonging to the chronograph and to which a pen is attached. At the bottom of its swing, the pendulum of the clock closes an electrical circuit for an instant and the action of the momentary current passing through the electro-magnet is to cause the pen to "kick". When there is no current, the pen traces out a straight line on the moving tape (or a uniform line on the rotating cylindrical paper), but when contact is made by the pendulum, the "kick" of the pen results in a distinctive mark being made on the paper. In this way, the seconds of the clock are mechanically registered. Usually, there is an automatic arrangement by which the pen misses the 60th second of every minute, and thus it is easy, by simple reference to the clock, to ascertain the hour and minute of any "blank"; this time can be written down on the paper and any particular second of the following minute deduced by counting the "kicks" of the pen from the "blank". This is illustrated in the upper trace of Fig. 43. The chronograph carries a second pen, attached to an electro-magnet, which makes a trace parallel to that of the clock pen, and this pen can be actuated by the observer closing a circuit by means of a tapping-key. When he sees, in the telescope, the star on a vertical wire he gives the key a sharp tap and the second pen makes a "kick" on the moving paper. The time of transit over this wire can be inferred later from the chronograph record (this is illustrated in the lower trace of Fig. 43).

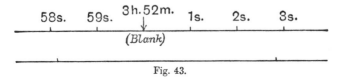

Fig. 43.

There are in use several modifications of the procedure just outlined. In one, the observer moves a vertical wire, by means of the attached micrometer, in such a way that the wire appears to bisect continuously the image of the star as it moves across the field of view; at definite points, corresponding to the positions of the system of vertical wires, the frame supporting the wire closes an electrical circuit and the chronograph pen makes the appropriate records. In other instruments, the moving wire is driven across by a small motor, at a rate adjusted according to the declination of the star, and the observer has only to make the necessary small corrections to the rate to ensure the accurate bisection of the stellar image by the moving wire.

53. *The measurement of declination.*

We have already mentioned that the rotation axis of the telescope carries two finely graduated circles, the graduations being generally at 5′ intervals from 0° to 360°. In the focal plane of the object-glass there is a horizontal wire and when observations of a star's declination are being made the telescope is adjusted so that, in the neighbourhood of the collimation axis, the star appears to travel along the horizontal wire. In this position, the circles indicate a particular reading R_1, the accurate determination of which is facilitated by the use of four microscopes (90° apart) on each circle. To obtain the zenith distance of the star we require to know the circle reading R when the telescope is pointed accurately to the zenith. We shall suppose for the moment that the horizontal wire is fixed in the focal plane of the object-glass. The reading R can then be determined by pointing the telescope downwards towards the basin of mercury and moving the telescope gradually (by slow-motion screws) until the horizontal wire and its image coincide when viewed in the special eye-piece of Fig. 41; the addition of 180° to the reading of the circles in this position gives the reading R, corresponding to the position of the telescope when pointing to the zenith. The difference between the readings R and R_1 is the meridian zenith distance of the star. This zenith distance of course contains the refraction and when the latter is removed we obtain the true zenith distance of the star. The latitude being presumed known, we finally deduce the declination of the star. But in practice it is inadvisable to attempt the delicate adjustment of the instrument just contemplated. As we invariably know the approximate declination of the star to be observed, the telescope can be pointed with sufficient accuracy to ensure that in due course the star will appear in the field of view. Generally, it will appear to move (disregarding a slight curvature in its path) parallel to and at some distance from the horizontal wire. In this position of the telescope the circles can be read either before or after the star's appearance. Let the reading be R_2. Now assume that the horizontal wire can be moved parallel to itself by means of a micrometer to such a position that the star appears to travel along the wire. Let the two micrometer readings be M_0 and M_1, the former the original reading (corresponding, say, to some definite reading on the micrometer head) and the latter the reading when the star travels along the wire. The value of one revolution of the micrometer can be found in terms of seconds of arc by observations of two stars of known declination which are close together in the sky; with the telescope fixed, the difference in declination can be measured in terms of the micrometer scale and hence the value of this scale (in seconds of arc) can be deduced. Now, corresponding to M_0, there is the particular reading R of the circles when the telescope is pointing accurately to the zenith; the zenith distance of the star is not quite the difference between R and R_2, as in the case of the fixed horizontal wire; we have to apply to this difference a correction given by $(M_1 - M_0)$ expressed in seconds of arc.

Thus the zenith distance of the star is determined and its declination can be deduced as before.

54. *The measurement of right ascension.*

The point on the celestial equator from which right ascensions are measured is the vernal equinox, whose position is given by the centre of the sun about March 21 when the sun passes from south declination to north declination. At the summer solstice (about June 21) the sun has reached its maximum northerly declination and then its declination is simply the obliquity of the ecliptic. Observations of its meridian zenith distances on several days before and after the solstice will lead to an accurate value of its declination exactly at the solstice; in other words the obliquity of the ecliptic can be found. It is to be noticed that in these considerations it is the sun's centre that is specified; in the actual observations the meridian zenith distance of the upper (or lower) limb is measured; to obtain the zenith distance of the centre, the sun's *semi-diameter* (the angle subtended at the observer by a radius of the sun) requires to be added (or subtracted).

Since the vernal equinox is so closely associated with the sun, it follows that the right ascension of any star is fundamentally related to the position of the sun on any day. Let us suppose that the complete observation of the sun on a particular day gives

(i) the clock time T (corrected for the instrumental errors a, b and c) of the transit of its centre, and

(ii) the zenith distance of its centre at transit.

Assuming that we know the latitude accurately we deduce, as in the previous section, the declination of the sun's centre [we omit in this discussion the correction due to "parallax" (see Chapter IX)]. From the declination and the value of the obliquity ϵ (which we shall assume known), we can calculate the right ascension α of the sun at the time of transit on the day in question. In the triangle ΥSA (Fig. 44), PSA is the meridian through the sun S, and $SA\Upsilon = 90°$, $\Upsilon A = \alpha$, $AS = \delta$, $S\Upsilon A = \epsilon$. By formula **D**,

$$\cos \alpha \cos 90° = \sin \alpha \cot \delta - \sin 90° \cot \epsilon,$$

or $\sin \alpha = \tan \delta \cot \epsilon$ ......(21).

From (21) we obtain α, the sun's right ascension at the moment of transit on the day concerned. Let E be the error of the sidereal clock at the time T of transit. Then the true sidereal time of transit is $T + E$. But this is precisely the sun's right ascension α. Hence

$$\alpha = T + E$$

......(22).

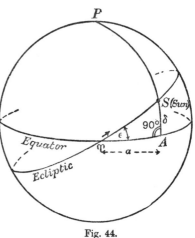

Fig. 44.

Suppose that the transit of a star is observed afterwards, as soon as practicable. Let the clock time of transit be T_1 (we suppose that both T_1 and T have been corrected for the instrumental errors a, b and c). The error of the clock may now be presumed to be slightly different; let it be E_1. Then, if α_1 is the star's right ascension,

$$\alpha_1 = T_1 + E_1 \qquad \qquad ......(23),$$

so that, from (22) and (23),

$$\alpha_1 - \alpha = (T_1 - T) + (E_1 - E) \qquad ......(24).$$

$(T_1 - T)$ is the interval by the clock between the times of transit of the sun and star and is therefore known. If the clock neither gains nor loses, $(E_1 - E) = 0$, so that α_1 can be found from (24). Generally, however, the mechanical perfection of the clock cannot be presumed and as a convenient hypothesis the clock may be supposed to gain (or lose)—over an interval of a few hours—at a uniform rate, so that $(E_1 - E)$ is proportional to the interval between observations; we can write

$$E_1 - E = r\,(T_1 - T),$$

so that (24) becomes

$$\alpha_1 - \alpha = (1 + r)\,(T_1 - T) \qquad ......(25).$$

Suppose that the observations are repeated next day. Then we shall have

$$\alpha_1 - \alpha' = (1 + r)\,(T_1' - T') \qquad ......(26),$$

in which α' is the sun's right ascension calculated from (21) and T', T_1' are the sidereal times of transit of the sun and star respectively. From (25) and (26) we have by division

$$\frac{\alpha_1 - \alpha'}{\alpha_1 - \alpha} = \frac{T_1' - T'}{T_1 - T} \qquad \ldots\ldots(27),$$

from which the right ascension α_1 of the star can be calculated.

This, in brief, is the fundamental principle involved in the measurement of the right ascensions of the stars. In practice, we may assume that the right ascensions of a sufficiently large number of stars have been accurately measured; such stars—"fundamental stars"—form a series of reference points on the celestial sphere and by means of them (or a selection of them) we are enabled to determine the right ascension of any other star by observing the interval between the transits of one or more fundamental stars and of the star in question, as in the case of the sun.

55. *The measurement of time.*

Stars which are selected for the determination of clock error and hence of the accurate sidereal time—and, by a further step, of the accurate U.T.—are called "clock stars". If α_1, α_2 are the known right ascensions of two clock stars, T_1, T_2 the observed times of transit (both corrected for the instrumental errors of azimuth, etc.), E_1, E_2 the clock errors at the respective times of transit, then

$$\alpha_1 = T_1 + E_1,$$
$$\alpha_2 = T_2 + E_2.$$

α_1, α_2 are both known, T_1 and T_2 are observed, hence E_1 and E_2 are found. The clock rate r is obtained from the equation

$$E_2 - E_1 = r\,(T_2 - T_1).$$

In practice several stars are observed to ensure greater accuracy in the values of the clock error and rate.

The modern clocks now installed in several important observatories are the most accurate time measurers ever devised. The time which they record is so nearly uniform that it may be considered as conforming to the Newtonian description of time as "equably flowing"—as for E.T.—and the normal behaviour of a modern clock is such that its error on any day can be almost exactly predicted several months in advance. The performance of a clock is of course checked by meridian observations of the stars. It may be added that the broadcasting of mean time from the principal observatories of the world has also reached a high degree of accuracy.

EXERCISES

[Symbols used:

ϕ = latitude, b = level error,

a = azimuth error, c = collimation error.]

1. If a and b are the only errors of a meridian circle, show that the time of transit of a star will be unaffected if its declination δ is given by

$$\delta = \phi + \tan^{-1}(b/a).$$

2. Prove that the error in the time of transit of a star due to the three instrumental errors is a minimum for a star whose declination is

$$\sin^{-1}\{(a\cos\phi - b\sin\phi)/c\}. \qquad [Coll.\ Exam.]$$

3. If two stars of declinations δ_1 and δ_2 can be found for which the three errors of adjustment produce no error in the time of transit, show that the correction to be added to the observed time of transit of a star of declination δ is

$$2c\sin\tfrac{1}{2}(\delta - \delta_1)\sin\tfrac{1}{2}(\delta - \delta_2)\sec\delta\sec\tfrac{1}{2}(\delta_1 - \delta_2). \qquad [M.T.]$$

4. If the observed time of transit of a star whose declination is 30° is found to be correct, while the observed times for stars of declinations 15° and 60° are found to be $-7^{s}\!\cdot\!4$ and $+31^{s}\!\cdot\!5$ in error, show that the error to be expected for a star in declination 45° is approximately 11 seconds. $\qquad [M.T.]$

5. In a transit instrument of 10 feet focal length, which is in correct adjustment except for collimation error, a star of declination 60° is observed to cross the meridian 2^{s} too soon. Show that to adjust the instrument the cross-wires must be moved a distance of 0·0087 inch. $\qquad [M.T.\ 1900.]$

6. The level constant b and the collimation constant c of a transit instrument are determined, in the usual way, with a possibility of errors Δb, Δc, respectively. The azimuth constant a is determined by observations of a polar star and of an equatorial star. Show that the resulting possible error in a is given by

$$\Delta a = \Delta b \tan\phi + \Delta c \sec\phi,$$

where ϕ is the latitude.

Prove also: $\qquad \Delta m = \Delta b \sec\phi + \Delta c \tan\phi,$

$$\Delta n = -\Delta c. \qquad [Lond.\ 1926.]$$

7. On the same day, the observed sidereal times of transits of two stars, corrected for the level and collimation errors, in latitude 51° 30′ N are $18^{h}\,51^{m}\,27^{s}\!\cdot\!2$ and $19^{h}\,7^{m}\,38^{s}\!\cdot\!9$. Calculate the azimuth error and the clock error, given the co-ordinates of the two stars as follows:

$$a_1 = 18^{\text{h}}\,51^{\text{m}}\,36^{\text{s}}{\cdot}5; \quad \delta_1 = +\,89^\circ\,2'.$$
$$a_2 = 7^{\text{h}}\,7^{\text{m}}\,11^{\text{s}}{\cdot}2; \quad \delta_2 = +\,87^\circ\,10'. \qquad [\textit{Lond.}\ 1926.]$$

8. Prove that a displacement of the pole of the earth's rotation by s feet in longitude L produces a change in the azimuth error of a meridian circle in latitude ϕ, longitude l, of $s/a \sin (l - L) \sec \phi \operatorname{cosec} 1''$, where a is the radius of the earth in feet. Estimate roughly the maximum value of this displacement for an observatory in latitude 60°, and discuss the possibility of detecting it in the routine of meridian circle observations. [*Lond.* 1922.]

CHAPTER V

PLANETARY MOTIONS

56. *Introduction.**

The nine major planets in the order of increasing distance from the sun are Mercury, Venus, the Earth, Mars, Jupiter, Saturn, Uranus, Neptune and Pluto. The laws according to which the planets move with reference to the sun were discovered by John Kepler (1571–1630), and half a century later Kepler's three laws were shown by Sir Isaac Newton (1642–1727) to be deducible from the law of universal gravitation which he stated in the *Principia* in 1687. A complete investigation into the motions of the planets is the province of Dynamical Astronomy and as such is outside the scope of this book. But certain principles and results relating to the planetary motions and, in particular, to the earth's motion are necessary if we are to understand clearly some of the problems with which Spherical Astronomy is more closely associated.

57. *Kepler's first law.*

Kepler's first law states that the path, or orbit, of a planet around the sun is an ellipse, the position of the sun being at a focus of the ellipse. Fig. 45 shows an ellipse of which S and F

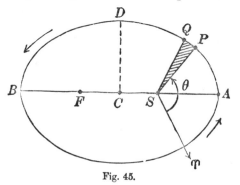

Fig. 45.

* In this chapter the time is assumed throughout to be expressed in terms of Ephemeris Time (E.T.); v. Appendix F (p. 424).

are the two foci, C is the centre (midway between S and F) and AB is the *major axis*. The sun will be supposed to be at S and the planet to move round the ellipse in the direction of the arrows. At A, the planet is nearest the sun; it is then said to be in *perihelion*. At B, it is furthest from the sun and it is then said to be in *aphelion*. CA is called the *semi-major axis*; its length is denoted by a. CD, drawn perpendicularly to CA, is the *semi-minor axis*, denoted by b. The ratio $CS : CA$ is called the *eccentricity* which we denote by e. The semi-minor axis b can be expressed in terms of a and e by the formula

$$b^2 = a^2 (1 - e^2) \qquad \qquad \dots\dots(1).$$

The perihelion distance SA is $a (1 - e)$ and the aphelion distance SB is $a (1 + e)$.

If P is any position of the planet in its orbit, SP is called the *radius vector* (denoted by r) and the distance SP is the *heliocentric distance*, that is, the distance of the planet from the sun. Let ST be a straight line in the plane of the ellipse. Then the position of P is specified by the radius vector r and the angle θ which SP makes with ST, θ being measured in the direction of the planet's motion. Let ω be the value of θ when the planet is in perihelion, that is, at A so that $T\hat{S}A = \omega$. Then $P\hat{S}A = \theta - \omega$. The equation of the ellipse is known to be

$$r = \frac{p}{1 + e \cos (\theta - \omega)} \qquad \qquad \dots\dots(2),$$

in which $\qquad\qquad p = b^2/a = a (1 - e^2) \qquad\qquad \dots\dots(3).$

The time required for the planet to describe its orbit is called the *period* which we denote by T. The earth's orbital period is the *year* which, for the present, may be taken to be $365\frac{1}{4}$ mean solar days.

58. *Kepler's second law.*

Kepler's second law states that the radius vector SP (Fig. 45) sweeps out equal areas in equal times. Let P correspond to the planet's position at time t and Q its position at time $t + \Delta t$. Let $r + \Delta r$ denote the radius vector SQ and $\theta + \Delta \theta$ the angle QST. Then $Q\hat{S}P = \Delta \theta$. If $\Delta \theta$ is sufficiently small, the arc PQ may be regarded as a straight line and the area swept out in the infinitesimal interval Δt is simply the area of the triangle QSP, that is, $\frac{1}{2}r\,(r + \Delta r)\sin\Delta\theta$, or, with sufficient accuracy, $\frac{1}{2}r^2\Delta\theta$. The rate of description of area is this last expression divided by Δt; as this rate is constant according to Kepler's second law, we can write

$$r^2\frac{d\theta}{dt} = h \qquad\qquad \dots\dots(4),$$

where h, a constant, is twice the rate of description of area by the radius vector.

Now the whole area of the ellipse is πab and this is described in the interval defined by the period T. Hence

$$\frac{2\pi ab}{T} = h \qquad\qquad \dots\dots(5),$$

or, using (1),

$$\frac{2\pi a^2\,(1 - e^2)^{\frac{1}{2}}}{T} = h \qquad\qquad \dots\dots(6).$$

In time T, the radius vector sweeps out an angle of 360° or 2π. Let n denote the average rate of description of angle by the radius vector. Then

$$n = 2\pi/T \qquad\qquad \dots\dots(7).$$

n is called the *mean angular motion* of the planet. In moving from SP to SQ the radius vector sweeps out the angle $\Delta\theta$ in the interval Δt. At P the angular velocity is thus $\dfrac{d\theta}{dt}$; thus n is the mean value of $\dfrac{d\theta}{dt}$ for all points in the orbit. By the help of (7), formula (6) can be written

$$na^2\,(1 - e^2)^{\frac{1}{2}} = h \qquad\qquad \dots\dots(8).$$

59. *Kepler's third law.*

Kepler's third law, expressed mathematically, is as follows. Let a, a_1 be the semi-major axes of two planetary orbits and T, T_1 the corresponding orbital periods. Then, by the third law,

$$\frac{a^3}{T^2} = \frac{a_1{}^3}{T_1{}^2} \qquad\qquad \dots\dots(9),$$

or, using (7),

$$n^2a^3 = n_1{}^2a_1{}^3 \qquad\qquad \dots\dots(10),$$

in which n and n_1 are the mean angular motions in the two orbits.

We have from (9)

$$\frac{a_1}{a} = \left(\frac{T_1}{T}\right)^{\frac{2}{3}} \qquad\qquad \dots\dots(11).$$

If a, T refer to the earth's orbit around the sun, the ratio of the semi-major axis of the orbit of any planet P_1 to the semi-major axis of the earth's orbit is given by (11) when the orbital period T_1 for the planet P_1 is known in years, since for the earth T is one year. The orbital periods of the planets are known from observations (see section 80) and hence their semi-major axes can be derived in terms of a as the unit of distance. The semi-major axis a of the earth's orbit is known as the *astronomical unit* of distance. Expressing T_1 in years, we have from (11), putting $T = 1$ year,

$$a_1 = (T_1)^{\frac{2}{3}} \text{ astronomical units.}$$

60. *Newton's law of gravitation.*

The statement of this law is as follows. Every particle of matter attracts every other particle of matter with a force proportional to the product of the masses of the two particles concerned and inversely proportional to the square of the distance between them. Stated mathematically, the law is:

$$F = G \frac{mm_1}{r^2} \qquad \ldots\ldots(12),$$

where m, m_1 are the masses of the particles, r the distance between them, F the gravitational force of attraction and G a constant—the *constant of gravitation*. In the c.g.s. system of units, the value of G is $6\cdot658 \times 10^{-8}$, which means that the force of attraction between two particles each of 1 gramme in mass and separated by a distance of 1 centimetre is $6\cdot658 \times 10^{-8}$ dynes. For our present purpose we can regard the sun and planets as "particles".

The application of the law of gravitation (12) to the motion of a planet around the sun leads to the three laws of Kepler. Let M, m denote the masses of the sun and a planet respectively and let μ be defined by

$$\mu = G (M + m) \qquad \ldots\ldots(13).$$

Then it is found that the constant h in formula (4) is given by

$$h^2 = \mu p = \mu a (1 - e^2) \qquad \ldots\ldots(14).$$

But, by (8), $\qquad h^2 = n^2 a^4 (1 - e^2).$

Hence $\qquad n^2 a^3 = \mu \equiv G (M + m) \qquad \ldots\ldots(15).$

For another planet,

$$n_1{}^2 a_1{}^3 = \mu_1 \equiv G\,(M + m_1) \qquad \ldots\ldots(16).$$

Hence from (15) and (16),

$$\frac{n^2 a^3}{n_1{}^2 a_1{}^3} = \frac{M + m}{M + m_1},$$

or, using (7),
$$\frac{a^3}{a_1{}^3} = \frac{M + m}{M + m_1} \cdot \frac{T^2}{T_1{}^2} \qquad \ldots\ldots(17).$$

The equation (17) gives the correct form of Kepler's third law which in our notation had been expressed by (9). Actually the mass of a planet is very small compared with that of the sun and so the quantity $\dfrac{M + m}{M + m_1}$ in (17) is very nearly equal to unity; consequently, Kepler's third law, although not strictly accurate, is nevertheless a very good approximation.

61. *The masses of the planets.*

The formulae of the preceding section enable us to calculate the mass of any planet, which is accompanied by one or more satellites, as a fraction of the sun's mass. If m, a, T refer to the earth, we have, by (7) and (15),

$$G\,(M + m) = 4\pi^2 \frac{a^3}{T^2} \qquad \ldots\ldots(18).$$

Now the motion of a satellite around a planet is given by the same laws as the motion of a planet around the sun. In the case of the satellite, the planet is the controlling body and if m_1, m' are the masses of the planet and the satellite respectively, a_1 the semi-major axis of the satellite's orbit around the planet, and T_1 the period of its orbital revolution, Newton's law of gravitation leads to an equation analogous to (18); it is

$$G\,(m_1 + m') = 4\pi^2 \frac{a_1{}^3}{T_1{}^2} \qquad \ldots\ldots(19).$$

Hence, by (18) and (19),

$$\frac{m_1 + m'}{M + m} = \left(\frac{a_1}{a}\right)^3 \left(\frac{T}{T_1}\right)^2 \qquad \ldots\ldots(20).$$

Now the mass m' of the satellite is small compared with the mass of the planet m_1; hence in (19) and (20) we can neglect m'.

Similarly we can neglect the earth's mass m in comparison with the sun's mass M, and (20) can be written

$$\frac{m_1}{M} = \left(\frac{a_1}{a}\right)^3 \left(\frac{T}{T_1}\right)^2 \qquad \ldots\ldots(21).$$

We shall suppose that the semi-major axis a_1 of the satellite's orbit is known in astronomical units and that its period of revolution T_1 is also known in terms of years. Then in (21), $a = 1$ astronomical unit, $T = 1$ year, and therefore the ratio of m_1 (the planet's mass) to M (the sun's mass) is determined.

As an example we shall find the mass of Mars (in terms of the sun's mass) from the relevant orbital elements of the satellite Deimos. The semi-major axis of the orbit of Deimos around Mars is $0\cdot00015695$ astronomical unit; its period of revolution is $1\cdot26244$ day or $\dfrac{1\cdot26244}{365\frac{1}{4}}$ year.

Hence, by (21),

$$\frac{m_1}{M} = (0\cdot00015695)^3 \times \frac{(365\frac{1}{4})^2}{(1\cdot26244)^2}$$

$$= \frac{1}{3093500},$$

that is to say, the sun's mass is rather more than three million times the mass of Mars.

62. *Perturbations of the elements.*

We have assumed so far that the path of a planet around the sun is determined by the mutual gravitational attraction of the planet and the sun only. But every other planet and body in the solar system exerts a gravitational attraction on the planet concerned and the effects are shown in small changes in its orbital elements such as the semi-major axis a and the eccentricity e. Such changes are known as *perturbations of the elements*; in general their values are small. The magnitude of any perturbation due to a particular disturbing planet depends, among other things, on the mass of the latter and it is thus possible to deduce from observations, in combination with the formulae of Dynamical Astronomy, the masses of planets which are unaccompanied by satellites. In this way, the masses of Mercury and Venus (which have no satellites) are derived. Later, we shall have occasion to refer more particularly to the perturbations of orbital elements, quoting the results of Dynamical Astronomy, in so far as they directly impinge on the particular problems with which we are more immediately concerned in this book.

63. *The dynamical principles of orbital motion.*

Let S and P be the positions of the sun and a planet at any time t and let their co-ordinates referred to unaccelerated rectangular axes OA, OB, OC in space be (X_1, Y_1, Z_1) and (X, Y, Z) respectively (Fig. 46). If the masses of the sun and planet are

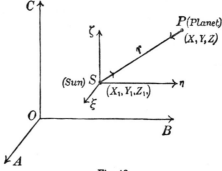

Fig. 46.

M and m, then by Newton's law of gravitation, the planet P is attracted towards S with the force GMm/r^2, where r is the distance SP. The component of this force in the positive direction of the axis OA is

$$\frac{GMm}{r^2} \cdot \frac{X_1 - X}{r} \quad \text{or} \quad - \frac{GMm\,(X - X_1)}{r^3}.$$

If $\ddot{X}\left(\equiv \dfrac{d^2 X}{dt^2}\right)$ denotes the acceleration of P parallel to OA, we have, by Newton's second law of motion,

$$m\ddot{X} = - \frac{GMm\,(X - X_1)}{r^3} \qquad \ldots\ldots(22).$$

Now gravitational attraction is mutual and the sun S will be attracted towards P with the force GMm/r^2 and the component of this force parallel to OA is

$$\frac{GMm}{r^2}\,\frac{(X - X_1)}{r}.$$

If $\ddot{X}_1$ is the acceleration of the sun parallel to OA, we have as before

$$M\ddot{X}_1 = \frac{GMm\,(X - X_1)}{r^3} \qquad \ldots\ldots(23).$$

Dividing (22) by m and (23) by M and subtracting the resulting equations, we obtain

$$\ddot{X} - \ddot{X}_1 = -\,G\,(M + m)\frac{(X - X_1)}{r^3} \qquad \ldots\ldots(24).$$

There are two similar equations in Y and Z.

Write $\quad \xi = X - X_1, \ \eta = Y - Y_1, \ \zeta = Z - Z_1.$

Then $(\xi,\ \eta,\ \zeta)$ are the co-ordinates of the planet P referred to rectangular axes passing through the sun, and we have from (24), writing μ for $G\,(M + m)$ as in (13),

$$\ddot{\xi} + \frac{\mu\xi}{r^3} = 0 \qquad \ldots\ldots(25).$$

Similarly

$$\ddot{\eta} + \frac{\mu\eta}{r^3} = 0 \qquad \ldots\ldots(26),$$

$$\ddot{\zeta} + \frac{\mu\zeta}{r^3} = 0 \qquad \ldots\ldots(27).$$

These are the equations of motion of the planet P with reference to the sun.

Multiply (26) by ζ and (27) by η and subtract. We obtain

$$\zeta\ddot{\eta} - \eta\ddot{\zeta} = 0,$$

that is,

$$\frac{d}{dt}\,(\zeta\dot{\eta} - \eta\dot{\zeta}) = 0.$$

Integrating, we have $\quad \zeta\dot{\eta} - \eta\dot{\zeta} = A \qquad \ldots\ldots(28),$

in which A is a constant of integration.

In a similar way, we derive from (25) and (27) and then from (25) and (26),

$$\xi\dot{\zeta} - \zeta\dot{\xi} = B \qquad \ldots\ldots(29),$$

and $\qquad \eta\dot{\xi} - \xi\dot{\eta} = C \qquad \ldots\ldots(30),$

in which B and C are constants of integration.

Multiply (28), (29) and (30) by ξ, η, ζ in order and add. Then

$$A\xi + B\eta + C\zeta = 0 \qquad \ldots\ldots(31),$$

which is the equation of a plane passing through the origin of the ξ, η, ζ co-ordinates, that is, passing through the sun S. The interpretation is that, as the co-ordinates (ξ, η, ζ) of the planet

P satisfy the relation given by (31), its motion with respect to the sun takes place in a plane. This plane is the orbital plane.

64. *The equation of the orbit.*

We can now refer the motion of the planet to two axes passing through the sun and lying in the orbital plane: let (x, y) be the co-ordinates of the planet referred to these axes. In Fig. 45, we shall suppose that ST is the x-axis, the y-axis being at right angles to ST and of course in the plane of the orbit. The equations of motion of the planet are then

$$\ddot{x} + \mu\frac{x}{r^3} = 0 \qquad \text{......(32)},$$

$$\ddot{y} + \mu\frac{y}{r^3} = 0 \qquad \text{......(33)}.$$

These equations, in rectangular co-ordinates, will now be transformed in terms of the polar co-ordinates r and θ. (In Fig. 45, $SP = r$ and $T\hat{S}P = \theta$.) We have

$$x = r\cos\theta \quad \text{and} \quad y = r\sin\theta \qquad \text{......(34)}.$$

Let α and β denote the components of the acceleration of the planet P along SP and at right angles to SP respectively. Then

$$\alpha = \ddot{x}\cos\theta + \ddot{y}\sin\theta \qquad \text{......(35)},$$

and $$\beta = \ddot{y}\cos\theta - \ddot{x}\sin\theta \qquad \text{......(36)}.$$

We have, from the first of (34),

$$\dot{x} = \dot{r}\cos\theta - r\sin\theta.\dot{\theta},$$

and $$\ddot{x} = \ddot{r}\cos\theta - 2\dot{r}\sin\theta.\dot{\theta} - r\cos\theta.\dot{\theta}^2 - r\sin\theta.\ddot{\theta} \quad \text{...(37)}.$$

The expression for $\ddot{y}$ can be derived in a similar way; it is

$$\ddot{y} = \ddot{r}\sin\theta + 2\dot{r}\cos\theta.\dot{\theta} - r\sin\theta.\dot{\theta}^2 + r\cos\theta.\ddot{\theta} \quad \text{...(38)}.$$

Inserting in (35) the values of $\ddot{x}$ and $\ddot{y}$ given by (37) and (38), we obtain, after some reduction,

$$\alpha = \ddot{r} - r\dot{\theta}^2 \qquad \text{......(39)}.$$

But by (32) and (33),

$$\alpha \equiv \ddot{x}\cos\theta + \ddot{y}\sin\theta = -\frac{\mu}{r^3}(x\cos\theta + y\sin\theta)$$

$$= -\frac{\mu}{r^2}.$$

Hence $$\ddot{r} - r\dot{\theta}^2 = -\frac{\mu}{r^2} \qquad \text{......(40)}.$$

By a similar procedure, we find that

$$\beta = 2\dot{r}\dot{\theta} + r\ddot{\theta} = 0.$$

But

$$2\dot{r}\dot{\theta} + r\ddot{\theta} = \frac{1}{r}\frac{d}{dt}(r^2\dot{\theta}).$$

Hence

$$\frac{d}{dt}(r^2\dot{\theta}) = 0,$$

and by integration

$$r^2\dot{\theta} = h \qquad \ldots\ldots(41),$$

where h is the constant of integration. Thus (41) is the mathematical expression of Kepler's second law which we considered previously in section 58.

The equation of the path of the planet around the sun can now be derived from (40) and (41); it is, of course, a relation between r and θ. The process consists in eliminating the time t from (40) by means of (41); in these equations the time occurs only in the differential coefficients. We shall write for convenience

$$u = \frac{1}{r} \qquad \ldots\ldots(42),$$

so that, by (41),

$$\dot{\theta} = hu^2 \qquad \ldots\ldots(43).$$

Now

$$\dot{r} = \frac{dr}{dt} = -\frac{1}{u^2}\frac{du}{dt} = -\frac{1}{u^2}\frac{du}{d\theta}\cdot\frac{d\theta}{dt}.$$

Hence, by (43),

$$\dot{r} = -h\frac{du}{d\theta}.$$

Again,

$$\ddot{r} = \frac{d}{dt}(\dot{r}) = \frac{d}{d\theta}(\dot{r})\cdot\frac{d\theta}{dt}$$

$$= -hu^2\frac{d}{d\theta}\left(h\frac{du}{d\theta}\right),$$

so that

$$\ddot{r} = -h^2u^2\frac{d^2u}{d\theta^2} \qquad \ldots\ldots(44).$$

Also, using (43),

$$r\dot{\theta}^2 = h^2u^3 \qquad \ldots\ldots(45).$$

Hence (40) becomes, by means of (44) and (45),

$$-h^2u^2\frac{d^2u}{d\theta^2} - h^2u^3 = -\mu u^2,$$

or

$$\frac{d^2u}{d\theta^2} + u = \frac{\mu}{h^2} \qquad \ldots\ldots(46).$$

The general solution of (46) is given by

$$u = \frac{\mu}{h^2}[1 + e\cos(\theta - \omega)] \qquad \ldots\ldots(47),$$

in which e and ω are the two essential constants of integration, or by (42),

$$r = \frac{h^2/\mu}{1 + e \cos (\theta - \omega)} \qquad \ldots\ldots(48).$$

This last equation is identical with (2)—the equation of an ellipse—if $p = h^2/\mu$, that is, making use of (3), if

$$h^2 = \mu a\,(1 - e^2) \qquad \ldots\ldots(49),$$

which is equation (14) previously considered. The constant of integration e is seen to be identified with the eccentricity.

It is to be remarked that (48) is the general equation of a conic section which may be

        (i)   an ellipse, if $e < 1$,

       (ii)   a parabola, if $e = 1$,

     (iii)   a hyperbola, if $e > 1$.

Although case (i) is that with which we are closely concerned here, the extension of the possibilities concerning the motion of a body under the gravitational attraction of the sun should be noted.

The equation (41) is simply the mathematical expression of Kepler's second law. Also, by defining the mean angular motion n as in (7), we are led—using (5), (7) and (48)—to the formula

$$n^2 a^3 = \mu \equiv G\,(M + m)$$

already mentioned. Thus from the single law of gravitation we have derived the mathematical equivalents of Kepler's three laws.

65. *Velocity of a planet in its orbit.*

Let V denote the velocity of the planet at the point P in its orbit (Fig. 47). V will be directed along the tangent PT. The components of V are (i) $\dot{r}$ along the radius vector in the direction PR, and (ii) $r\dot{\theta}$ along PL at right angles to the radius vector. We thus have

$$V^2 = \dot{r}^2 + r^2\dot{\theta}^2 \qquad \ldots\ldots(50).$$

Now

$$\dot{r} = - h\,\frac{du}{d\theta},$$

and from (47)

$$\frac{du}{d\theta} = - \frac{\mu}{h^2}\,e \sin (\theta - \omega),$$

so that

$$\dot{r} = \frac{\mu}{h}\,e \sin (\theta - \omega) \qquad \ldots\ldots(51).$$

Also, since $r^2\dot\theta = h$, we have

$$r\dot\theta = hu = \frac{\mu}{h}[1 + e\cos(\theta - \omega)] \qquad \ldots\ldots(52).$$

Hence, squaring (51) and (52), we have from (50)

$$V^2 = \frac{\mu^2}{h^2}[1 + 2e\cos(\theta - \omega) + e^2],$$

which can be written

$$V^2 = \frac{\mu^2}{h^2}[2 + 2e\cos(\theta - \omega) - (1 - e^2)],$$

or, using (47), $\qquad V^2 = 2\mu u - \frac{\mu^2}{h^2}(1 - e^2).$

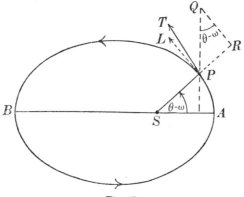

Fig. 47.

But from (49) $\qquad h^2 = \mu a(1 - e^2).$

Hence we obtain, writing $1/r$ for u,

$$V^2 = \mu\left(\frac{2}{r} - \frac{1}{a}\right) \qquad \ldots\ldots(53).$$

This formula gives the velocity V as a function of the radius vector r.

It is seen from (53) that V is greatest when r is least, that is, when the planet is in perihelion. Then $r = a(1 - e)$, and if V_1 denotes the velocity at perihelion we have

$$V_1{}^2 = \frac{\mu}{a}\cdot\frac{1 + e}{1 - e} \qquad \ldots\ldots(54).$$

Similarly, the velocity is a minimum when r is a maximum, that is, when the planet is in aphelion. If V_2 denotes the velocity at aphelion, we obtain

$$V_2{}^2 = \frac{\mu}{a} \cdot \frac{1-e}{1+e} \qquad \ldots\ldots(55).$$

From (54) and (55), $\qquad V_1 V_2 = \frac{\mu}{a}.$

Thus the product of the linear velocities at perihelion and aphelion is independent of the eccentricity of the orbit.

66. *Components of the linear velocity perpendicular to the radius vector and to the major axis.*

We now derive a theorem which will be used later in the investigation of certain problems. In Fig. 47 let PR represent the velocity $\dot{r}$. Draw PQ perpendicular to the major axis AB and draw RQ perpendicular to PR. Then the velocity $\dot{r}$ is equivalent to (i) a velocity along PQ and represented in magnitude by PQ, and (ii) a velocity parallel to $\overrightarrow{QR}$ and represented in magnitude by QR. Now $P\hat{S}A = \theta - \omega$ and by the construction, $P\hat{Q}R = \theta - \omega$. Hence $PQ = PR \operatorname{cosec}(\theta - \omega)$ and $QR = PR \cot(\theta - \omega)$. We thus find that the velocity $\dot{r}$ is equivalent to

   (i) $\dot{r} \operatorname{cosec}(\theta - \omega)$ along PQ,

   (ii) $\dot{r} \cot(\theta - \omega)$ parallel to $\overrightarrow{QR}$.

Now the velocity V is equivalent to $\dot{r}$ along PR and $r\dot{\theta}$ along PL, PL being perpendicular to SP. Hence V is equivalent to

   (i) $\dot{r} \operatorname{cosec}(\theta - \omega)$ along PQ, that is, perpendicular to the major axis,

and (ii) $r\dot{\theta} - \dot{r} \cot(\theta - \omega)$ along PL, that is, perpendicular to the radius vector.

From (51), we obtain for (i),

$$\dot{r} \operatorname{cosec}(\theta - \omega) = \frac{e\mu}{h} \qquad \ldots\ldots(56),$$

and, similarly, from (51) and (52) for (ii),

$$r\dot{\theta} - \dot{r} \cot(\theta - \omega) = \frac{\mu}{h} \qquad \ldots\ldots(57).$$

Hence (57) and (56) express the result that the velocity V of a planet at any point of its orbit can be resolved into a constant velocity μ/h perpendicular to the radius vector and a constant velocity $e\mu/h$ perpendicular to the major axis.

67. *The true and eccentric anomalies.*

Theoretically, Kepler's second law enables us to calculate the position of a planet in its orbit at any time provided we know the semi-major axis a, the eccentricity e, the time τ at which the planet passed through perihelion and the orbital period T. In Fig. 48 let P be the position of the planet at time t. In the

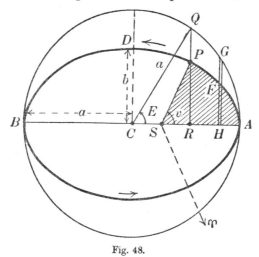

Fig. 48.

interval $(t - \tau)$ the radius vector moving from SA to SP sweeps out the shaded area SPA. Now by the second law

$$\text{Area } SPA : \text{Area of ellipse} = t - \tau : T,$$

that is,
$$\text{Area } SPA = \frac{\pi ab\,(t - \tau)}{T},$$

or, introducing the mean angular motion n ($n = 2\pi/T$), we can write

$$\text{Area } SPA = \tfrac{1}{2}\,nab\,(t - \tau) \qquad \ldots\ldots(58),$$

in which b is given by $b^2 = a^2\,(1 - e^2)$. Thus, at a given instant t, the right-hand side of (58) can be calculated when a, e, τ and n are known. Hence the area SPA is found and the position of the planet in its orbit determined.

However simple this method appears theoretically, in practice it is inconvenient and we shall now derive the formulae generally applied for determining the position of a planet in its orbit.

Let the radius vector SP make the angle v with SA; v is called the *true anomaly* and it is clearly related to θ and ω by the formula

$$v = \theta - \omega \qquad \dots\dots(59).$$

Let a circle be described on the major axis AB as diameter; its radius is thus a. Let RP, the perpendicular from P to AB, be produced to meet this circle at Q. Then the angle QCA is called the *eccentric anomaly* which we denote by E.

By a well-known property of the ellipse

$$PR : QR = b : a \qquad \dots\dots(60),$$

where b is the semi-minor axis CD.

Now $PR = r \sin v$ and $QR = CQ \sin E = a \sin E$. Hence, from (60),

$$r \sin v = b \sin E \qquad \dots\dots(61).$$

Again, $SR = r \cos v$. Also $SR = CR - CS = a \cos E - ae$. Hence

$$r \cos v = a (\cos E - e) \qquad \dots\dots(62).$$

Square (61) and (62) and add. Then, putting $b^2 = a^2 (1 - e^2)$, we obtain, after a little reduction,

$$r = a (1 - e \cos E) \qquad \dots\dots(63).$$

Again, $\qquad 2r \sin^2 \dfrac{v}{2} = r (1 - \cos v)$

$$= a (1 - e \cos E) - a (\cos E - e),$$

applying (63) and (62); hence

$$2r \sin^2 \frac{v}{2} = a (1 + e) (1 - \cos E) \qquad \dots\dots(64).$$

Similarly, $\qquad 2r \cos^2 \dfrac{v}{2} = a (1 - e) (1 + \cos E) \qquad \dots\dots(65).$

Divide (64) by (65). Then

$$\tan^2 \frac{v}{2} = \frac{1 + e}{1 - e} \cdot \frac{1 - \cos E}{1 + \cos E},$$

from which $\qquad \tan \dfrac{v}{2} = \left(\dfrac{1 + e}{1 - e}\right)^{\frac{1}{2}} \tan \dfrac{E}{2} \qquad \dots\dots(66).$

Thus, by means of (63) and (66), we can express the radius vector r and the true anomaly v in terms of the eccentric anomaly E. We now derive Kepler's equation by means of which the eccentric anomaly E can be expressed in terms of known quantities. The eccentric anomaly thus enters into our theoretical discussion as an intermediary angle.

68. *Kepler's equation.*

From our definition of the mean angular motion n we see that the product $n\,(t-\tau)$, which occurs in formula (58), represents the angle described in an interval $(t-\tau)$ by a radius vector rotating about S with constant angular velocity n. We define $n\,(t-\tau)$ as the *mean anomaly*, which we denote by M, so that

$$M = n\,(t-\tau) \qquad \dots\dots(67).$$

Hence, from (58), the area SPA in Fig. 48 is given by

$$\text{Area } SPA = \tfrac{1}{2}abM \qquad \dots\dots(68).$$

We now express this area in terms of the eccentric anomaly E.

The shaded area SPA is equal to the area of the triangle PSR together with the area RPA. Take first the triangle PSR. Its area is $\tfrac{1}{2}SR.PR$. But $SR = CR - CS$ or $SR = a\cos E - ae$. Also $PR = \dfrac{b}{a}\,QR$ by (60), and therefore $PR = b\sin E$. Hence the area of the triangle PSR is $\tfrac{1}{2}ab\sin E\,(\cos E - e)$. Consider now the area RPA. Divide it up into strips perpendicular to AB and extend the strips to meet the circle on AB as diameter. Consider the strip FH. Since $FH = \dfrac{b}{a}\,GH$, then the sum of all the strips forming the area RPA is equal to $\dfrac{b}{a}$ times the sum of the strips forming the area QRA; hence

$$\text{Area } RPA = \frac{b}{a}.\,\text{Area } QRA \qquad \dots\dots(69).$$

But the area QRA is the area of the sector CQA minus the area of the triangle QCR. But since $Q\hat{C}A = E$, the area of CQA is $\tfrac{1}{2}a^2E$ and the area of the triangle QCR is $\tfrac{1}{2}a^2\sin E\cos E$. Hence (69) becomes

$$\text{Area } RPA = \frac{b}{a}\,[\tfrac{1}{2}a^2E - \tfrac{1}{2}a^2\sin E\cos E]$$

$$= \tfrac{1}{2}ab\,(E - \sin E\cos E).$$

Add to RPA the area of the triangle PSR previously found and we obtain

$$\text{Area } SPA = \tfrac{1}{2}ab\,(E - e\sin E) \qquad \ldots\ldots(70).$$

From (68) and (70), we then have

$$E - e\sin E = M \equiv n\,(t - \tau) \qquad \ldots\ldots(71).$$

This is Kepler's equation. It is a relation between the eccentric anomaly E and the mean anomaly M. If M and e are known, it is then possible to determine the corresponding value of E. It is to be noted, from the manner in which (71) has been derived, that both E and M are supposed to be expressed in circular measure.

69. *Solution of Kepler's equation.*

The general method of solution depends on deriving an approximate value of E, nearly satisfying the equation, either by inspection or by special tables or by one of the many graphical processes devised at different times. Let this approximate value be denoted by E_0 and let the true value be $E_0 + \Delta E_0$. Then, rigorously, $\quad (E_0 + \Delta E_0) - e\sin(E_0 + \Delta E_0) = M,$

or $\quad E_0 + \Delta E_0 - e\sin E_0\cos\Delta E_0 - e\cos E_0\sin\Delta E_0 = M.$

Since ΔE_0 is supposed to be small, we can write as an approximation: $\cos\Delta E_0 = 1$, $\sin\Delta E_0 = \Delta E_0$, so that

$$(E_0 - e\sin E_0) + \Delta E_0\,(1 - e\cos E_0) = M.$$

As E_0 and e are known we can compute a quantity M_0 given by

$$M_0 = E_0 - e\sin E_0 \qquad \ldots\ldots(72),$$

so that we then obtain

$$\Delta E_0 = \frac{M - M_0}{1 - e\cos E_0} \qquad \ldots\ldots(73),$$

from which ΔE_0 can be calculated. Then $(E_0 + \Delta E_0)$ is a more accurate value of E. The process can then be repeated, if necessary, with $(E_0 + \Delta E_0)$ as a new approximation to E.

If the eccentricity is small, say less than 0·1, we obtain by inspection of Kepler's equation an approximate value of the eccentric anomaly. For, neglecting the term $e\sin E$, we obtain for the first approximation E_0 the simple result

$$E_0 = M,$$

and by the application of (73) a more accurate value of the eccentric anomaly can be derived.

When the eccentricity is large, it is not so easy to obtain an approximate value of the eccentric anomaly satisfying Kepler's equation. In these circumstances the use of special tables, such as Bauschinger's[*] or Astrand's[†], or of a graphical construction, greatly facilitates the computations. In Bauschinger's tables, the values of E are tabulated for different values of the eccentricity e and the mean anomaly M; then, by inspection or by an easy interpolation, a very good approximate value E_0 can be obtained. The application of (73) follows as already indicated.

Example. Required to calculate the eccentric anomaly of Mars 200 days after perihelion passage, given: $e = 0 \cdot 09334$, $T = 1 \cdot 8809$ years.

The mean angular motion n is defined by $n = 2\pi/T$ or $360/1 \cdot 8809$ degrees per year or $1886 \cdot 52$ seconds of arc per mean solar day. Hence

$$M = 200 \times 1886'' \cdot 52 = 377304'' = 104° \, 48' \, 24''.$$

If we neglect the term $e \sin E$ in Kepler's equation we can take, as a rough value of the eccentric anomaly, $E_0 = 105°$ (the value of M to the nearest degree).

Compute now the value of M_0 given by (72), namely

$$M_0 = E_0 - e \sin E_0.$$

We first notice that, in this theoretical formula, M_0 and E_0 are necessarily expressed in circular measure. If we now suppose them expressed in seconds of arc, we have

$$M_0 \sin 1'' = E_0 \sin 1'' - e \sin E_0,$$

or $\qquad\qquad M_0 = E_0 - e \sin E_0 \operatorname{cosec} 1''.$

$$\log e = \bar{2} \cdot 97007$$

$$\log \sin E_0$$

$$(\equiv \log \sin 105°) = \bar{1} \cdot 98494$$

$$\log \operatorname{cosec} 1'' = 5 \cdot 31443$$

$$\overline{\qquad 4 \cdot 26944 \qquad} \text{ which is } \log 18597.$$

Hence $\qquad M_0 = E_0 - 18597'' = 105° - 5° \, 9' \, 57'',$

or $\qquad\quad M_0 = 99° \, 50' \, 3''.$

---

[*] J. Bauschinger, *Tafeln zur theoretischen Astronomie* (Leipzig, 1901).
[†] J. J. Astrand, *Hülfstafeln* (Leipzig, 1890).

Then the formula (73) for ΔE_0 becomes

$$\Delta E_0 = \frac{104° \ 48' \ 24'' - 99° \ 50' \ 3''}{1 - e \cos 105°} \ .$$

But $e \cos 105° = -0\cdot09334 \sin 15° = -0\cdot0242.$

Hence $\Delta E_0 = \dfrac{4° \ 58' \ 21''}{1\cdot0242} = \dfrac{17901}{1\cdot0242}$ seconds of arc

$$= 17478''$$
$$= 4° \ 51' \ 18''.$$

Hence $E_0 + \Delta E_0 = 109° \ 51' \ 18'',$

which is a more accurate value of the eccentric anomaly satisfying Kepler's equation. We ought to repeat the process again, taking now 109° 51′ 18″ as an approximation to the value of E, and to continue until the required accuracy is obtained. We leave the step thus indicated to the student.

From Bauschinger's tables we extract the following:

	$e = 0\cdot0$	$e = 0\cdot1$
$M = 104°$	$E = 104°$	$E = 109°\cdot40$
$M = 105°$	$E = 105°$	$E = 110°\cdot37$

Hence, by a rough interpolation for $M = 104° \ 48'$ and $e = 0\cdot093$, we derive $E_0 = 109°\cdot8$, which is close to the result of our first computation. The tables thus save, in this instance, the computation of our approximate value of E. A more accurate value of E is obtained by making the calculations, based on (73), in the manner already indicated, taking for E_0 the value 109°·8.

70. *Summary of the formulae of elliptic motion.*

We now collect the more important formulae of elliptic motion which have been derived in the previous pages:

$$
\begin{array}{ll}
\text{(i)} & r = \dfrac{a\,(1 - e^2)}{1 + e \cos v} \\[2mm]
\text{(ii)} & n^2 a^3 = \mu \equiv G\,(M + m) \\[2mm]
\text{(iii)} & h^2 = \mu a\,(1 - e^2) \\[2mm]
\text{(iv)} & E - e \sin E = M \equiv n\,(t - \tau) \\[2mm]
\text{(v)} & r = a\,(1 - e \cos E) \\[2mm]
\text{(vi)} & \tan \dfrac{v}{2} = \left(\dfrac{1 + e}{1 - e}\right)^{\frac{1}{2}} \tan \dfrac{E}{2}
\end{array}
\qquad \Bigg\} \ \ \dots\dots(74).
$$

In these formulae a, e and τ are *elements of the elliptic orbit.*

To obtain the position of the planet in its orbit at time t, the elements a, e, τ and the period being given, the procedure is as follows:

(a) Calculate the eccentric anomaly E by means of (iv), in the manner outlined in section 69, or otherwise.

(b) Calculate the radius vector r by means of (v).

(c) Calculate* the true anomaly v by means of (vi).

(d) As a check, calculate the radius vector r by means of (i), using the value of the true anomaly v found in (c).

71. *The eccentric anomaly expressed as a series in terms of* e *and the mean anomaly.*

We now express Kepler's equation in a different form, in which we shall derive E as a series in terms of the eccentricity e and the mean anomaly M. We have Kepler's equation

$$E = M + e \sin E \qquad \ldots\ldots(75).$$

We shall regard e as a small fraction and the first approximation to E—denoted by E_1—will evidently be given by neglecting $e \sin E$, so that
$$E_1 = M.$$

A more accurate value of E—the second approximation denoted by E_2—will clearly be given by writing E_1 (or M) on the right of (75). Thus

$$E_2 = M + e \sin M \qquad \ldots\ldots(76).$$

In the same way, denoting the third approximation to E by E_3, we write
$$E_3 = M + e \sin E_2,$$
which by (76) becomes

$$E_3 = M + e \sin [M + e \sin M],$$

or $\quad E_3 = M + e \sin M \cos [e \sin M] + e \cos M \sin [e \sin M]$.

But, since e is small, we can write this last equation

$$E_3 = M + e \sin M + e \cos M . e \sin M \qquad \ldots\ldots(77),$$

which is correct up to terms in e^2. (77) is equivalent to

$$E_3 = M + e \sin M + \tfrac{1}{2} e^2 \sin 2M \qquad \ldots\ldots(78).$$

We can proceed in this way as far as we like. A further approximation will give

$$E = M + \left(e - \frac{e^3}{8}\right) \sin M + \tfrac{1}{2} e^2 \sin 2M + \tfrac{3}{8} e^3 \sin 3M \ldots(79).$$

---

* In the method to be described in section 77 the quantities E and r are alone calculated.

5   **S A**

When terms containing e^4 and higher powers can be neglected, formula (79) enables us to calculate the value of E when e and M are given.

72. *The true anomaly expressed as a series in terms of* e *and the eccentric anomaly.*

We begin with the formula

$$\tan \frac{v}{2} = \left(\frac{1+e}{1-e}\right)^{\frac{1}{2}} \tan \frac{E}{2} \qquad \ldots \ldots (80).$$

Let us define an angle ϕ, lying between 0 and $\frac{\pi}{2}$, by

$$\sin \phi = e \qquad \ldots \ldots (81).$$

Then we can write (80) in the form

$$\tan \frac{v}{2} = \frac{1 + \tan \dfrac{\phi}{2}}{1 - \tan \dfrac{\phi}{2}} \tan \frac{E}{2},$$

or, putting $\tan \dfrac{\phi}{2} = x$,

$$\tan \frac{v}{2} = \frac{1+x}{1-x} \tan \frac{E}{2} \qquad \ldots \ldots (82).$$

Since $2 \cos \theta = e^{i\theta} + e^{-i\theta}$ and $2i \sin \theta = e^{i\theta} - e^{-i}$, where i is defined by $i^2 = -1$, we can write

$$\tan \frac{v}{2} = \frac{e^{iv/2} - e^{-iv/2}}{i \left(e^{iv/2} + e^{-iv/2}\right)} = \frac{e^{iv} - 1}{i \left(e^{iv} + 1\right)},$$

with a similar formula for $\tan \dfrac{E}{2}$. Hence (82) can be written

$$\frac{e^{iv} - 1}{e^{iv} + 1} = \frac{1+x}{1-x} \cdot \frac{e^{iE} - 1}{e^{iE} + 1},$$

from which $e^{iv} = \dfrac{e^{iE} - x}{1 - x e^{iE}},$

or $e^{iv} = e^{iE} (1 - x e^{-iE})/(1 - x e^{iE}),$

from which, by taking logarithms of both sides,

$$iv = iE + \log (1 - x e^{-iE}) - \log (1 - x e^{iE}).$$

Now $x = \tan \dfrac{\phi}{2}$ and $\phi = \sin^{-1}(e)$; hence $x < 1$ numerically, since

$e < 1$. Applying the formula for the logarithmic series we derive, after some reduction,

$$v = E + 2\left(x \sin E + \frac{x^2}{2}\sin 2E + \frac{x^3}{3}\sin 3E + \dots\right)\dots(83).$$

Now

$$x \equiv \tan\frac{\phi}{2} = \frac{\sin^2\dfrac{\phi}{2}}{\sin\dfrac{\phi}{2}\cos\dfrac{\phi}{2}} = \frac{1-\cos\phi}{e} = \frac{1-(1-e^2)^{\frac{1}{2}}}{e},$$

or
$$x = \tfrac{1}{2}e + \tfrac{1}{8}e^3 + \dots \qquad \dots\dots(84).$$

Hence from (83) and (84) we obtain finally

$$v = E + (e + \tfrac{1}{4}e^3)\sin E + \tfrac{1}{4}e^2\sin 2E + \tfrac{1}{12}e^3\sin 3E \ \dots(85),$$

correct to the third power of e.

73. *The equation of the centre.*

By means of (79) and (85) we can express v as a series in e and M. We shall keep only terms up to e^3. Now in (85), E occurs in the form $\sin E$, $\sin 2E$ and $\sin 3E$, and we first find from (79) the values of these quantities expressed in terms of e and M. Thus, by (79),

$$\sin E = \sin\left[M + \left(e - \frac{e^3}{8}\right)\sin M + \tfrac{1}{2}e^2\sin 2M + \tfrac{3}{8}e^3\sin 3M\right]$$
$$\dots\dots(86),$$

but, as in (85) $\sin E$ is multiplied by the factor e, we need only keep terms in (86) up to e^2, so that

$$\sin E = \sin\left[M + e\sin M + \tfrac{1}{2}e^2\sin 2M\right],$$

which can be expanded in the form

$$\sin E = \sin M\cos\left[e\sin M + \tfrac{1}{2}e^2\sin 2M\right]$$
$$\qquad\qquad + \cos M\sin\left[e\sin M + \tfrac{1}{2}e^2\sin 2M\right]$$
$$= \sin M\left[1 - \frac{e^2}{2}\sin^2 M\right] + \cos M\left[e\sin M + \tfrac{1}{2}e^2\sin 2M\right]$$
$$= \sin M + \tfrac{1}{2}e\sin 2M + \tfrac{1}{2}e^2(\sin 2M\cos M - \sin^3 M)$$
$$= (1 - \tfrac{1}{8}e^2)\sin M + \tfrac{1}{2}e\sin 2M + \tfrac{3}{8}e^2\sin 3M.$$

Similarly, keeping only the necessary terms, we derive

$$\sin 2E = \sin 2M + e(\sin 3M - \sin M),$$
$$\sin 3E = \sin 3M.$$

Hence from (79) and the expressions for sin E, sin $2E$ and sin $3E$ just found, the formula (85) becomes

$$v - M = (2e - \tfrac{1}{4}e^3) \sin M + \tfrac{5}{4}e^2 \sin 2M + \tfrac{13}{12}e^3 \sin 3M$$

$$\dots\dots(87).$$

This last formula is known as the *equation of the centre.* Its importance lies in the fact that the true anomaly v is expressed directly in terms of the eccentricity e and the mean anomaly M. When e and M are given, v can thus be calculated. We shall make use of the equation of the centre in a later chapter.

74. *The orbit in space.*

We shall take the plane of the ecliptic as the fundamental plane. In general, the orbital plane of any planet, other than the earth, will be inclined at some particular angle to the ecliptic. Suppose a sphere drawn with the sun as centre (Fig. 49).

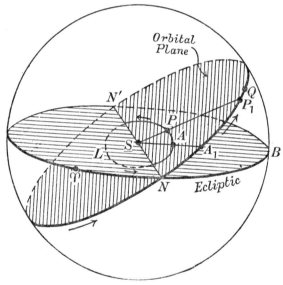

Fig. 49.

As the planet's orbital plane (shown shaded in the figure) passes through the sun, it will therefore cut the sphere in a great circle NQN'. This great circle intersects the ecliptic in two points N and N', called the *nodes.* The orbital ellipse APL is drawn in the figure, A being perihelion; SA produced cuts the great circle NQN' in A_1 and, if P is the position of the planet in its orbit at any time t, the radius vector SP when produced will cut the great circle NQN' in P_1. Thus, since $A\hat{S}P$ is the true anomaly v, the angle A_1SP_1 or the great circle arc A_1P_1 is also v. In the figure, v increases in the direction of the arrow between A_1 and P_1. N is then called the *ascending node* and N' the *descending node*.

The angle P_1NB (denoted by i) defines the inclination of the orbital plane to the ecliptic; i is simply called the *inclination*. The arc NA_1, measured from the ascending node to the point A_1, will be denoted by ω.

Let Υ be the position of the vernal equinox. The arc ΥN is the *longitude of the ascending node*, which we denote by θ. The sum of the arcs ΥN and NA_1 is called the *longitude of perihelion*, denoted by ϖ. Hence

$$\varpi = \theta + \omega \qquad \ldots\ldots(88).$$

It is to be noted that it is only the part θ of the longitude of perihelion that is measured along the ecliptic.

θ, i, ϖ are three elements of the orbit. The longitude of the node (θ) determines the points N and N' on the celestial sphere at which the plane of the orbit intersects the ecliptic; the element i specifies the angle at which the orbital plane is inclined to the ecliptic; the longitude of perihelion (ϖ) determines the direction of perihelion with reference to the ecliptic and the vernal equinox.

As the three elements a, e and τ considered in section 70 refer to the orbital ellipse only, we now see that the complete specification of a planetary orbit requires six elements; they are a, e, τ, θ, i and ϖ.

In Fig. 49 the sum of the arcs ΥN and NP_1 is called the *true longitude of the planet* in its orbit. Denote it by L. Then

$$L = \theta + \omega + v,$$

or, by (88), $\qquad L = \varpi + v \qquad \ldots\ldots(89).$

Now consider a radius vector coinciding with SA at time τ and moving in the plane of the orbit with mean angular velocity n. At time t, the radius vector has described the angle $n(t - \tau)$—the mean anomaly M—and we shall suppose that it then intersects the sphere in Q. Thus the arc $A_1 Q$ is M. Let l denote the sum of the arcs ΥN and NQ. Then

$$l = \theta + \omega + n(t - \tau),$$

or
$$l = \varpi + n(t - \tau) \qquad \ldots\ldots(90).$$

l is called the *mean longitude* of the planet.

The formula (90) is generally written in the form

$$l = nt + \epsilon \qquad \ldots\ldots(91),$$

where ϵ is given by
$$\epsilon = \varpi - n\tau \qquad \ldots\ldots(92).$$

From (91) it is seen that ϵ is the mean longitude when $t = 0$. This instant is called the "epoch" and ϵ is thus the *mean longitude at the epoch*.

From (90) we have

$$M \equiv n(t - \tau) = l - \varpi,$$

so that, by (91), $M = nt + \epsilon - \varpi.$

Hence Kepler's equation can be written

$$E - e \sin E = nt + \epsilon - \varpi \qquad \ldots\ldots(93).$$

Since ϵ is defined by (92) in terms of the two elements ϖ and τ, it may be regarded as one of the six elements of the orbit in place of τ.

As an illustration, the elements of the orbit of Mars for the epoch 1929 January 0, Greenwich Mean Noon, are given below:

$a = 1 \cdot 52369$ astronomical units $\qquad \theta = 49° \, 0' \, 36'' \cdot 0$

$e = 0 \cdot 09334 \qquad\qquad\qquad\qquad i = 1° \, 51' \, 0'' \cdot 5$

$\epsilon = 84° \, 42' \, 33'' \cdot 9 \qquad\qquad\qquad \varpi = 334° \, 45' \, 7'' \cdot 3$

75. *The heliocentric ecliptic rectangular co-ordinates of a planet.*

In Fig. 50 let $S\Upsilon$, SB, SK be rectangular axes through the sun, forming a right-handed system, Υ being the vernal equinox, B the point on the ecliptic $90°$ from Υ and K the pole of the ecliptic. Let P be the position of the planet in its orbit at time t so that $SP = r$. Let P_1 be the point of intersection of SP with the celestial sphere. Denote by (x_1, y_1, z_1) the rectangular co-ordinates of P with respect to the axes $S\Upsilon$, SB and SK. Then

$$\frac{x_1}{r} = \cos PS\Upsilon = \cos P_1 S\Upsilon,$$

or, writing the arc $P_1 \Upsilon$ for $P_1 \hat{S} \Upsilon$, we have

$$x_1 = r \cos P_1 \Upsilon \qquad \qquad ......(94).$$

Similarly, $\qquad y_1 = r \cos P_1 B \qquad \qquad ......(95),$

$$z_1 = r \cos P_1 K \qquad \qquad ......(96).$$

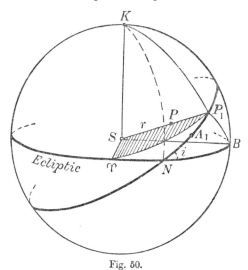

Fig. 50.

Consider now the spherical triangle $P_1 \Upsilon N$ in which $\Upsilon N = \theta$, $NP_1 = \omega + v$ and $P_1 \hat{N} \Upsilon = 180° - i$. Then, by the cosine formula **A**,

$$\cos P_1 \Upsilon = \cos \theta \cos (\omega + v) - \sin \theta \sin (\omega + v) \cos i.$$

Hence

$$x_1 = r \left[\cos \theta \cos (\omega + v) - \sin \theta \sin (\omega + v) \cos i \right] ...(97).$$

Similarly, from the triangle $P_1 NB$ in which $NB = 90° - \theta$ and $P_1 \hat{N} B = i$,

$$y_1 = r \left[\sin \theta \cos (\omega + v) + \cos \theta \sin (\omega + v) \cos i \right] ...(98).$$

In the triangle $KP_1 N$, $KN = 90°$, $K \hat{N} P_1 = 90° - i$; hence by applying formula **A** we derive

$$z_1 = r \sin (\omega + v) \sin i \qquad \qquad ......(99).$$

Thus from the formulae (97), (98) and (99) the heliocentric co-ordinates (x_1, y_1, z_1) can be calculated when r and v and the elements θ, i and ω are given.

It is convenient in practice to calculate auxiliary angles a, A, b_1 and B_1 defined by

$$\left.\begin{array}{ll} \sin a \sin A = \cos \theta; & \sin a \cos A = -\sin \theta \cos i \\ \sin b_1 \sin B_1 = \sin \theta; & \sin b_1 \cos B_1 = \cos \theta \cos i \end{array}\right\} \quad (100),$$

in which θ and i are supposed known. Then

$$\left.\begin{array}{l} x_1 = r \sin a \sin (A + \omega + v) \\ y_1 = r \sin b_1 \sin (B_1 + \omega + v) \\ z_1 = r \sin (\omega + v) \sin i \end{array}\right\} \quad \ldots\ldots(101).$$

76. *The heliocentric equatorial co-ordinates of a planet.*

In Fig. 51 let ΥC be the equator on the celestial sphere. Consider the right-handed system of equatorial rectangular axes $S\Upsilon$, SC and SQ, where C is the point on the equator 90° from Υ

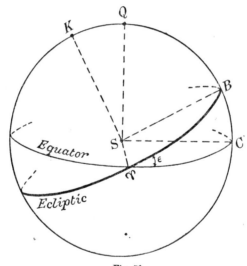

Fig. 51.

and Q is the north pole of the equator. Let (x, y, z) be the co-ordinates of the planet referred to these axes. Now $B\hat{\Upsilon}C = KQ = \epsilon$, the obliquity of the ecliptic. The axes SC and SQ can be obtained from SB and SK by rotating the latter about ST through the angle ϵ. Hence we obtain the following relations between (x, y, z) and (x_1, y_1, z_1):

$$\left.\begin{aligned} x &= x_1 \\ y &= y_1 \cos \epsilon - z_1 \sin \epsilon \\ z &= y_1 \sin \epsilon + z_1 \cos \epsilon \end{aligned}\right\} \qquad \ldots\ldots(102).$$

Inserting the values of x_1, y_1 and z_1 given by (97), (98) and (99) we derive the expressions for x, y and z in terms of r and v, the elements θ, i and ω, and ϵ.

Define auxiliary angles b, B, c and C by the following:

$$\left.\begin{aligned} \sin b \sin B &= \sin \theta \cos \epsilon; & \sin b \cos B &= \cos \theta \cos i \cos \epsilon - \sin i \sin \epsilon \\ \sin c \sin C &= \sin \theta \sin \epsilon; & \sin c \cos C &= \cos \theta \cos i \sin \epsilon + \sin i \cos \epsilon \end{aligned}\right\}$$

$$\ldots\ldots(103).$$

Then we have
$$\left.\begin{aligned} x &= r \sin a \sin (A + \omega + v) \\ y &= r \sin b \sin (B + \omega + v) \\ z &= r \sin c \sin (C + \omega + v) \end{aligned}\right\} \qquad \ldots\ldots(104).$$

The convenience of these formulae is seen when the rectangular co-ordinates are required for several positions of the planet. For when the auxiliary quantities $\sin a$, $\sin b$, $\sin c$, A, B and C have been computed by means of (100) and (103) from the known values of the elements θ, i and ω, and of ϵ they can be used in (104) for determining the several values of (x, y, z). For each position of the planet the radius vector r and the true anomaly v are supposed to be calculated in advance by the methods already described.

An alternative method of deriving the co-ordinates (x, y, z) is described in the following section.

77. *The heliocentric equatorial co-ordinates of a planet (alternative method).*

The heliocentric equatorial co-ordinates can be put in another form* which is now favoured by computers. The first equation of (104) can be written

$$x = r \cos v \left[\sin a \sin (A + \omega)\right] + r \sin v \left[\sin a \cos (A + \omega)\right].$$

Put $P_x = \sin a \sin (A + \omega)$ and $Q_x = \sin a \cos (A + \omega)$.

* *Journal of the Brit. Astron. Association*, vol. XXXII, p. 231 (C. E. Adams).

Then $\qquad P_x = \sin a \sin A \cos \omega + \sin a \cos A \sin \omega,$

which by (100) becomes

$$P_x = \cos \theta \cos \omega - \sin \theta \sin \omega \cos i.$$

Similarly, $\quad Q_x = - \cos \theta \sin \omega - \sin \theta \cos \omega \cos i.$

Now by (62), $r \cos v = a (\cos E - e)$, and by (61),

$$r \sin v = b \sin E,$$

where E is the eccentric anomaly and b is the semi-minor axis of the orbit. We then have, adding the formulae for y and z which can be derived in a similar manner,

$$\left. \begin{array}{l} x = aP_x \cos E + bQ_x \sin E - aeP_x \\ y = aP_y \cos E + bQ_y \sin E - aeP_y \\ z = aP_z \cos E + bQ_z \sin E - aeP_z \end{array} \right\} \quad \ldots\ldots(104\,a),$$

where

$$\left. \begin{array}{l} P_x = \cos \theta \cos \omega - \sin \theta \sin \omega \cos i \\ Q_x = - \cos \theta \sin \omega - \sin \theta \cos \omega \cos i \\ P_y = (\sin \theta \cos \omega + \cos \theta \sin \omega \cos i) \cos \epsilon - \sin \omega \sin i \sin \epsilon \\ Q_y = (-\sin \theta \sin \omega + \cos \theta \cos \omega \cos i) \cos \epsilon - \cos \omega \sin i \sin \epsilon \\ P_z = \sin \omega \sin i \cos \epsilon + (\sin \theta \cos \omega + \cos \theta \sin \omega \cos i) \sin \epsilon \\ Q_z = \cos \omega \sin i \cos \epsilon - (\sin \theta \sin \omega - \cos \theta \cos \omega \cos i) \sin \epsilon \end{array} \right\}$$

$$\ldots\ldots(104\,b).$$

For each position of the planet, the eccentric anomaly E is calculated from the elements, supposed known. The quantities P_x, etc., are also found from the elements. The co-ordinates (x, y, z) can then be easily calculated from (104 a) for as many positions of the planet as are desired.

78. *The heliocentric rectangular co-ordinates of the earth.*

In the case of the earth, the equations (104) assume a simple form as the inclination of the earth's orbit to the ecliptic is zero. Writing (x', y', z') for the heliocentric equatorial co-ordinates of the earth at any time t, we obtain easily the following formulae:

$$\left. \begin{array}{l} x' = r' \cos (\theta' + \omega' + v') \\ y' = r' \sin (\theta' + \omega' + v') \cos \epsilon \\ z' = r' \sin (\theta' + \omega' + v') \sin \epsilon \end{array} \right\} \quad \ldots\ldots(105),$$

in which dashes refer to the respective quantities associated with the earth's orbit. But $(\theta' + \omega')$ is the longitude of perihelion for the earth's orbit; denote it by ϖ'. Then

$$\left.\begin{array}{l} x' = r' \cos (\varpi' + v') \\ y' = r' \sin (\varpi' + v') \cos \epsilon \\ z' = r' \sin (\varpi' + v') \sin \epsilon \end{array}\right\} \qquad \dots\dots(106).$$

The values of r' and v' can be calculated for any time t from the elements of the earth's orbit; hence x', y', z' can be obtained.

An alternative method is based on the formulae (104 a) and (104 b); in the latter, the corresponding values of P_x, etc., are obtained by putting the inclination i equal to zero. As before, the values of the eccentric anomaly for the earth are calculated and the co-ordinates derived by means of (104 a).

79. The planet's geocentric right ascension and declination.

Consider now rectangular axes drawn through the earth (Fig. 52) and parallel to the equatorial axes through the sun illustrated in Fig. 51. Let X, Y, Z denote the co-ordinates of the sun with reference to the axes drawn through E. Then it is clear that

$$X = -x', \quad Y = -y', \quad Z = -z'$$
$$\dots\dots(107).$$

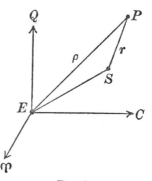

Fig. 52.

X, Y and Z are the *geocentric* equatorial rectangular coordinates of the sun and are calculable by means of (106) as already indicated. Their values are tabulated in the almanacs for every day throughout the year.

Let (ξ, η, ζ) denote the co-ordinates at time t of the planet P with respect to E as origin (Fig. 52). Then since (x, y, z) are the co-ordinates of P with respect to S as origin, we have

$$\xi = X + x; \quad \eta = Y + y; \quad \zeta = Z + z \quad \dots\dots(108).$$

Consider in Fig. 53 the celestial sphere centred at E (the earth's centre). The straight line joining E and the planet cuts the sphere in R. Denote the distance EP by ρ. Then ρ is the *geocentric distance* of the planet at time t. Draw the meridian QRT intersecting the equator ΥC in T. Denote by α, δ the geocentric right ascension and declination of the planet so that $\Upsilon T = \alpha$, $TR = \delta$. Then

$$\xi/\rho = \cos P\hat{E}\Upsilon = \cos R\hat{E}\Upsilon,$$

or $\qquad\qquad \xi = \rho \cos R\Upsilon.$

Similarly, $\qquad \eta = \rho \cos RC,$

$$\zeta = \rho \cos RQ.$$

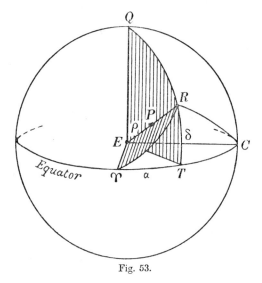

Fig. 53.

But formula **A** gives, for the triangle $R\Upsilon T$ (right-angled at T),

$$\cos R\Upsilon = \cos \alpha \cos \delta.$$

Hence $\qquad\qquad \xi = \rho \cos \alpha \cos \delta.$

Similarly, $\qquad \eta = \rho \sin \alpha \cos \delta \quad \Big\}\qquad ......(109).$

$$\zeta = \rho \sin \delta$$

Hence from (104), (108) and (109) we obtain

$$\rho \cos \alpha \cos \delta = X + r \sin a \sin (A + \omega + v) \quad ......(110),$$

$$\rho \sin \alpha \cos \delta = Y + r \sin b \sin (B + \omega + v) \quad ......(111),$$

$$\rho \sin \delta \qquad = Z + r \sin c \sin (C + \omega + v) \quad ......(112).$$

These equations in the form stated were first given by Gauss.

Given the elements of the planet's orbit, the right-hand sides of (110)–(112) can be calculated for any time t, as we have already explained (X, Y and Z are taken from the *Nautical Almanac*). Thus dividing (111) by (110) we obtain

$$\tan \alpha = \frac{Y + r \sin b \sin (B + \omega + v)}{X + r \sin a \sin (A + \omega + v)},$$

from which α can be derived. Similarly, by dividing (112) by (111), we obtain an expression for $\tan \delta$ cosec α, from which δ is easily calculated. Thus the geocentric right ascension and declination of the planet can be derived for any time t. The observed position will then be the geocentric position as calculated, modified by the correction for refraction and certain other corrections to be discussed in later chapters.

We have thus indicated how the position of a planet in the sky at any time can be derived from a knowledge of its orbit.

The converse problem—*the determination of the elements of a planetary orbit* from an adequate number of observations—is one of much greater difficulty. A vast amount of astronomical literature is devoted to this problem alone and the subject is outside the scope of this book. The reader is referred, for the general principles of the methods of determining orbits from observation, to *Dynamical Astronomy* by H. C. Plummer (Cambridge, 1918) or to *Bahnbestimmung der Planeten und Kometen* by G. Stracke (Berlin, 1929).

80. *The orbital and synodic periods of a planet.*

We have assumed throughout this chapter that the orbital period of a planet is known and we now indicate how this period can be obtained from observation. We shall take the simple case when the planet's orbital plane is assumed to coincide with the plane of the ecliptic and the orbits of the earth and planet are assumed to be circular (Fig. 54). In this figure, the planet considered is taken to be an *outer* planet, for which the heliocentric distance is greater than the heliocentric distance of the earth. At some moment, let the sun, the earth and the planet be collinear as at S, E_1 and P_1 respectively. The earth and planet are then said to be in heliocentric conjunction. As viewed from the earth at E_1 (we neglect the dimensions of the earth in comparison with the dimensions of the orbits concerned), the planet's position P_1 will be diametrically opposite in the sky to the position of the sun; the planet is then said to be in *opposition*. If heliocentric conjunction occurred at a particular place—say, Greenwich—at apparent midnight (when the sun's hour angle is 12^h) the planet would then be on the meridian of the place; this instant could be determined accurately from observation of the planet with the meridian circle. This particular assumption, however, is unlikely to be realised in practice and so observations must be made on a few nights before and after the expected heliocentric conjunction. On each night the difference between

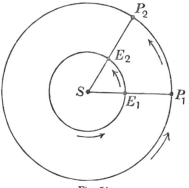

Fig. 54.

the right ascensions of the planet and the sun, at the moment of transit of the planet, can be determined; on some nights this difference will be a little less than 12^h and on other nights a little more than 12^h. By interpolation the instant at which the difference is precisely 12^h can be derived. This gives the time of heliocentric conjunction. Let this time be denoted by t_1.

Let t_2 be the time at which the earth and planet are next in heliocentric conjunction and let their respective positions be E_2 and P_2. Since the planet is at a greater distance from the sun than the earth, by Kepler's third law its mean angular velocity is less than that of the earth. Hence at the end of the interval $(t_2 - t_1)$ the earth has described $360° + E_2\hat{S}E_1$ or $2\pi + \phi$, where $\phi = E_2\hat{S}E_1$, and the planet has described the angle P_2SP_1 or ϕ. Let T_1, T denote the orbital periods of the earth and planet. Then their mean angular motions are $2\pi/T_1$ and $2\pi/T$ respectively. Hence

$$2\pi + \phi = \frac{2\pi}{T_1}(t_2 - t_1),$$

and

$$\phi = \frac{2\pi}{T}(t_2 - t_1).$$

Hence, eliminating ϕ,

$$(t_2 - t_1)\left(\frac{1}{T_1} - \frac{1}{T}\right) = 1 \qquad \ldots\ldots(113).$$

The interval between two successive heliocentric conjunctions in longitude is called the *synodic period* of the planet, which we denote by S. Hence $S = t_2 - t_1$ and therefore, by (113),

$$\frac{1}{S} = \frac{1}{T_1} - \frac{1}{T}.$$

Now, for the earth, T_1 is one year. Also S is obtained from observations and if it is expressed in terms of the year as unit we derive the formula for the planet's orbital period T, in the same unit, as

$$\frac{1}{T} = 1 - \frac{1}{S},$$

or
$$T = \frac{S}{S-1} \qquad \qquad \ldots\ldots(114).$$

For an *inner* planet, the period T' is given by

$$T' = \frac{S'}{1 + S'}.$$

where S' is the corresponding synodic period.

For the outer planets, the synodic periods are evidently greater than one year. For Mercury and Venus the synodic periods, found in this way, are 116 and 584 days.

Owing to the assumptions made in this section with reference to the orbits of the earth and the planet under consideration, the orbital periods derived in this way are approximate; accurate values can only be obtained from a more thorough study of the planetary orbits.

81. *The earth's orbit.*

The eccentricity of the earth's orbit is 0·016739 and the longitude of perihelion is 101° 45′ 8″·1; these quantities refer to the beginning of 1931.

The *sidereal year* is defined to be the interval between two successive returns of the earth to the same point among the stars as viewed from the sun. The sidereal year is 365·2564 mean solar days.

82. *The sun's apparent orbit.*

For many of our purposes, it will be convenient to consider the sun's apparent orbit, that is, the orbit it appears to describe relative to the earth. The eccentricity is, of course, the same as that given in the previous section.

The point of the orbit at which the sun is nearest the earth is *perigee* and the point most remote is *apogee*. As the direction of perihelion (as viewed from the sun) is exactly opposite to the direction of perigee (as viewed from the earth), it follows that

$$\varpi = \varpi_1 + 180° \qquad \ldots\ldots(115),$$

where ϖ denotes the longitude of perigee and ϖ_1 denotes the longitude of perihelion.

It is clear that the sidereal year is the interval between two returns of the sun (moving in the apparent orbit) to the same point among the stars, as viewed from the earth.

The *tropical year* is defined to be the interval between two successive passages of the sun through the vernal equinox. As we shall see in a later chapter, the vernal equinox is not a fixed point on the ecliptic, as we have hitherto assumed it to be, and consequently the tropical year differs somewhat in length from the sidereal year. The tropical year is 365·2422 mean solar days. When the term "year" is used without any qualifying adjective it is the tropical year that is referred to.

Owing to the gravitational attraction of the planets on the earth, the elements of the earth's orbit are not quite constant. In particular, the longitude of perihelion (or, by (115), the longitude of perigee) undergoes small changes. The interval between two successive passages of the earth, in its orbit, through perihelion—or the interval between two successive passages of the sun, in the apparent orbit, through perigee—is called the *anomalistic year*, which is 365·2596 mean solar days.

83. *The moon's orbit.*

The formulae which we have derived for the motion of a planet around the sun are applicable to the motion of the moon around the earth. The moon's orbit in space is defined with reference to the ecliptic as in the case of a planet and the appropriate figure, corresponding to Fig. 49, would be identical in character with the latter, with this exception, that the centre of the celestial sphere is now E (the earth) instead of S (the sun). The longitude of the moon's node, the longitude of *perigee* (the point of the moon's orbit nearest the earth—the furthest point is *apogee*) and the inclination of the orbital plane are defined in the same way as in section 74.

So far as the gravitational attraction of the earth alone is concerned, the moon's orbit would be an ellipse; but, owing to the gravitational influence of the sun and—to a lesser degree—of the planets, the elements of the orbit undergo considerable changes —*perturbations*—which have to be taken into account in many problems with which the moon is associated and, in particular, in the theory of precession and nutation, to be discussed in Chapter x.

The *synodical month* is the interval between two successive "new moons". *New moon* occurs when the geocentric longitudes of the sun and moon are the same. The average value of the synodical month is 29·5306 mean solar days.

The *sidereal month* is defined to be the interval given by the moon's complete circuit of the stars as seen from the earth; its mean value is 27·3217 mean solar days.

Owing to the perturbations, the direction of perigee is altering and the interval required by the moon to move in its path around the earth from perigee to perigee is called the *anomalistic month*; its value is 27·5546 mean solar days.

Again, owing to perturbations, the moon's ascending node has a backward movement along the ecliptic, the longitude of the node decreasing at the rate of nearly 20° per annum (the period required by the moon's node to make a circuit of the ecliptic is 18·6 years). The *nodical month* is defined to be the interval between two successive passages of the moon through the ascending node; its value is 27·2122 mean solar days.

EXERCISES

[Symbols used:

a = semi-major axis of orbit, E = eccentric anomaly,

e = eccentricity, M = mean anomaly.]

v = true anomaly,

1. If the semi-major axes of the orbits of Mercury and Jupiter are 0·387 and 5·203 astronomical units and Jupiter's orbital period is 11·862 years, show that Mercury's orbital period is 0·2406 year.

2. Neglecting the mass of the first satellite of Jupiter, calculate the mass of this planet in terms of that of the earth from the following data:

Period of first satellite: 1^d 18^h 28^m.

Mean distance of first satellite from Jupiter's centre: 267,000 miles.

Radius of the earth: 3960 miles.

Acceleration of gravity at earth's surface: 32·2 feet per second per second.

[*Lond.* 1926.]

3. The orbital period of Jupiter is 4333 mean solar days and Jupiter's mass is 1/1048 times the sun's mass. Show that the period of a small body, of negligible mass, moving in an elliptic orbit round the sun with the same major axis as that of Jupiter, is $4335\frac{1}{16}$ days. [*Lond.* 1930.]

4. If T is the orbital period of a planet, show that a small increase Δa in the semi-major axis a will produce an increase $\dfrac{3T\Delta a}{2a}$ in the period.

5. If V_1 and V_2 are the linear velocities of a planet at perihelion and aphelion respectively, prove that $(1-e)\,V_1 = (1+e)\,V_2.$

6. If $e = \sin \phi$, prove that
$$\tan \frac{v}{2} = \tan (45° + \tfrac{1}{2}\phi)\tan \tfrac{1}{2}E.$$

7. If $e = \sin \phi$, show that, when powers of e above the second are neglected, the value of E satisfying Kepler's equation is given by
$$\tan E = \sec \phi \tan 2\chi,$$
where
$$\tan \chi = \tan (45° + \tfrac{1}{2}\phi)\tan \tfrac{1}{2}M,$$
and
$$\tan E = \sin M/(\cos M - e).$$

8. Prove that
$$\cos v = \frac{\cos E - e}{1 - e \cos E},$$
$$\sin v = \frac{(1-e^2)^{\frac{1}{2}}\sin E}{1 - e \cos E}.$$

9. Prove that, if the fourth and higher powers of e are neglected,
$$E = M + \frac{e \sin M}{1 - e \cos M} - \frac{1}{2}\left(\frac{e \sin M}{1 - e \cos M}\right)^3$$
is a solution of Kepler's equation. [*Ball.*]

10. The relevant elements of an orbit are as follows: $e = 0{\cdot}961733$, T (period) $= 76{\cdot}085$ years; time of perihelion passage, 1910 May 24. If $E = 101°{\cdot}3$ is an approximate solution of Kepler's equation when $M = 47°{\cdot}3$ and $e = 0{\cdot}96$, show that the value of E for 1900 May 24 is $101°\,20'\,33''{\cdot}1$. [*Ball.*]

11. A comet describes a hyperbolic orbit around the sun; prove that the velocity V is given by
$$V^2 = \mu \left(\frac{2}{r} + \frac{1}{a}\right).$$

If its minimum heliocentric distance is k astronomical units and its maximum linear velocity l times the earth's velocity, show that the angle between the asymptotes of the hyperbola is
$$2 \sin^{-1}\left(\frac{1}{l^2 k - 1}\right).$$
[*Edin.* 1921.]

12. If ψ is the angle between the direction of a planet's motion and the direction perpendicular to the radius vector, prove that

$$\tan \psi = \frac{e \sin E}{(1 - e^2)^{\frac{1}{2}}}.$$

13. A planet moves in an orbit inclined at a small angle i to the ecliptic; show that, if its declination is a maximum, either the motion in latitude vanishes or the longitude is approximately $90° + i \cot \epsilon \sin \theta$, where θ is the longitude of the ascending node and ϵ is the obliquity of the ecliptic.

14. The mean distance of Venus from the sun is 0·72 of that of the earth. Determine the greatest altitude at which Venus, supposed to have a circular orbit in the plane of the ecliptic, can be visible after sunset in a given latitude, and the time of year at which this may occur.   [*M.T.*]

15. If (λ_1, β_1), (λ_2, β_2) and (λ_3, β_3) are the heliocentric longitudes and latitudes of a planet at three different points of its orbit, prove that

$$\tan \beta_1 \sin (\lambda_2 - \lambda_3) + \tan \beta_2 \sin (\lambda_3 - \lambda_1) + \tan \beta_3 \sin (\lambda_1 - \lambda_2) = 0.$$

16. Prove that the equation of the centre is given in terms of the true anomaly v by the expression

$$\sum_{p=1}^{\infty} (-)^{p-1} \frac{2\lambda^p \{(p+1) - (p-1) \lambda^2\}}{p (1 + \lambda^2)} \sin pv,$$

where $\lambda = e/(1 + \sqrt{1 - e^2})$, e being the eccentricity of the orbit.

Show that the maximum value of the equation of the centre occurs approximately when $v = \frac{1}{2}\pi + \sin^{-1} (3e/4)$, e being small.   [*Lond.* 1930.]

17. The perihelion distance of a parabolic comet is a astronomical units $(a < 1)$. Assuming that the earth's orbit is circular and that the comet moves in the ecliptic, show that, if t (measured in years) is the interval during which the comet is within the earth's orbit,

$$t = \frac{1}{3\pi} (1 + 2a) (2 - 2a)^{\frac{1}{2}}.$$

18. (*Euler's Theorem.*) If r and r_1 are the radii vectores of two points C and C_1 in a parabolic orbit and if k is the distance CC_1, prove that the time in the orbit between C and C_1 is

$$\frac{T_0}{12\pi} \left\{ \left(\frac{r + r_1 + k}{a} \right)^{\frac{3}{2}} - \left(\frac{r + r_1 - k}{a} \right)^{\frac{3}{2}} \right\},$$

where T_0 is the length of the sidereal year and a is the semi-major axis of the earth's orbit.

19. (*Lambert's Theorem.*) If r and r_1 are the radii vectores of two points C and C_1 in an elliptic orbit, k the distance CC_1, t the time required by the planet to move from C to C_1 and T the orbital period, prove that

$$\frac{2\pi t}{T} = \eta - \sin \eta - (\eta_1 - \sin \eta_1),$$

where $\sin \frac{1}{2} \eta = \frac{1}{2} \left(\frac{r + r_1 + k}{a} \right)^{\frac{1}{2}}$; $\sin \frac{1}{2} \eta_1 = \frac{1}{2} \left(\frac{r + r_1 - k}{a} \right)^{\frac{1}{2}}.$

CHAPTER VI

TIME

84. *Sidereal time.*

In Chapter II we briefly discussed the subject of time and now it is necessary to consider the problem in greater detail. Sidereal time at any instant at a given place is the hour angle of the vernal equinox. Now in Chapter II we regarded the ecliptic and celestial equator as fixed great circles on the celestial sphere, so that the vernal equinox was also regarded as a fixed point whose direction, for convenience, we might visualise as that of a particular star. But owing to the phenomena of precession and nutation the celestial equator can no longer be regarded as a fixed great circle and, consequently, the vernal equinox must be treated as a point on the celestial sphere moving slowly, according to well-ascertained laws, with reference to the background of the stars.

Precession and nutation will be fully considered in Chapter x but for our present purpose we shall assume certain results. In our discussion of sidereal time it will be sufficient to regard the ecliptic as a fixed great circle. Owing to precession, the north celestial pole P (defined by the direction of the earth's axis) describes a small circle around the pole K of the ecliptic in a period of about 25,000 years (Fig. 55). At present the north celestial pole P is a little over 1° from the second magnitude star α Ursae Minoris (Polaris or the pole-star), but their

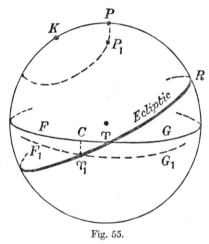

Fig. 55.

relative positions are altering from day to day and from year to year. Two thousand years ago the pole P was 12° from Polaris and 12,000 years hence it will be within a few degrees of the first magnitude star Vega. It is the direction of the earth's axis that is altering continuously with reference to the background of the stars.

Now consider Fig. 55, in which P is the north celestial pole, say, at the beginning of 1900 (denoted by 1900·0) and P_1 its position one year later. PP_1 is the arc of a small circle of which K is the pole. FTG is the celestial equator corresponding to the position P, and $F_1T_1G_1$ is the equator one year later corresponding to the position P_1 of the pole. T is the vernal equinox for 1900·0 and T_1 the vernal equinox for 1901·0. T and T_1 are called the *mean equinoxes* at the dates in question and the corresponding celestial equators are called the *mean equators*. We assume that owing to precession the north celestial pole moves uniformly along the small circle arc PP_1 and that the mean equinox moves uniformly backwards along the ecliptic from T to T_1. It is found that the motion of T along the ecliptic is at the rate of 50″·2 per annum.

Let us return to our definition of sidereal time. Suppose for simplicity that there is a star in the direction of T which we shall assume for the moment to be fixed in direction. The sidereal time would then be defined as the hour angle of T or the hour angle of the star, and the period of rotation of the earth is then simply the interval between two successive transits of this star (or of T) across the meridian of any particular observatory. But when we define sidereal time in relation to the moving equinox, we can no longer regard the earth's rotational period to be the interval between two successive transits of the equinox. In Fig. 55 let CT_1 be a great circle arc drawn through T_1 perpendicularly to the equator FTG. Then the equinox at any given date is separating, in right ascension, from the equinox T for 1900·0 at the annual rate measured by TC. But from the small triangle TCT_1,

$$TC = TT_1 \cos \epsilon,$$

where ϵ is the obliquity of the ecliptic. Hence

$$TC = 50''·2 \cos \epsilon,$$

and inserting the value of ϵ (23° 27′) we easily find that, in time measure, the mean equinox is separating, in right ascension, from

Υ at the rate of $0^s\cdot008$ per sidereal day. Now the direction of motion of the equinox is westwards in the sky—opposite to that in which right ascension increases—and it follows that the interval between two successive transits of the moving equinox over any meridian is $0^s\cdot008$ less than the interval given by a fixed equinox or star. The first interval is a *sidereal day* (defined with reference to the moving equinox) and the second interval is the rotational period of the earth.

Owing to nutation, the true equator at any instant is slightly different from the mean equator at that instant. Consequently the true equinox is displaced slightly along the ecliptic relative to the mean equinox; these small displacements are periodic in character, the period being about 18 years. The difference in right ascension between the true equinox at any instant and the mean equinox at that instant is evidently periodic in character also and may amount numerically to $1^s\cdot2$.

We define *uniform sidereal time* to be associated with the moving mean equinox (precession only being involved) and *true sidereal time* to be associated with the true equinox. As the motion of the true equinox along the ecliptic can be regarded as compounded of (*a*) the uniform motion due to precession ($50''\cdot2$ per annum), and (*b*) the small oscillatory motion, with respect to the mean equinox, due to nutation, it is evident that the interval between two successive transits of the true equinox over a meridian will differ from the interval between two successive transits of the mean equinox by a small amount which is periodic in character. But the differences from day to day are so small that in practice, generally, the sidereal day is taken to mean the interval between two successive transits of the mean equinox. As we have seen, the adopted sidereal day is $0^s\cdot008$ shorter than the earth's rotational period. Sidereal clocks are regulated according to uniform sidereal time.

When a star's position is observed at any instant, that position is referred to the true equator and equinox at that instant. For example, when the transit of a star is observed, the star's right ascension is the true sidereal time of transit; or if we suppose that the right ascension is known (referred to the true equator) we obtain the true sidereal time at the moment of transit. This, however, will not be the time shown by the sidereal clock, which, as we have said, is regulated (we assume, correctly) to keep uniform sidereal time; the difference is the small effect due to nutation. The magnitude of this quantity, which has a principal period of 18 years, can be derived from other considerations and it is tabulated for every day in the almanacs under the heading "nutation in R.A." in the sense:

True sid. time = uniform sid. time + nutation in R.A. ...(1).

For example, at 1931 October 1, U.T. 0^h, the nutation in R.A. is $-0^s.136$ and on November 1 it is $-0^s.151$, a numerical increase of $0^s.015$ in a month or, on the average, $0^s.0005$ per day approximately. This last small quantity represents the daily difference during October between a sidereal day as shown by a clock keeping uniform sidereal time and a sidereal day as defined by the true equinox; even for the most accurate clocks and observations this difference is at present too small to be detected within an interval of a sidereal day or two. It is only when comparatively long intervals are concerned that nutation has to be taken into account. For example, suppose we have a clock keeping uniform sidereal time accurately. Then on October 1 the clock is fast, by (1), on true sidereal time by $0^s.136$; on November 1 it is fast by $0^s.151$. Suppose further that on these dates the true sidereal time is determined accurately. Then during the interval concerned the effect of nutation is to give the clock an apparent "gaining rate" of $0^s.015$ per month. The performances of the modern crystal clocks are such that this minute "rate" cannot be neglected in the delicate problem of the measurement of time.

85. *The mean sun.*

We shall begin by ignoring precession and nutation, thus assuming for the present that the equinox Υ and the equator FTG are fixed (Fig. 56). Let A be the direction of the sun as seen from the earth when the sun is nearest the earth. Since the sun appears to describe an ellipse relative to the earth in the course of a year, A is the direction of perigee. Let n denote the mean angular velocity of the sun in its apparent orbit around the earth. When the sun's position on the celestial sphere is at A, suppose a fictitious body to start from A and to move along the ecliptic with the mean angular velocity n. When the sun reaches the

point S on the ecliptic we shall suppose the fictitious body to be
at M_1. If t and τ are the times corresponding to the position of
the sun at S and at A respec-
tively, the arc AM_1 is the mean
anomaly M which is given by

$$M = n\,(t - \tau) \ \ldots(2).$$

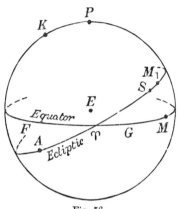

Fig. 56.

When the fictitious body is
at T (at time τ_1) suppose a second
fictitious body, which we call
the *mean sun*, to start from T
and to move along the equator
with the sun's mean angular
velocity n. When the first fictitious body is at M_1, the mean
sun is at M, so that $TM_1 = TM$.
But TM_1 is the sun's mean longi-
tude l and TM is the right ascension of the mean sun (R.A.M.S.);
hence

$$\text{R.A.M.S.} = l = n\,(t - \tau_1) \qquad \ldots\ldots(3).$$

It is evident that according to this equation the R.A.M.S. increases at a uniform rate.

When the movements of the equator and equinox due to
precession and nutation are taken into account, the mean sun is
defined to travel along the true equator (which is itself in motion)
so that its true right ascension at any moment is equal to the
sun's mean longitude. Owing to the non-uniform motion of the
true equinox along the ecliptic due to nutation, it is evident that
the right ascension of the mean sun (R.A.M.S.) is not quite a
uniformly increasing quantity, but in many problems the minute
fluctuations concerned may be disregarded; they are, however,
included in the tabulated values of the R.A.M.S. in the almanacs.

As in Chapter II we define a *mean solar day* to be the interval
between two consecutive transits of the mean sun over an
observer's meridian.

Now the true sidereal time at any instant at a given place is
the hour angle of the true equinox and, as in Chapter II (denoting
the hour angle of the mean sun by H.A.M.S. at the instant concerned), we have

$$\text{True sid. time} = \text{H.A.M.S.} + \text{R.A.M.S.} \qquad \ldots\ldots(4).$$

The mean sun is the basis of civil time reckoning.

We have also, for a star X,

$$\text{True sid. time} = \text{H.A. } X + \text{R.A. } X \qquad \ldots\ldots(5),$$

where R.A. X denotes the right ascension of the star referred to the true equinox and equator at the particular instant under consideration. From (4) and (5),

$$\text{H.A. } X + \text{R.A. } X = \text{H.A.M.S.} + \text{R.A.M.S.} \qquad \ldots\ldots(6),$$
$$\text{or} \qquad \text{H.A.} X + \text{R.A.} X = (\text{U.T.} - 12^{\text{h}} \pm \lambda) + \text{R.A.M.S.} \qquad \ldots\ldots(7),$$

where λ is the longitude of the observer, $+$ for east longitudes and $-$ for west longitudes. Formula (7) enables us to calculate, for example, a star's hour angle when its R.A., the longitude of the observer and the U.T. of the observation are all known.

We remind the reader that, on any day,

R.A.M.S. at U.T. 0^{h}

   $=$ the tabulated sidereal time at U.T. 0^{h} for that day

   -12^{h} (see Chapter II, p. 46).

86. *The sidereal year and the tropical year.*

The time required by the sun to make a complete circuit of the ecliptic is called a *sidereal year*. Thus, it is the interval between its passage through any fixed point on the ecliptic and its next passage through the same point.

The *tropical year* is the average interval between two consecutive passages of the sun through the vernal equinox (which must now be regarded as in motion owing to precession and nutation). When the mean of a large number of such intervals is taken, the periodic effects due to nutation will be eliminated and the average value is the tropical year. Thus we need only consider the motion of the mean equinox due to precession. In Fig. 55, p. 136, we shall now suppose that Υ is the mean equinox when the sun's R.A. and declination are both zero—the sun is then in the direction of Υ. As the sun moves along the ecliptic in the direction ΥR, the mean equinox moves slowly in the opposite direction. Let Υ_1 denote the position of the mean equinox when it and the sun are again coincident. Then the tropical year is the time taken by the sun to describe $360°$ less $\Upsilon_1\Upsilon$. From observations it is found that

The tropical year $= 365 \cdot 2422$ mean solar days.

The relation between the tropical and sidereal years is evidently given by

$$\text{Sid. year : trop. year} = 360° : (360° - 50''\cdot2)$$

(the precessional motion of the equinox, i.e. TT_1, is $50''\cdot2$ per *tropical* year). It is then found that

$$\text{The sidereal year} = 365\cdot2564 \text{ mean solar days.}$$

For civil purposes and in most astronomical relations the tropical year is used, and when we employ the term "year" in succeeding pages it is the tropical year that is implied.

87. *Relation between mean and sidereal time.*

We have defined a tropical year as the average interval between two successive passages of the sun through the moving mean equinox, and it is evident that this interval is the same as that between two successive passages of the mean sun through the mean equinox, nutation being ignored. Hence in the course of a tropical year the R.A.M.S. increases from $0°$ to $360°$, so that the increase in the R.A.M.S. is at the rate of $360° \div 365\cdot2422$ or $59'\ 8''\cdot33$ per mean solar day. Let t_1 denote the sidereal time when the hour angle of the mean sun at a given place is H_1 and let R_1 denote the corresponding value of the R.A.M.S. Then by (4),

$$t_1 = H_1 + R_1 \qquad \qquad \text{......(8).}$$

Let t_2 be the sidereal time one mean solar day later. Then the hour angle of the mean sun has increased by $360°$ and the R.A.M.S. by $59'\ 8''\cdot33$, or in time measure by 24^h and $3^m\ 56^s\cdot556$ respectively. Hence

$$t_2 = (H_1 + 24^h) + (R_1 + 3^m\ 56^s\cdot556) \qquad \text{......(9),}$$

so that, by (8), $t_2 - t_1 = 24^h\ 3^m\ 56^s\cdot556.$

But $t_2 - t_1$ is the interval of sidereal time corresponding to 24^h of mean solar time. Hence

$$24^h \text{ mean solar time} = 24^h\ 3^m\ 56^s\cdot556 \text{ sid. time} ...(10).$$

It is easily calculated from (10) that

$$24^h \text{ sid. time} = (24^h - 3^m\ 55^s\cdot910) \text{ mean solar time} ...(11).$$

The relation (10) may also be derived from the following consideration. At a particular instant the mean sun and the mean equinox are coincident and after a tropical year they are together again. In this interval, the earth has rotated about its axis 365·2422 times with respect to the mean sun and once more with respect to the mean equinox. Hence

$$365·2422 \text{ mean solar days} = 366·2422 \text{ sidereal days} \quad …(12),$$

from which

$$24^h \text{ mean solar time} = \left(1 + \frac{1}{365·2422}\right) 24^h \text{ sid. time}$$
$$……(13).$$

This last relation is evidently the same as (10).

To facilitate the conversion of any interval of mean time into its equivalent, expressed in sidereal time, and *vice versa*, the following tables are given; the entries are easily derived from (10) and (11).

Table I. *Conversion of mean time into sidereal time*

$$24^h \text{ mean solar time} \equiv (24^h + 3^m 56^s·556) \text{ sid. time}$$

1^h	„	„	$\equiv (1^h + \quad 9^s·8565)$ „
1^m	„	„	$\equiv (1^m + \quad 0^s·1643)$ „
1^s	„	„	$\equiv (1^s + \quad 0^s·0027)$ „

Table II. *Conversion of sidereal time into mean time*

$$24^h \text{ sid. time} \equiv (24^h - 3^m 55^s·910) \text{ mean solar time}$$

1^h	„	$\equiv (1^h - \quad 9^s·8296)$	„ „
1^m	„	$\equiv (1^m - \quad 0^s·1638)$	„ „
1^s	„	$\equiv (1^s - \quad 0^s·0027)$	„ „

In the almanacs there are more extensive tables which facilitate the problem of time-conversion.

Example 1. To find the sidereal time at Greenwich on 1931 February 24 at u.t. $8^h 47^m 38^s·52$.

From the almanacs, the Greenwich sidereal time on February 24 at u.t. 0^h (midnight) is $10^h 11^m 37^s·67$. The mean solar interval concerned is $8^h 47^m 38^s·52$ and by means of Table I this can be expressed in sidereal time (taking the hours, minutes, seconds separately) as follows:

Mean solar time		Sidereal time	
8^h	$\equiv$	$8^h + 8 \times 9^s·8565$ or 8^h	$+ 1^m 18^s·85$
47^m	$\equiv$	$47^m + 47 \times 0^s·1643$ or $47^m +$	$7^s·72$
38^s	$\equiv$	$38^s + 38 \times 0^s·0027$ or $38^s +$	$0^s·10$
$0^s·52$	$\equiv$	$0^s·52 +$	$0^s·00$

The sum of the quantities on the right is

$$8^h 47^m 38^s·52 + 1^m 26^s·67 \text{ or } 8^h 49^m 5^s·19.$$

Hence the interval of 8^h 47^m $38^s \cdot 52$ mean solar time is equivalent to 8^h 49^m $5^s \cdot 19$ sidereal time.

But at U.T. 0^h the sidereal time is 10^h 11^m $37^s \cdot 67$. Hence at U.T. 8^h 47^m $38^s \cdot 52$ the sidereal time is 19^h 0^m $42^s \cdot 86$.

Example 2. To find the U.T. on 1931 April 5 when the Greenwich sidereal time is 18^h 31^m $52^s \cdot 38$.

From the almanac, the sidereal time at U.T. 0^h on April 5 is 12^h 49^m $19^s \cdot 83$, which subtracted from 18^h 31^m $52^s \cdot 38$ gives the sidereal interval concerned. This interval is thus 5^h 42^m $32^s \cdot 55$ sidereal time. We use Table II as follows:

Sidereal time		Mean solar time		
5^h	$\equiv$	$5^h - 5 \times 9^s \cdot 8296 = 5^h$		$- 49^s \cdot 15$
42^m	$\equiv$	$42^m - 42 \times 0^s \cdot 1638 = 42^m$		$- 6^s \cdot 88$
32^s	$\equiv$	$32^s - 32 \times 0^s \cdot 0027 = 32^s$		$- 0^s \cdot 09$
$0^s \cdot 55$			$= 0^s \cdot 55 -$	$0^s \cdot 00$

The sum of the quantities on the right is

$$5^h \; 42^m \; 32^s \cdot 55 - 56^s \cdot 12 \quad \text{or} \quad 5^h \; 41^m \; 36^s \cdot 43.$$

Hence when the Greenwich sidereal time is as stated in the problem, the U.T. is 5^h 41^m $36^s \cdot 43$.

The computations can be considerably curtailed by means of the special tables in the almanacs already mentioned.

88. *The calendar.*

We have already mentioned that the tropical year is the unit on which civil reckoning is based. For obvious reasons, the *civil year* contains an integral number of mean solar days and, as we have seen, the tropical year is equal to $365 \cdot 2422$ mean solar days.

In the Julian calendar, introduced by Julius Caesar, the tropical year was taken to be $365\frac{1}{4}$ days; three years out of every four were each given 365 days while the fourth—called a *leap-year*—was given 366 days. The leap-year was chosen to be that one which was divisible by 4 and the extra day was added to February. Thus, according to the Julian rule, the years 1928, 1932, 1936, etc., are all leap-years in each February of which there are 29 days. If the assumed length of the tropical year had been accurate, then there would have been an exact accordance between cycles of four tropical years and cycles of four civil years.

In 1582, Pope Gregory introduced the calendar now in use; it was not, however, till 1752 that it was adopted in English-speaking countries. The Gregorian calendar introduces the necessary correction, based on the accurate value of the tropical year, to the Julian system. According to the Julian system, the years 1700, 1800, 1900 and 2000 would all be leap-years; in the Gregorian calendar only 2000 is defined to be a leap-year, the rule being that, when a year ends in two "noughts", it is not a leap-year unless it is divisible by 400. In a cycle of 400 years, there are 100 leap-years according to the Julian calendar and 3 less, or 97, in the Gregorian calendar. Hence, according to the Gregorian calendar,

400 civil years = (400 × 365 + 97) mean solar days,

from which the average civil year is 365·2425 mean solar days. This is so near the value of the tropical year that no appreciable discrepancy can arise for many centuries.

There is one small matter relating to the civil year that we can conveniently consider here. The first day in the year is, according to civil usage, denoted January 1 and an event that happens, say, at 6 a.m. at Greenwich on January 1 is said to occur on January 1, 6^h 0^m U.T. or on Jan. $1^d·25$. The event thus occurs 1·25 days after an epoch, which is denoted astronomically by Jan. $0^d·0$ which, in effect, is the midnight that ushers in the day December 31. Thus the instant given by 1931 December 31, 18^h 0^m U.T. can be written as 1932 Jan. $0^d·75$. A similar procedure is applicable to the other months.

89. *The Besselian year.*

We have defined the length of the tropical year but not the instant, according to civil reckoning, at which it is assumed to begin. It is the general astronomical practice to define the beginning of the tropical year (or solar year, as it is sometimes called) as the instant when the right ascension of the mean sun —or the sun's mean longitude—is exactly 18^h 40^m or 280°. This instant falls near the beginning of the civil year. The year defined in this way is usually called the Besselian year, after the German astronomer Bessel who first introduced it into astronomical practice. Thus the beginning of the Besselian year in 1931 is 1931 Jan. $1^d·322$ (civil time at Greenwich) or 1931 January 1,

$7^h\ 43^{m}\cdot68$ U.T. According to Washington civil time this instant is Jan. $1^{d}\cdot108$, the difference of $0^{d}\cdot214$ or $5^h\ 8^{m}\cdot16$ being simply the difference of longitude between Washington and Greenwich. The beginning of the next Besselian year is clearly obtained by adding $365\cdot242$ days. Hence in 1932, the beginning of the Besselian year is 1932 Jan. $1^{d}\cdot564$ U.T. or 1932 Jan. $1^{d}\cdot350$ (Washington Civil Time).

It is the general practice to denote the beginning of any Besselian year by the notation $1931\cdot0$, $1932\cdot0$, $1933\cdot0$, *etc.*

The year defined in this way is used in calculations and observations relating to the heavenly bodies. For example, if the R.A. and declination of a star are observed at a particular instant, the co-ordinates refer to the actual or true equinox and equator at that instant. By means of certain principles which we shall elaborate in Chapter x, the star's position can be deduced with respect to the mean equinox and equator at the beginning of the Besselian year. Thus for example, the positions of stars observed at different dates during 1931 can all be referred to the mean equinox and equator for $1931\cdot0$ and by a further process to some standard equinox such as that for $1900\cdot0$ or $1950\cdot0$.

90. *The Julian date.*

In certain observations (such as observations of variable stars) it is found convenient to express the instant of observation as so many days and fraction of a day after a definite fundamental epoch. The epoch chosen is mean noon on January 1 of the year 4713 B.C., and for any given date the number of days which have elapsed since this epoch defines the *Julian date* (J.D.) of the date in question. For example, for mean noon 1931 January 1, the Julian date is denoted by J.D. 2426343; the time of an observation made say on 1931 January 3, 18^h U.T. (that is 6 hours after mean noon on January 3) is denoted by J.D. $2426345\cdot25$. It is essential to remember that the Julian day is measured from mean noon. In the almanacs the Julian date is given for every day in the year, and there are also tables which enable the astronomer to find the Julian date for any day in any year.

91. *The equation of time.*

As in Chapter II (p. 43) the equation of time (E) is defined by

$$E = \text{H.A. } \odot - \text{H.A.M.S.} \qquad \qquad \ldots\ldots(14),$$

in which H.A. ⊙ denotes the hour angle of the sun and H.A.M.S. the hour angle of the mean sun at a given time and place. We have also

Sid. time = H.A. ⊙ + R.A. ⊙ = H.A.M.S. + R.A.M.S.   ......(15),

and E = R.A.M.S. − R.A. ⊙   ......(16).

But the R.A.M.S. is defined to be the sun's mean longitude l and if α denotes the sun's R.A., we have

$$E = l - \alpha \qquad ......(17).$$

If α is obtained from an observation of the sun, it is then easy to determine the value of E, for the corresponding value of l can be calculated from the known mean angular motion of the sun and the longitude of perigee, or taken directly from the almanacs where it is tabulated as R.A.M.S.

We now show how E can be expressed in terms of the elements of the orbit and of the time.

In Fig. 57 the celestial sphere is drawn with the earth as centre. Relative to the earth, the sun will appear to describe an elliptic orbit lying in the plane of the ecliptic. The principles and formulae of Chapter V apply to the sun's apparent orbit around the earth.

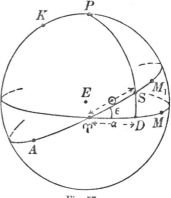

Fig. 57.

Let ⊙ denote the sun's true longitude at a given instant— the direction of the sun defining the point S on the ecliptic; then the arc ΥS = ⊙. We can write

$$E = (⊙ - \alpha) - (⊙ - l) \qquad ......(18).$$

In Fig. 57 let A denote perigee. Then the arc AS is the sun's true anomaly v. Let M_1 be the position of a fictitious body coinciding with the true sun at perigee and moving in the ecliptic with the sun's mean angular motion n; then the arc AM_1 is the mean anomaly M and the arc ΥM_1 is l. Thus,

$$⊙ - l = \Upsilon S - \Upsilon M_1$$
$$= AS - AM_1,$$

that is $⊙ - l = v - M$ ......(19).

Hence $E = -(\alpha - ⊙) - (v - M)$ ......(20).

The equation of time thus consists of two parts: (1) the quantity $(\alpha - \odot)$ called the *reduction to the equator*, and (2) the quantity $(v - M)$ which is the *equation of the centre* discussed in section 73.

We first express the quantity $(\odot - \alpha)$ in the form of a series. From the right-angled triangle TSD (PSD is the meridian through S) in which

$$TS = \odot, \quad TD = \alpha, \quad S\hat{T}D = \epsilon \quad \text{and} \quad S\hat{D}T = 90°,$$

we have by the four-parts formula **D**

$$\cos \alpha \cos \epsilon = \sin \alpha \cot \odot,$$

or
$$\tan \alpha = \cos \epsilon \tan \odot,$$

which can be written

$$\tan \alpha = \frac{1 - \tan^2 \dfrac{\epsilon}{2}}{1 + \tan^2 \dfrac{\epsilon}{2}} \tan \odot \qquad \ldots\ldots(21).$$

Since ϵ is about $23\frac{1}{2}°$, $\tan^2 \dfrac{\epsilon}{2}$ is approximately $1/25$. Write y for $\tan^2 \dfrac{\epsilon}{2}$; then

$$\tan \alpha = \frac{1 - y}{1 + y} \tan \odot \qquad \ldots\ldots(22),$$

in which we regard y as a small quantity.

(22) is essentially of the same form as formula (82) of Chapter v, p. 118, which is given in the form of a series by (83), p. 119. Hence writing y for $-x$ in (83) and α for $\dfrac{v}{2}$ and $\odot$ for $\dfrac{E}{2}$, we have

$$\odot - \alpha = y \sin 2\odot - \tfrac{1}{2} y^2 \sin 4\odot + \tfrac{1}{3} y^3 \sin 6\odot \ \ \ldots\ldots(23).$$

In this formula, $\odot$ and α are expressed in circular measure. Expressing the difference $(\odot - \alpha)$ in seconds of time we obtain

$$\odot - \alpha = \operatorname{cosec} 1^s \left[y \sin 2\odot - \tfrac{1}{2} y^2 \sin 4\odot \ldots \right].$$

Now $\operatorname{cosec} 1^s = 206265/15$ and $y \left(\equiv \tan^2 \dfrac{\epsilon}{2} \right) = 0 \cdot 0430687$, using the value $23° \, 26' \, 54''$ for the obliquity ϵ in 1931. We then obtain

$$\odot - \alpha = 592^s \cdot 24 \sin 2\odot - 12^s \cdot 75 \sin 4\odot + 0^s \cdot 36 \sin 6\odot$$
$$\ldots\ldots(24).$$

Formula (24) gives the part of the equation of time, depending on the obliquity of the ecliptic, expressed as a series involving the sun's true longitude $\odot$.

We now consider the part of the equation of time depending on the eccentricity, that is, the equation of the centre. Quoting formula (87) of Chapter v we write, keeping only terms up to e^2,

$$v - M \equiv \odot - l = 2e \sin M + \tfrac{5}{4}e^2 \sin 2M \quad \ldots \ldots (25).$$

Expressing $(v - M)$ in seconds of time and using the value of e for 1931·0, namely 0·0167381, we find

$$v - M = 460^{s}\cdot 33 \sin M + 4^{s}\cdot 82 \sin 2M \quad \ldots \ldots (26).$$

Now the right-hand side of (23) is expressed in terms of $\odot$, the sun's true longitude, which by (25) is given in terms of the mean anomaly M by

$$\odot = l + 2e \sin M + \tfrac{5}{4}e^2 \sin 2M \quad \ldots \ldots (27).$$

We substitute the value of $\odot$ given by (27) on the right-hand side of (23). As we have pointed out, y is about $1/25$ and e is about $1/60$; regarding y and e as small quantities of the same order of smallness and keeping only terms up to the second order in the value of $\odot - \alpha$, we can write with the accuracy indicated

$$\sin 2\odot = \sin (2l + 4e \sin M)$$
$$= \sin 2l + 4e \sin M \cos 2l$$

(we require the development of $\sin 2\odot$ up to the first order only, since in (23) $\sin 2\odot$ is multiplied by y). Similarly we have, with the limitations imposed,

$$\sin 4\odot = \sin 4l.$$

Hence (23) becomes—up to the second order—

$$\odot - \alpha = y \sin 2l + 4ey \sin M \cos 2l - \tfrac{1}{2}y^2 \sin 4l \quad \ldots (28).$$

Combining (28) with (25) we have, for the equation of time,

$$E = y \sin 2l - 2e \sin M + 4ey \sin M \cos 2l$$
$$- \tfrac{1}{2}y^2 \sin 4l - \tfrac{5}{4}e^2 \sin 2M \quad \ldots (29).$$

Inserting the numerical values of y and e, we obtain

$$E = 592^{s}\cdot 2 \sin 2l - 460^{s}\cdot 3 \sin M + 39^{s}\cdot 6 \sin M \cos 2l$$
$$- 12^{s}\cdot 7 \sin 4l - 4^{s}\cdot 8 \sin 2M \quad \ldots (30).$$

Now in Fig. 57, the mean anomaly M is the arc AM_1 which is equal to $A\Upsilon + \Upsilon M_1$, so that

$$M = A\Upsilon + l.$$

Also, if ϖ denotes the longitude of perigee, we have $\varpi = 360° - A\Upsilon$.

Hence $$M = 360° + (l - \varpi) \quad \ldots \ldots (31).$$

In the formula (30) we can thus write $(l - \varpi)$ for M on the right-hand side. For 1931 the value of ϖ is 281° 45′ 14″. After some reduction, we then obtain

$$E = -97^{\text{s}} \cdot 8 \sin l - 431^{\text{s}} \cdot 3 \cos l + 596^{\text{s}} \cdot 6 \sin 2l - 1^{\text{s}} \cdot 9 \cos 2l$$
$$+ 4^{\text{s}} \cdot 0 \sin 3l + 19^{\text{s}} \cdot 3 \cos 3l - 12^{\text{s}} \cdot 7 \sin 4l \dots (32).$$

Up to the order of approximation adopted, this formula gives the equation of time in terms of the sun's mean longitude.

Let t_0 be the time when the sun's mean longitude is zero. If n is the sun's mean angular motion, the mean longitude l at any time t is given by $\qquad l = n \, (t - t_0).$

Thus the equation of time can be expressed in terms of the time and the constants associated with the sun's apparent orbit.

92. *The seasons.*

Consider the apparent path of the sun during the year (Fig. 58). The earth E is the centre of the celestial sphere and relative to E the sun appears to describe the great circle $CTAB$ on the sphere —the ecliptic—which in-tersects the celestial equa-tor in the two points T and B, the vernal and autum-nal equinoxes (sometimes called the First Point of Aries and the First Point of Libra respectively). P is the north pole of the equator. The sun's decli-nation increases from 0° at T (about March 21) to a maximum of 23° 27′ at A (about June 21). The

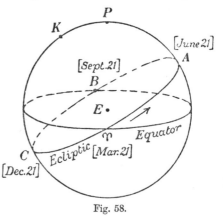

Fig. 58.

point A is called the *summer solstice*. Similarly, the point C where the sun (about December 21) has its greatest southerly declination $(-23° 27′)$ is called the *winter solstice*.

Astronomically, the four parts of the year, or seasons, during which the sun is successively in the quadrants TA, AB, BC and CT are called, *spring, summer, autumn* and *winter* respectively. As these terms are used rather loosely in ordinary speech, we shall consider them only in their astronomical significance.

The general characteristics of the seasons at any place depend on the relative amounts of heat received from the sun from day to day. The two astronomical factors are (a) the interval during which the sun is above the horizon on any day, and (b) the general sequence of altitudes which the sun attains during this interval. It is convenient to use the sun's altitude or zenith distance at apparent noon as a criterion of (b).

Refraction being neglected, the number of hours during which the sun (strictly, the sun's centre) is above the horizon is given by $2H$, where H is the angle between 0^h and 12^h given by

$$\cos H = - \tan \phi \tan \delta \qquad \ldots\ldots(33),$$

where ϕ is the latitude of the place; this is formula (24), p. 47.

If the declination δ is positive, then in northern latitudes H lies between 6^h and 12^h and therefore the sun is above the horizon for more than 12 hours out of the 24. This occurs between March 21 (approximately), when the sun is at Υ, and September 21 (approximately), when the sun is at B. Thus as far as (a) is concerned, the days increase in warmth from March 21 to June 21 corresponding to the increase in δ from $0°$ to $+ 23° 27'$; from the latter date, a decrease ensues. If δ is negative, then in northern latitudes the hour angle of setting is by (33) less than 6^h, and consequently the sun is above the horizon for less than 12^h; this refers to the two seasons autumn and winter. The minimum number of hours of daylight occurs when the sun attains its greatest southerly declination, that is at the winter solstice (December 21). It is easily seen that the number of hours of daylight increases from December 21 to June 21 and decreases from June 21 to December 21.

The sun's zenith distance z at apparent noon on any day is given by $$z = \phi - \delta \qquad \ldots\ldots(34).$$

It is clear that at any place whose north latitude lies between $0°$ and $23° 27'$, the sun is in the zenith (or practically so) twice during spring and summer, that is, between March 21 and September 21. For latitude $23° 27'$ N, the sun is in the zenith only on June 21. This northern parallel of latitude ($23° 27'$ N) is called the *Tropic of Cancer*. The corresponding south parallel ($23° 27'$ S) is the *Tropic of Capricorn*, and for places between this parallel and the equator, the sun is in the zenith twice during

the interval between September 21 and March 21. The zone of
the earth's surface bounded by the Tropics of Cancer and
Capricorn is the *Torrid Zone*. The parallels of latitude 66° 33′ N
and 66° 33′ S are called the *Arctic Circle* and *Antarctic Circle*
respectively. The zone between the parallels of 23° 27′ N and
66° 33′ N is called the *North Temperate Zone*, and the corresponding zone in the southern hemisphere is the *South Temperate
Zone*.

From (33), it is seen that when $\phi = 66° 33′$ N and $\delta = + 23° 27′$,
the hour angle of setting is 12^{h}; that is to say, at the Arctic Circle
on June 21 the sun is above the horizon during the whole
24 hours. In latitude 70°, for example, the sun is above the
horizon continuously—by (33)—during the period for which
the sun's declination is between $+ 20°$ and $+ 23° 27′$, that is to
say (referring to the almanac for 1931), between May 22 and
July 24. At the north pole the sun is above the horizon continuously between March 21 and September 21 and below the
horizon for the remainder of the year.

Summarising for the northern hemisphere, we have these
characteristics of the seasons. During spring, the number of
hours of daylight increases from 12 hours at March 21 until
June 21 when the maximum is reached; this maximum can be
24 hours on particular days for places on or north of the Arctic
Circle; during spring, the sun's altitude at noon increases, for
places north of the Tropic of Cancer, and reaches a maximum
on June 21; for places between the equator and the parallel of
23° 27′ N, it reaches a maximum of 90° on some date between
March 21 and June 21.

During summer, the number of hours of daylight decreases
from the maximum on June 21 to 12 hours on September 21.

During autumn, the number of hours of daylight decreases
from 12 hours on September 21 to a minimum on December 21,
and during this period the sun's altitude at noon progressively
decreases.

During winter, the number of hours of daylight increases from
the minimum on December 21 to 12 hours on March 21, and
during this season the sun's altitude at noon progressively
increases.

The relation between the seasons and the sun's declination in the southern hemisphere can readily be inferred from the preceding discussion.

The respective lengths (in mean solar days) of the seasons can be obtained by considering the equation of the centre which, for our present purpose, we write simply in the form—from (25)—

$$l = \odot - 2e \sin M$$

or, using (31),

$$l = \odot - 2e \sin (l - \varpi) \qquad \ldots\ldots(35).$$

As l is approximately equal to $\odot$, we can replace l on the right of (35) by $\odot$ and then

$$l = \odot - 2e \sin (\odot - \varpi) \qquad \ldots\ldots(36).$$

Let l_1, l_2, l_3 and l_4 denote the values of the sun's mean longitude at the beginning of spring, summer, autumn and winter respectively. At the beginning of spring $\odot = 0$, therefore

$$l_1 = 2e \sin \varpi \qquad \ldots\ldots(37).$$

At the beginning of summer, $\odot = 90°$ or $\dfrac{\pi}{2}$ in radian measure; hence

$$l_2 = \frac{\pi}{2} - 2e \cos \varpi \qquad \ldots\ldots(38).$$

Similarly

$$l_3 = \pi - 2e \sin \varpi \qquad \ldots\ldots(39),$$

$$l_4 = \frac{3\pi}{2} + 2e \cos \varpi \qquad \ldots\ldots(40).$$

Let t_1, t_2, t_3 and t_4 denote the instants when the mean longitude of the sun has the values l_1, l_2, l_3 and l_4. Then

$$l_2 - l_1 = n (t_2 - t_1) \qquad \ldots\ldots(41),$$

where n is the mean angular motion. If $t_2 - t_1$ is expressed in mean solar days, $n = 2\pi/T$, where $T = 365 \cdot 2422$ mean solar days. Hence by (41)

$$t_2 - t_1 = \frac{T}{2\pi} (l_2 - l_1) \quad \text{mean solar days} \quad \ldots\ldots(42).$$

But $(t_2 - t_1)$ is the number of mean solar days in spring, and if we denote this interval by I_1 we have from (37), (38) and (42),

$$I_1 = \frac{T}{2\pi} \left\{ \frac{\pi}{2} - 2e (\sin \varpi + \cos \varpi) \right\},$$

or

$$I_1 = 91 \cdot 31 - \frac{eT}{\pi} (\sin \varpi + \cos \varpi) \qquad \ldots\ldots(43).$$

Similarly if I_2, I_3 and I_4 denote the number of mean solar days in summer, autumn and winter respectively, we obtain

$$I_2 = 91\cdot31 - \frac{eT}{\pi}(\sin \varpi - \cos \varpi) \qquad \ldots\ldots(44),$$

$$I_3 = 91\cdot31 + \frac{eT}{\pi}(\sin \varpi + \cos \varpi) \qquad \ldots\ldots(45),$$

$$I_4 = 91\cdot31 + \frac{eT}{\pi}(\sin \varpi - \cos \varpi) \qquad \ldots\ldots(46).$$

Putting $T = 365\cdot2422$, $e = 0\cdot01674$, $\varpi = 281° \ 45'$, we find that, for the northern hemisphere,

Spring	contains	92 days	20·2 hours
Summer	,,	93 ,,	14·4 ,,
Autumn	,,	89 ,,	18·7 ,,
Winter	,,	89 ,,	0·5 ,,

93. *Time of the sun's transit over any meridian.*

We shall conclude this chapter by considering two problems involving time. An example of the first problem is the following: to calculate, to the nearest second, the U.T. and standard time of the transit of the sun's centre on 1931 January 4, at the Dominion Observatory, Victoria, B.C. of which the longitude is $8^h 13^m 40^s$ W.

From the 1931 almanac, we have the following:

At U.T. 0^h, Jan. 4, the equation of time (E.T.) is $-4^m 30^s\cdot7$.

,, ,, Jan. 5, ,, ,, ,, ,, $-4^m 58^s\cdot1$.

When the sun's centre is on the meridian of Victoria, the sun's hour angle is 0^h, and therefore the hour angle of the sun at Greenwich at this instant is $8^h 13^m 40^s$. Neglecting at first the equation of time, the approximate time of transit is U.T. $20^h 14^m$ January 4, and thus from the data concerning the equation of time given above we have, to the nearest minute, E.T. $= -5^m$. But E.T. at any instant is given by

$$\text{E.T.} = \text{H.A. } \odot - \text{H.A.M.S.} \qquad \ldots\ldots(47).$$

Hence, approximately, we can write

$$-5^m = 8^h \ 13^m \ 40^s - \text{H.A.M.S. (Greenwich)},$$

which gives H.A.M.S. at Greenwich $= 8^h \ 18^m \ 40^s$. Hence the approximate U.T. of transit at Victoria is January 4, $20^h\cdot3$.

This U.T. is now sufficiently accurate to enable us to obtain the correct E.T. at transit. In 24 hours, the E.T. has increased numerically by $27^s \cdot 4$; therefore in $20^h \cdot 3$, it has increased numerically by $\dfrac{20 \cdot 3}{24} \times 27^s \cdot 4$ or $23^s \cdot 2$. Thus at the moment of transit, the E.T. is $-4^m 54^s$ (to the nearest second). Inserting this value in (47), we obtain:

$$\text{H.A.M.S. at Greenwich} = 8^h 13^m 40^s + 4^m 54^s.$$

Thus the U.T. of transit at Victoria is $20^h 18^m 34^s$.

If we require greater accuracy, we make a closer interpolation for the equation of time with the U.T. just derived.

The standard time kept at Victoria is "Pacific time" corresponding to the meridian of 8^h west. Hence the standard time of the sun's transit at Victoria on January 4 is $12^h 18^m 34^s$ (Pacific time).

94. *Time of the moon's transit over any meridian.*

This is the second problem and, as an example, we shall find the U.T. and standard time of the moon's transit at Victoria on 1931 May 24.

We first find the approximate time of transit as follows. From the almanac, the U.T. of transit of the moon's centre at Greenwich on May 24 is $18^h 14^m \cdot 9$; on May 25, it is $19^h 4^m \cdot 3$. Thus the transit on May 25 occurs $24^h 49^m \cdot 4$ (mean solar time) after the transit on May 24. During this interval the moon's hour angle has increased by 24^h. When the moon transits at Victoria, its hour angle at Greenwich is $8^h 13^m 40^s$ (the longitude of Victoria) and consequently this occurs at an interval of $\dfrac{24^h 49^m \cdot 4}{24^h} \times 8^h 13^m 40^s$ (expressed in mean solar time) after the transit at Greenwich. This interval can be written

$$\left(1 + \frac{49 \cdot 4}{24 \times 60}\right) \times 8^h 13^m 40^s \quad \text{or} \quad (8^h 13^m 40^s + 16^m \cdot 9)$$

or $8^h 30^m \cdot 6$ approximately. Hence when the moon transits at Victoria, the corresponding U.T. is $(18^h 14^m \cdot 9 + 8^h 30^m \cdot 6)$, May 24 approximately or $2^h 45^m \cdot 5$, May 25.

A more accurate result can now be derived as follows. With the U.T. and date just obtained, we find from the almanac that the moon's R.A. (correct to 1^s) is $10^h\,38^m\,55^s$, and that the R.A.M.S. is $4^h\,6^m\,55^s$. Since the hour angle at transit is 0^h, we have for Victoria

$$\text{R.A. moon} = \text{H.A.M.S.} + \text{R.A.M.S.},$$

so that H.A.M.S. (Victoria) is $(10^h\,38^m\,55^s - 4^h\,6^m\,55^s)$ or $6^h\,32^m\,0^s$. Adding the longitude of Victoria ($8^h\,13^m\,40^s$) and 12^h, we find the U.T. of transit at Victoria to be $2^h\,45^m\,40^s$, May 25. Accordingly the standard time of transit is $18^h\,45^m\,40^s$ (Pacific time) on May 24.

EXERCISES

[Symbols used:

    ϕ = latitude of observer, ϵ = obliquity of the ecliptic.]

    e = eccentricity of earth's orbit,

1. For a place within the tropics, prove that the hour angle H of the sun (a, δ) when the ecliptic is vertical is given by

$$H = \sin^{-1}(\sin a \cot \delta \tan \phi) - a.$$

2. A star (a, δ) rises at the same time as the sun at a place in north latitude ϕ when the sun's right ascension is a_1. Prove that

$$\alpha_1 - \sin^{-1}(\sin a_1 \tan \phi \tan \epsilon) = a - \sin^{-1}(\tan \delta \tan \phi).$$

3. Prove that at a place on the Arctic Circle the daily displacement of the point of sunset is equal to the sun's change in longitude during the same interval.   [*M.T.*]

4. Neglecting the eccentricity of the earth's orbit, prove that at a place within the Arctic Circle the sun will be above the horizon for

$$\frac{365\frac{1}{4}}{\pi} \cos^{-1}\left(\frac{\cos \phi}{\sin \epsilon}\right) \text{ days.}$$

5. If S is the sun's semi-diameter in minutes of arc, show that at a solstice the time taken by the sun's disc to cross the prime vertical at a place in latitude ϕ ($> \epsilon$) is, in minutes,

$$\frac{2S}{15\,(\sin^2 \phi - \sin^2 \epsilon)^{\frac{1}{2}}}.$$

6. Observations of equal altitude of ζ Persei are taken with a theodolite on 1931 January 3, at U.T. $18^h\,45^m\,8^s$ and $22^h\,44^m\,19^s$. Calculate the longitude of the observer.   [*Lond.* 1925.]

7. The longitude of Columbia University, New York, is $4^h\,55^m\,50^s$ west of Greenwich. The sidereal time at mean noon at Greenwich on a certain day is $17^h\,23^m\,8^s$. Show that on the same day when the sidereal time at Columbia University is $20^h\,8^m\,4^s$ the hour angle of the mean sun at the same place is $2^h\,43^m\,41^s$.

8. If δ is the sun's declination, S its semi-diameter in minutes of arc, I the interval in minutes between the disappearance of its lower and of its upper limb at sunset for a place in latitude ϕ, prove that ϕ is given approximately by

$$\sin^2 \phi = \cos^2 \delta - \frac{4}{225} \frac{S^2}{I^2}.$$ [*Lond.* 1930.]

9. Two stars, of equal R.A. and of declinations δ and δ', are observed when their altitudes are the same; the sidereal interval h between the observations is noted. Prove that H, the hour angle of the first star at the moment of observation, is given by

$$\cos(\psi + H) = \frac{2 \sin \psi \tan \phi}{\cos \delta' \sin h} \cdot \sin \tfrac{1}{2}(\delta - \delta') \cos \tfrac{1}{2}(\delta + \delta'),$$

where

$$\cot \psi = \frac{\cos \delta' \cos h - \cos \delta}{\cos \delta' \sin h}.$$

10. The right ascension and declination of the sun near the summer solstice are $(90° - 2\beta)$ and δ respectively. Prove that $(\epsilon - \delta)$ expressed in seconds of arc is given by

$$\epsilon - \delta = \operatorname{cosec} 1'' [\tan^2 \beta \sin 2\epsilon - \tfrac{1}{2} \tan^4 \beta \sin 4\epsilon + \ldots].$$

11. Let T and T' be the times shown by a sidereal clock when a star is at the same altitude, first on one side of the meridian and then on the other, and let a be the right ascension of the star. Show that $a - \tfrac{1}{2}(T + T')$ is the correction to be applied to clock time to obtain true sidereal time.

12. If the sidereal clock times when the sun arrives at equal altitudes on each side of the meridian are T and T', and if the change of declination δ of the sun in the interval is $d\delta$, and the right ascension of the sun at culmination is a, show that the correction to be applied to clock time to obtain the true sidereal time is

$$a - \tfrac{1}{2}(T + T') - \tfrac{1}{2} \left(\frac{\tan \delta}{\tan \tfrac{1}{2}(T' - T)} - \frac{\tan \phi}{\sin \tfrac{1}{2}(T' - T)} \right) d\delta.$$

13. From the side of a church running due east a tall buttress projects southwards forming a rectangular corner in which the afternoon sun casts a triangle of shadow on the ground. Show that in winter the total amount of shade during the day (ground-area integrated with respect to time) is proportional to

$$\frac{1}{\sin \phi} \log \frac{\sin(-\delta)}{\cos \phi \sin(\phi - \delta)},$$

where δ is the declination of the sun. [*M.T.* 1925.]

14. If the sun's right ascension increases by Δa while its longitude increases by a small amount ΔL, show that

$$\Delta a = \Delta L \cos \epsilon \sec^2 \delta,$$

where ϵ is the obliquity of the ecliptic and δ is the sun's declination.

Hence find the approximate dates when that part of the equation of time which is known as the reduction to the equator is a maximum.

[*Lond.* 1929.]

15. If $\odot$ is the longitude of the sun and a its right ascension, show that the greatest value of $(a - \odot)$ occurs when

$$\tan \odot = (\sec \epsilon)^{\frac{1}{2}} \text{ and } \tan a = (\cos \epsilon)^{\frac{1}{2}}.$$

16. Prove that, if the eccentricity of the earth's orbit were zero, the equation of time in minutes would be

$$\frac{720}{\pi} \tan^{-1} \frac{(1 - \cos \epsilon) \tan \odot}{1 + \cos \epsilon \tan^2 \odot},$$

where ϵ is the obliquity of the ecliptic and $\odot$ the sun's longitude.

[*M.T.* 1912.]

17. Show that the part of the equation of time due to the orbital eccentricity is stationary in value when the sun's true anomaly is

$$\cos^{-1}\left[\frac{(1 - e^2)^{\frac{3}{4}} - 1}{e}\right],$$

and that the part due to the obliquity is stationary in value when the sun's longitude is

$$\tan^{-1}\{(\sec \epsilon)^{\frac{1}{2}}\}.$$

18. Show that when the equation of time is a maximum or minimum the sun's longitude $\odot$ is given by

$$(1 - e^2)^{\frac{3}{2}} (\cos^2 \odot + \cos^2 \epsilon \sin^2 \odot) = \cos \epsilon \{1 + e \cos (\odot - \varpi)\}^2,$$

where ϖ is the longitude of perigee.

19. Show that in latitude 45° the difference between the times from sunrise to apparent noon and from apparent noon to sunset is

$$\frac{D}{365\frac{1}{4}} \tan \delta \sec \delta (\sec 2\delta)^{\frac{1}{2}} \cot (360° \, T/365\frac{1}{4}),$$

where D is the length of the day, δ the sun's declination, and T the number of days since the vernal equinox, the earth's orbit being supposed circular.

[*Ball.*]

20. Assuming the sun's semi-diameter at mean distance to be 961″, show that the time t, expressed in sidereal seconds, required for the semi-diameter to cross the meridian is expressed by the relation

$$\frac{961}{r \cos \delta} = 15 \left(1 - \frac{\cos \epsilon}{r^2 (\tau + 1) \cos^2 \delta}\right) t,$$

where τ is the length of the year in mean solar days and r the sun's distance in astronomical units.

[*Coll. Exam.*]

21. Prove that, for a ship steaming V nautical miles per hour on a course S $\theta°$ W, the interval I between the passage of the sun over the meridian of the ship and the moment when the sun's altitude is a maximum, is given by

$$I = 15^{s}\cdot 3 \, (V \cos \theta + \Delta\delta) \, (\tan \phi - \tan \delta) \Big/ \left(1 - \frac{V \sin \theta \sec \phi}{900}\right)^2,$$

where ϕ is the ship's latitude at apparent noon and $\Delta\delta$ is the northerly change (in minutes of arc) of the sun's declination per hour at apparent noon.

[*Lond.* 1926.]

22. Given that:

the tropical year   = 365·2422 mean solar days

the sidereal year   = 365·2564   ,,   ,,

the anomalistic year = 365·2596   ,,   ,,

determine the amount and sign of annual precession and the motion of perihelion.

CHAPTER VII

PLANETARY PHENOMENA AND HELIOGRAPHIC CO-ORDINATES

95. *The geocentric motion of a planet.*

We shall assume for the present that the orbits of the earth and a planet P are circular and in the plane of the ecliptic. In Fig. 59 let E and P denote the positions of the earth and planet at any given moment. The plane of the paper represents the plane of the ecliptic. Let $S\Upsilon$ denote the direction of the vernal equinox from the sun S and $E\Upsilon$ its direction from the earth E. The

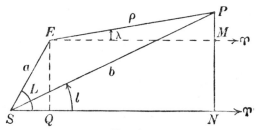

Fig. 59.

angles $ES\Upsilon$ and $PS\Upsilon$ are accordingly the heliocentric longitudes of the earth and planet (denoted by L and l respectively) and $P\hat{E}\Upsilon$ is the geocentric longitude of the planet (denoted by λ). Let a and b be the radii of the orbits so that $SE = a$ and $SP = b$; also let the planet's geocentric distance EP be denoted by ρ. Our object is to express the changes in the geocentric longitude λ in terms of the heliocentric longitudes L and l. Draw perpendiculars PMN, EQ to $S\Upsilon$. Then we have at once

$$\rho \sin \lambda = b \sin l - a \sin L \qquad \ldots\ldots(1),$$

and
$$\rho \cos \lambda = b \cos l - a \cos L \qquad \ldots\ldots(2).$$

In these equations ρ, λ, l and L vary with the time. Differentiating, we obtain

$$\rho \cos \lambda \frac{d\lambda}{dt} + \sin \lambda \frac{d\rho}{dt} = b \cos l \frac{dl}{dt} - a \cos L \frac{dL}{dt} \qquad \ldots\ldots(3),$$

$$\rho \sin \lambda \frac{d\lambda}{dt} - \cos \lambda \frac{d\rho}{dt} = b \sin l \frac{dl}{dt} - a \sin L \frac{dL}{dt} \qquad \ldots\ldots(4).$$

Multiply (3) by $\rho \cos \lambda$ and (4) by $\rho \sin \lambda$ and add; we obtain

$$\rho^2 \frac{d\lambda}{dt} = b\rho \cos(l - \lambda) \frac{dl}{dt} - a\rho \cos(L - \lambda) \frac{dL}{dt} \quad \ldots\ldots(5).$$

Multiply (1) by $\sin l$ and (2) by $\cos l$ and add. Then

$$\rho \cos(l - \lambda) = b - a \cos(L - l) \quad \ldots\ldots(6).$$

Multiply (1) by $\sin L$ and (2) by $\cos L$ and add. Then

$$\rho \cos(L - \lambda) = b \cos(L - l) - a \quad \ldots\ldots(7).$$

Using (6) and (7) we obtain from (5),

$$\rho^2 \frac{d\lambda}{dt} = \{b^2 - ab \cos(L - l)\} \frac{dl}{dt} + \{a^2 - ab \cos(L - l)\} \frac{dL}{dt}$$
$$\ldots\ldots(8).$$

Now, since the planet's orbit is supposed to be circular, $\dfrac{dl}{dt}$ is the mean angular motion n which is related to the radius b by

$$n^2 b^3 = G(M + m),$$

where M, m are the masses of the sun and planet and G is the constant of gravitation. We can neglect m in comparison with M in this problem and writing $G . M = \mu$, we obtain

$$n \equiv \frac{dl}{dt} = \mu^{\frac{1}{2}} b^{-\frac{3}{2}} \quad \ldots\ldots(9).$$

Similarly,
$$\frac{dL}{dt} = \mu^{\frac{1}{2}} a^{-\frac{3}{2}} \quad \ldots\ldots(10).$$

Insert these values of $\dfrac{dl}{dt}$ and $\dfrac{dL}{dt}$ in (8) and there results,

$$\rho^2 \frac{d\lambda}{dt} = \mu^{\frac{1}{2}} \{(b^{\frac{1}{2}} + a^{\frac{1}{2}}) - (ab^{-\frac{1}{2}} + ba^{-\frac{1}{2}}) \cos(L - l)\}$$
$$\ldots\ldots(11).$$

Also, from the triangle PES, we obtain

$$\rho^2 = a^2 + b^2 - 2ab \cos(L - l) \quad \ldots\ldots(12).$$

Thus the value of $\dfrac{d\lambda}{dt}$ given by (11) can be expressed completely in terms of L and l.

The formula (11) can be written as

$$\rho^2 \frac{d\lambda}{dt} = \mu^{\frac{1}{2}} (ab^{-\frac{1}{2}} + ba^{-\frac{1}{2}}) \left\{ \frac{(b^{\frac{1}{2}} + a^{\frac{1}{2}}) a^{\frac{1}{2}} b^{\frac{1}{2}}}{b^{\frac{3}{2}} + a^{\frac{3}{2}}} - \cos(L - l) \right\}$$
$$\ldots\ldots(13).$$

Now $(b^{\frac{1}{2}} + a^{\frac{1}{2}})\, a^{\frac{1}{2}} b^{\frac{1}{2}}$ is less than $(b^{\frac{3}{2}} + a^{\frac{3}{2}})$ if $a^{\frac{1}{2}} b^{\frac{1}{2}} < b + a - a^{\frac{1}{2}} b^{\frac{1}{2}}$, that is, if $(a^{\frac{1}{2}} - b^{\frac{1}{2}})^2 > 0$, which is evidently always true. Hence it is possible to define an angle a between $0°$ and $90°$ such that

$$\cos a = \frac{(b^{\frac{1}{2}} + a^{\frac{1}{2}})\, a^{\frac{1}{2}} b^{\frac{1}{2}}}{b^{\frac{3}{2}} + a^{\frac{3}{2}}} \qquad \ldots\ldots(14).$$

The angle a can be calculated from the known values of a and b. Hence, from (13),

$$\rho^2 \frac{d\lambda}{dt} = A\left[\cos a - \cos(L - l)\right] \qquad \ldots\ldots(15),$$

where A is evidently a positive quantity. It is clear from (15) that $\dfrac{d\lambda}{dt}$ can be positive or negative or zero according to the values taken by $(L - l)$.

When $\dfrac{d\lambda}{dt}$ is positive, that is when the geocentric longitude of the planet is increasing, the geocentric motion is said to be *direct*; when $\dfrac{d\lambda}{dt}$ is negative, that is when the geocentric longitude is decreasing, the geocentric motion is said to be *retrograde*; when $\dfrac{d\lambda}{dt} = 0$, the planet is said to be *stationary*.

For a stationary point, (15) gives $\cos(L - l) = \cos a$, whence $L - l = a$ or $360° - a$. Hence when $(L - l)$ is within the range defined by $(360° - a) \to 0° \to a$, $\cos(L - l) > \cos a$ and the geocentric motion is accordingly retrograde. When $(L - l)$ is outside the range indicated, the geocentric motion is direct.

The interval of time during which the planet's geocentric longitude λ increases from $0°$ to $360°$ is the synodic period (p. 131) which we denote by S days. Hence during each synodic period the geocentric motion is retrograde for $\dfrac{a \cdot S}{180}$ days and direct for $\dfrac{180 - a}{180} \cdot S$ days, a being expressed in degrees.

For Jupiter, $b = 5\cdot20a$ and from (14) it is easily found that $a = 54\frac{1}{2}°$. As the synodic period of Jupiter is 399 days, it follows that the motion of the planet is retrograde for 121 days and direct for 278 days.

In Fig. 60, the geocentric motion of Jupiter against the background of the stars is shown between 1931 January 15 and 1931 May 15. Between January 15 and March 7 the motion is retrograde and the right ascension of the planet is decreasing between these dates. On March 7 the planet is stationary, after which the

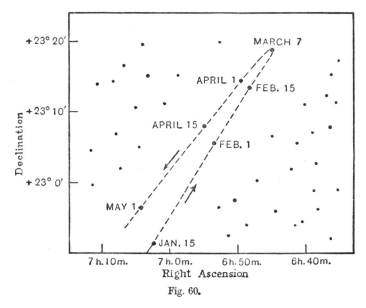

Fig. 60.

motion is direct and the right ascension is increasing. The separation of the two parts of the path is due to the effect of the inclination of the planet's orbit to the ecliptic. The effect of the inclination on the determination of the stationary points will be discussed later in section 97.

96. *The heliocentric distance of a planet, when stationary, in terms of its elongation.*

In Fig. 59, let P now denote the planet's position at a stationary point so that $E\hat{S}P = \alpha$, defined by (14). Let E denote the angle SEP which is called the *elongation* of the planet from the sun, as viewed from the earth. We shall suppose that the right ascension and declination of the planet at the stationary point are observed; the co-ordinates of the sun being taken from the almanac, the calculation of E can then be easily effected.

From the triangle ESP, we have
$$a \sin E = b \sin (E + \alpha),$$
from which $$\tan E = \frac{b \sin \alpha}{a - b \cos \alpha}$$ ......(16).

We consider the case when b is greater than a.

From (14), since by definition $0 < \alpha < 90°$, we have
$$\sin \alpha = + \frac{(b - a)(b + a)^{\frac{1}{2}}}{b^{\frac{3}{2}} + a^{\frac{3}{2}}}.$$

Hence (16) becomes, after a little reduction,
$$\tan E = - \frac{b}{(ab + a^2)^{\frac{1}{2}}}$$ ......(17),

from which it is evident that in the case concerned the value of E lies between $90°$ and $180°$.

As a is the earth's heliocentric distance and E is supposed known, this last formula enables the value of b, the planet's heliocentric distance, to be calculated. We must remember, of course, that the orbital eccentricities and the inclination to the ecliptic of the planet's orbital plane have been neglected in deriving (17) and consequently, in the general case, this formula can be expected to give only an approximate value of the planet's heliocentric distance. The minor planet Pallas was stationary soon after its discovery and its approximate heliocentric distance was first derived by means of the relation (17), the assumption of a circular orbit being made.

When b is less than a, the procedure is similar.

97. *The stationary points, the inclination being taken into account.*

We consider in this section the circumstances in which a planet is stationary in longitude when its orbital plane is inclined at an angle i to the plane of the ecliptic. We shall assume that the orbits of the earth and the planet are circular and of radii a and b respectively.

In Fig. 61, let NP define the plane of the planet's orbit about the sun S, the plane of the ecliptic being defined by NJ; the angle $PNJ = i$. Through K, the pole of the ecliptic, draw a great circle KPJ. Let NP be denoted by ψ, NJ by l and JP by β. The projection of the planet's radius vector b on the plane of the ecliptic is $b \cos \beta$ which we shall denote by B. From the triangle PJN (in which $P\hat{J}N = 90°$), we have

$$\sin \psi \cos i = \cos \beta \sin l \qquad \ldots\ldots(18),$$

$$\cos \psi = \cos \beta \cos l \qquad \ldots\ldots(19).$$

Consider now Fig. 62 in which SQ is the projection of the planet's radius vector on the plane of the ecliptic. Let SN be the direction of a node and let ϕ denote the angular distance of

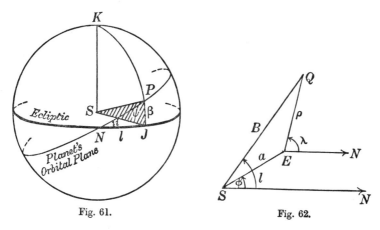

Fig. 61.   Fig. 62.

the earth E from the node. Let EQ be denoted by ρ and $N\hat{E}Q$ by λ (EN gives the geocentric direction of the node). We have also $N\hat{S}Q = l$ and $SQ = B$. As in section 95 we have

$$\rho \sin \lambda = B \sin l - a \sin \phi,$$

$$\rho \cos \lambda = B \cos l - a \cos \phi.$$

Now $B = b \cos \beta$; hence, using (18) and (19), we have

$$\left.\begin{array}{l} \rho \sin \lambda = b \cos i \sin \psi - a \sin \phi \\ \rho \cos \lambda = b \cos \psi - a \cos \phi \end{array}\right\} \qquad \ldots\ldots(20),$$

and differentiating,

$$\rho \cos \lambda \frac{d\lambda}{dt} + \frac{d\rho}{dt} \sin \lambda = b \cos i \cos \psi \frac{d\psi}{dt} - a \cos \phi \frac{d\phi}{dt} \, ,$$

$$\rho \sin \lambda \frac{d\lambda}{dt} - \frac{d\rho}{dt} \cos \lambda = b \sin \psi \frac{d\psi}{dt} - a \sin \phi \frac{d\phi}{dt} \, .$$

Multiplying these equations, in order, by $\rho \cos \lambda$ and $\rho \sin \lambda$, adding and making use of (20), we obtain

$$\rho^2 \frac{d\lambda}{dt} = \left(b \cos i \cos \psi \frac{d\psi}{dt} - a \cos \phi \frac{d\phi}{dt} \right) (b \cos \psi - a \cos \phi)$$

$$+ \left(b \sin \psi \frac{d\psi}{dt} - a \sin \phi \frac{d\phi}{dt} \right)(b \cos i \sin \psi - a \sin \phi)$$

$$= b^2 \cos i \frac{d\psi}{dt} + a^2 \frac{d\phi}{dt} - ab \frac{d\psi}{dt} [\cos \psi \cos \phi \cos i + \sin \psi \sin \phi]$$

$$- ab \frac{d\phi}{dt} [\cos \psi \cos \phi + \sin \psi \sin \phi \cos i].$$

If the planet is stationary in longitude we must have $\dfrac{d\lambda}{dt} = 0$. Also by Kepler's law

$$\frac{d\psi}{dt} = \mu^{\frac{1}{2}} b^{-\frac{3}{2}} \quad \text{and} \quad \frac{d\phi}{dt} = \mu^{\frac{1}{2}} a^{-\frac{3}{2}}.$$

We thus have for stationary points

$$b^{\frac{1}{2}} \cos i + a^{\frac{1}{2}} = ab^{-\frac{1}{2}} [\cos \psi \cos \phi \cos i + \sin \psi \sin \phi]$$
$$+ a^{-\frac{1}{2}} b [\cos \psi \cos \phi + \sin \psi \sin \phi \cos i] \dots (21).$$

Writing $\cos i = \cos^2 \dfrac{i}{2} - \sin^2 \dfrac{i}{2}$, we finally obtain, after multiplying throughout by $a^{\frac{1}{2}} b^{\frac{1}{2}}$,

$$a^{\frac{1}{2}} b^{\frac{1}{2}} (a^{\frac{1}{2}} + b^{\frac{1}{2}} \cos i) = (a^{\frac{3}{2}} + b^{\frac{3}{2}}) \cos (\psi - \phi) \cos^2 \frac{i}{2}$$

$$- (a^{\frac{3}{2}} - b^{\frac{3}{2}}) \cos (\psi + \phi) \sin^2 \frac{i}{2} \dots (22).$$

This is the relation sought. It is to be remembered that ψ and ϕ denote the angular distances of the planet and of the earth, measured in the plane of their respective orbits, from the line of nodes.

98. *The phases of the planets and the moon.*

We shall assume the heavenly bodies concerned to be spherical. In Fig. 63, let P be the centre of a planet (or the moon) and let the straight lines joining P to the earth and the sun cut the planet's surface at E and S. The hemisphere illuminated by the sun will be bounded by the great circle $BCKD$, of which S is the pole. The hemisphere towards the earth is bounded by the great circle $ACLD$ of which E is the pole. Hence the only part of the planet visible from the earth is the area consisting of the spherical triangles ABC and ABD. The plane $ACLD$ is the plane

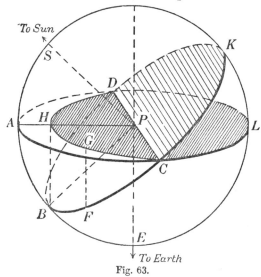

Fig. 63.

of the planet's *disc* as viewed from the earth. The great circle arc CAD is the visible *limit* of the planet. Any point on CB such as F will be seen on the disc at the point G, FG being parallel to the line of sight EP. The projection of all points on the semicircle CBD thus leads to a curve CHD in the plane of the great circle $ACLD$, and this curve is an ellipse of which CD is the major axis; the semi-minor axis is HP, where H is the projection of B on the plane of the disc. The planet is seen as shown in Fig. 64. The visible area of the planet's disc is the area bounded by the semi-circle CAD and the semi-circumference CHD of the ellipse. If r denotes the linear radius of the planet (or moon) this visible area A is given by

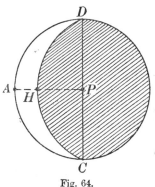

Fig. 64.

$$A = \tfrac{1}{2}\pi r^2 - \tfrac{1}{2}\pi r \cdot PH.$$

Let d denote the angle SPE; then d is the elongation of the earth from the sun as seen from P. Now $PH = PB \cos A\hat{P}B$, and since $S\hat{P}B = 90°$, it follows that the angle $APB = 180° - d$. Hence

$$A = \frac{\pi r^2}{2}(1 + \cos d) \qquad \dots\dots(23).$$

The *phase* is measured by the fraction of the diameter, perpendicular to the line of cusps, lying in the visible portion of the disc; the phase is thus $AH/2r$ or $(1 + \cos d)/2$. From (23) it is seen that the phase is also represented by the fraction of the area of the disc illuminated.

We now consider in greater detail the phases of the moon. In Fig. 65, let M denote the moon and E the earth. MS is the direction of the sun from M; $S\hat{M}E$ is the angle d. The moon's phase is then given by (23). Now the moon's distance from the earth is very small compared with the sun's distance from the earth, and in Fig. 65 we may assume ES parallel to MS without any serious loss of accuracy. Hence, denoting $S\hat{E}M$ by E, we have $E = 180° - d$ or $d = 180° - E$. This angle E is the elongation of the moon from the sun as viewed from the earth.

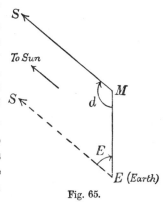

Fig. 65.

At new moon $E = 0$, for the moon is then directly, or almost directly, between the earth and the sun; thus $d = 180°$ and the phase is zero. At *quadrature*, the elongation is $90°$ and thus $d = 90°$, so that the phase is $\frac{1}{2}$; the moon then presents one-half of its illuminated surface towards the earth. When the moon is in opposition, the elongation is $180°$, so that $d = 0$; hence the phase is unity, that is, the complete disc of the moon is visible; the moon is then "full". After full moon, the phases are repeated in reverse order until the next new moon occurs. When more than one-half of the disc is visible, the moon is said to be *gibbous*.

99. *The brightness of the planets.*

The amount of light which reaches the earth from a planet depends (i) on the phase, and (ii) on the geocentric distance ρ of the planet, varying inversely as ρ^2. Thus if B denotes the apparent brightness, we can write

$$B = \frac{c\,(1 + \cos d)}{\rho^2} \qquad \text{......(24)},$$

where c is a constant depending on the surface illumination and reflective power. But from Fig. 66

$$\cos d = \frac{\rho^2 + b^2 - a^2}{2\rho b} \qquad \text{......(25)},$$

where a and b are the heliocentric distances of the earth and planet. Hence

$$B = c\,[\rho^2 + 2\rho b + b^2 - a^2]/2b\rho^3$$
$$\text{......(26)}.$$

When B is a maximum, the orbits being assumed circular, we must have $\dfrac{dB}{d\rho} = 0$

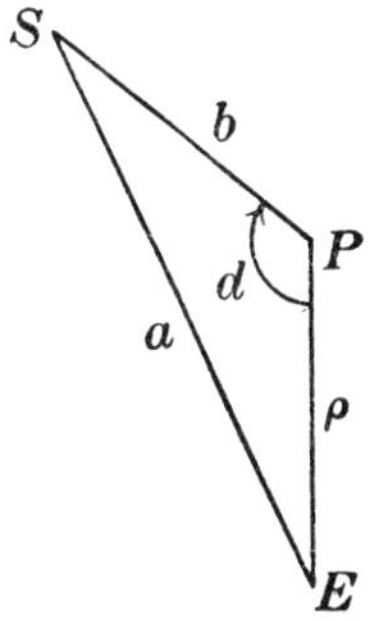

or $\qquad \rho^2 + 4b\rho + 3\,(b^2 - a^2) = 0$,

from which $\qquad \rho = (b^2 + 3a^2)^{\frac{1}{2}} - 2b \qquad \text{......(27)}.$

Fig. 66.

For the planet Venus, we have $b = 0\cdot723a$ and for maximum brightness (27) gives $\rho = 0\cdot430$. From (25), the corresponding value of d is $117°\cdot9$. The elongation from the sun—the angle SEP—is then found to be $39°\cdot7$.

It should be remarked that in the case of the planet Mercury, for which the orbital eccentricity is large (about $0\cdot2$), the symbol b in (26) represents the radius vector in the orbit. The brightness of Mercury is thus dependent on the planet's position in its orbit as well as on its geocentric distance.

100. *Heliographic co-ordinates.*

In studying sun-spots and the solar phenomena associated with them, it is of importance to know their co-ordinates on the solar surface; these co-ordinates are defined in much the same way as that in which points on the surface of the earth are expressed in terms of longitude and latitude. In Fig. 67, let the sphere represent the solar globe. The great circle TEN, of which K is the pole, is the intersection of the plane of the ecliptic with the sphere. The straight line from the sun's centre C, drawn in the direction of the vernal equinox, cuts the sphere in T. Since the earth is in the ecliptic, the straight line joining C to the earth cuts the solar surface in E. Any plane perpendicular to EC (which is the line of sight) we can regard as the plane of the sun's disc; for convenience we shall take the shaded plane defined by the radii CK and CT to be the plane of the disc (E is the pole of the great circle UKT). Any point X on the solar surface will be seen on the disc at X_1, such that XX_1 is perpendicular to the plane of the disc and therefore parallel to EC. The

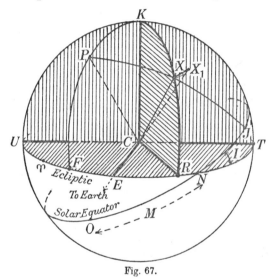

Fig. 67.

position of X_1 can be ascertained, from visual or photographic observations, with reference to certain rectangular axes in the plane of the disc. The problem is to deduce the solar or heliographic longitude and latitude of X from the observed position X_1.

The sun rotates about an axis, the northern extremity of which is the point P in Fig. 67, and the plane perpendicular to this axis cuts the sun's surface in the great circle ONJ, which is called the *solar equator*. The inclination JNT of the solar equator to the ecliptic is usually denoted by I. The point N is the ascending node of the solar equator on the ecliptic and its longitude TN (measured along the ecliptic from T) is denoted by Ω. The value of Ω is found to be given by

$$\Omega = 73° \, 40' + 50''{\cdot}25 \, (t - 1850{\cdot}0),$$

where t is expressed in years, so that for 1931·0 the value of Ω is 74° 47'·8.

The *heliographic co-ordinates* of a point X on the sun's surface are defined with reference to the solar equator. As the sun is a gaseous body, there is no definitely recognisable point on the solar equator from which to measure longitudes. A reference point O is chosen as follows. At mean noon on 1854 January 1, the node N defined a particular point on the solar surface. Owing to the solar rotation this point is carried round the equator and at any subsequent time we suppose that it is at O; assuming that the equatorial rotation occurs in a period of 25·38 days, the position of O with reference to N can be calculated for the instant concerned; in particular the arc ON which we denote by M can be found. The rotation occurs in the direction $\overrightarrow{ON}$. In Fig. 67, let PXJ be the solar meridian through X. Then OJ is the *heliographic longitude* of X (denoted by L) and the arc JX is the *heliographic latitude* of X (denoted by B).

101. *The heliographic co-ordinates of the centre of the disc.*

Referring to Fig. 67, we see that the point E on the solar surface will be seen as the centre of the disc. Let the heliographic co-ordinates of E be L_0 and B_0. Then in the spherical triangle PEN we have:

$$PN = 90°, \;\; P\hat{N}E = 90° - I, \;\; PE = 90° - B_0, \;\; \text{and} \;\; E\hat{P}N = M - L_0.$$

We require still the arc EN expressed in terms of known quantities. Now TE is the angle between the direction of the vernal equinox and the direction of the earth as seen from the sun's centre C, so that TE is the earth's heliocentric longitude. If $\odot$ denotes the geocentric longitude of the sun, we have

$$TE = \odot + 180° \qquad \dots\dots(28).$$

Also since $TN = \Omega$, we see that

$$EN = \Omega - \odot - 180° \qquad \dots\dots(29).$$

From the triangle PEN we derive the following formulae, using $\mathbf{D}$ and $\mathbf{A}$ in succession,

$$\tan (L_0 - M) = \tan (\odot - \Omega) \cos I \qquad \ldots\ldots(30),$$
$$\sin B_0 = \sin (\odot - \Omega) \sin I \qquad \ldots\ldots(31).$$

Assuming that M has been calculated for the time of observation, these formulae enable the heliographic co-ordinates L_0 and B_0 of the centre of the disc to be found; L_0 and B_0 are tabulated in the almanacs for every day during the year.

102. *Position angle of the sun's axis of rotation.*

Consider rectangular axes CE, CT and CK (Fig. 67). Let KXR be a great circle through K (the pole of the ecliptic) meeting the ecliptic in R. The radius of the sphere being taken as unity, the co-ordinates of X referred to these axes are (cos ER cos RX, sin ER cos RX, sin RX). Now X_1 is the projection of X on the plane KCT—the plane of the disc—and the co-ordinates of X_1 referred to CT and CK as axes are therefore (sin ER cos RX, sin RX) or, since $KR = 90°$, (sin ER sin KX, cos KX). Let ψ denote the angle between CX_1 and CK. Then we have

$$\tan \psi = \sin ER \tan KX \qquad \ldots\ldots(32).$$

Suppose that X now represents the point on the solar surface intersected by the radius drawn parallel to the *earth's axis of rotation*. The projection CX_1 on the disc defines the *north direction*, from which the position angle of any point on the disc is measured in practice. Now KX is the angle between the pole of the ecliptic and the pole of the earth's equator; hence $KX = \epsilon$, where ϵ is the obliquity of the ecliptic. Also the vernal equinox is 90° from both K and X. Hence it is 90° from the great circle KXR; hence $\Upsilon R = 90°$. But by (28), $\Upsilon E = 180° + \odot$; hence $ER = 90° - (180° + \odot)$ or

$$ER = 270° - \odot \qquad \ldots\ldots(33).$$

Denoting the angle X_1CK by x, we have from (32), writing x for ψ,

$$\tan x = - \cos \odot \tan \epsilon \qquad \ldots\ldots(34).$$

Again, the projection of CP on the plane KCT will make an angle, say, y with CK. Hence, by (32),

$$\tan y = \sin EF \tan KP \qquad \ldots\ldots(35),$$

where F is the point on the ecliptic cut by the great circle KP. Now since K is the pole of ΥRT and P is the pole of ONJ, it follows that N is the pole of KPF; hence $FN = 90°$. But $\Upsilon N = \Omega$ and therefore $\Upsilon F = \Omega - 90°$. Also, by (28), $\Upsilon E = 180° + \odot$; hence $EF = 180° + \odot - (\Omega - 90°)$ or

$$EF = 270° + (\odot - \Omega) \qquad \text{......(36)}.$$

Also KP is the inclination I. Hence (35) becomes, using (36),

$$\tan y = -\cos(\odot - \Omega)\tan I \qquad \text{......(37)}.$$

For convenience, we shall assume that we are dealing with a photograph of the solar disc. Let CN (Fig. 68) be the radius in the plane of the disc correspond-
ing to the projection of the solar radius parallel to the earth's axis. This radius CN can be inferred from the diurnal motion of the sun across the sky; if the telescope is at rest, the image of the sun will move along a parallel of declination, that is, perpendicular to CN; hence by suitable means the radius CN can be drawn on the photograph.

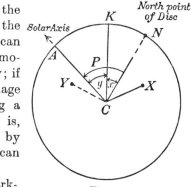

Fig. 68.

The *position angle* of any marking such as Y on the disc is measured from CN eastwards, that is, towards the left as we view the sun. Thus the position angle of Y is the angle NCY and of X it is $(360° - N\hat{C}X)$. Now if CA is the projection on the disc of the sun's axis

$$NCK = x \quad \text{and} \quad KCA = y,$$

where x and y are given by (34) and (37). Thus the position angle P of the sun's axis CA is given by

$$P = x + y \qquad \text{......(38)}.$$

Thus P can be calculated; it is tabulated in the almanacs.

If χ denotes the angle ACX (Fig. 68), X being any marking on the disc, and θ is the position angle of X we have, since $NCX = 360° - \theta$, that $\chi = P + 360° - \theta$ or

$$\chi = P - \theta \qquad \text{......(39)}.$$

As θ and P may be supposed known, the former by means of the photograph and the latter by means of the almanac, the angle χ can be found.

103. *The heliographic co-ordinates of a sun-spot.*

Let R and d denote in linear measure the radius of the sun and the distance of the earth from the sun respectively. If S is the angular semi-diameter of the sun—this can be found for the day of observation from the almanac—we have $\sin S = R/d$ or, since S is about $16'$, we can write with sufficient accuracy

$$S = \frac{R}{d} \operatorname{cosec} 1' \qquad \ldots\ldots(40),$$

in which S is expressed in minutes of arc. Let r_0 and r denote the measures of the radius of the disc and of the distance of the spot X from the centre C of the disc (Fig. 68); r_0 and r can be measured in any convenient unit on the photograph. If ρ_1 (in minutes of arc) is the angle between the direction of the sun's centre and the direction of the spot, both viewed from the earth, we have, with sufficient accuracy

$$\frac{\rho_1}{S} = \frac{r}{r_0} \qquad \ldots\ldots(41),$$

from which ρ_1 can be determined.

In Fig. 69, let S be the position of the spot on the solar surface

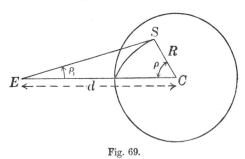

Fig. 69.

and let $S\hat{C}E$ be denoted by ρ, E being the earth. Then $ESC = 180° - (\rho + \rho_1)$. But from the plane triangle ESC,

$$\sin ESC = \frac{d}{R} \sin \rho_1$$

or, since ρ_1 is small (less than 16′) and is expressed in minutes of arc,

$$\sin (\rho + \rho_1) = \frac{d}{R} \rho_1 \sin 1'$$

or, using (40), $\sin (\rho + \rho_1) = \dfrac{\rho_1}{S}$ ......(42).

Since ρ_1 can be found by means of (41), this last formula (42) enables ρ to be calculated.

Consider now Fig. 70 in which S is the spot on the solar surface, CE gives the direction of the earth from the sun's centre C, and P is the north pole of the sun's axis. The great circle arc ES is thus the angle ρ given by (42). Denote the heliographic longitude and latitude of S by L and B; the corresponding co-ordinates of E are L_0 and B_0. In the spherical triangle PES we have: $PS = 90° - B$, $PE = 90° - B_0$, $ES = \rho$ and $E\hat{P}S = L - L_0$. Also, since the point E is seen as the centre C of the disc and

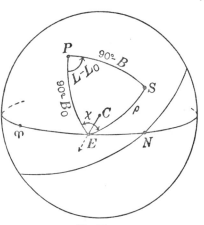

Fig. 70.

since the great circles EP and ES project into the straight lines CA and CX of Fig. 68, the angle PES is equal to the angle ACX or χ. Hence, by (39), $P\hat{E}S = P - \theta$, where θ is the position angle of the spot on the disc. Hence from the triangle PES we have the formulae, using **A** and **B**,

$$\sin B = \sin B_0 \cos \rho + \cos B_0 \sin \rho \cos (P - \theta) \quad ...(43),$$
$$\sin (L - L_0) = \sin \rho \sin (P - \theta) \sec B \qquad ......(44).$$

Since B_0, ρ, P, θ and L_0 are all supposed now to be known, the formulae (43) and (44) determine the heliographic latitude and longitude of the spot.

EXERCISES

1. Two planets P_1 and P_2 revolve in circular orbits at distances b_1, b_2 from the sun. Prove that when they appear stationary to one another,

$$\frac{b_1}{b_2} = \tfrac{1}{2} \tan \frac{\theta}{2} \tan \theta,$$

where $\tan \theta = 2 \cot E$, E being the elongation of P_2 as seen from P_1.

2. If θ is the angle subtended at the earth by the sun and a stationary point of a planet's orbit, and ϕ is the maximum elongation of the planet, prove that

$$2 \cot \theta = \sec \tfrac{1}{2}\phi + \operatorname{cosec} \tfrac{1}{2}\phi.$$

3. Show that, if the earth and a planet be supposed to describe coplanar, circular orbits, and the difference in longitude of the sun and the planet be θ, the rate of change of θ is, numerically,

$$\frac{2\pi}{S} \left(1 - \frac{a}{\rho} \cos \theta \right),$$

where S is the synodic period of the planet, a the radius of the earth's orbit, and ρ the distance of the planet from the earth at the moment.

[Coll. Exam.]

4. If the line joining two planets to one another subtends an angle of $60°$ at the sun when the planets appear to each other to be stationary, show that $a^2 + b^2 = 7ab$, where a, b are the distances of the planets from the sun.

[M.T.]

5. If u and v are the velocities of two planets in circular and coplanar orbits, show that the period of direct motion is to the period of retrograde motion as $(180° - \theta) : \theta$, where $\cos \theta = uv/(u^2 - uv + v^2)$.   *[Coll. Exam.]*

6. If a and b are the radii of the orbits (assumed to be circular and coplanar) of the earth E and a superior planet P, and u and v are their respective linear velocities, prove that the square of the velocity of P relative to E at a stationary point is

$$\frac{(u^2 - v^2)(bu - av)}{bu + av}.$$

7. One planet whose mean distance from the sun is a appears to have a phase E to another planet whose mean distance from the sun is b, and the latter appears to have a phase V. Prove that, if the inclinations of the orbits to each other and their eccentricities be neglected,

$$b^2 V (1 - V) = a^2 E (1 - E).$$

If the distance of Venus from the sun is $0 \cdot 72$ astronomical units, find what part of the earth's surface appears illuminated as seen from Venus.

8. The heliocentric distance of an inferior planet P, moving in a circular orbit in the ecliptic, is b astronomical units; the orbit of the earth (E) is also assumed circular. If the heliocentric co-ordinates of P and E are $(b \cos f\theta,\ b \sin f\theta)$ and

($\cos \theta$, $\sin \theta$) respectively, θ being measured from inferior conjunction, show that for a stationary point

$$\cos (f-1)\, \theta = \frac{1 + fb^2}{b(1+f)}.$$

9. Prove that when a planet whose orbit does not coincide with the ecliptic is absolutely stationary as seen from the earth, its direction of motion and that of the earth must intersect on the line of nodes, and that its projection on the plane of the ecliptic is also stationary. [*M.T.*]

10. Neglecting the inclination of the plane of a planet's orbit to the ecliptic, prove that, in latitude ϕ ($> \epsilon$), the altitude of an inferior planet at sunrise or sunset never exceeds $\sin^{-1} (\sin E \cos \overline{\phi - \epsilon})$,

where E is the maximum elongation of the planet; and that this most favourable case can only occur when maximum elongation coincides with an equinox.

How is this result modified if the observer is in the tropical zone?

[*Lond.* 1925.]

11. If the orbit of an outer planet is an ellipse of eccentricity e and semi-axis a, the inclination to the ecliptic being zero, and if it is in opposition at perihelion, show that its motion will appear direct if a, in astronomical units, is less than $(1 + e)/(1 - e)$. [*Ball.*]

12. Show that the phase of a superior planet, as seen from the earth, is least when the earth appears half illuminated to the planet, but that the apparent brightness of the planet is a maximum at opposition and a minimum at conjunction. [*Coll. Exam.*]

13. Assuming that Venus and the earth describe circular orbits in the ecliptic, show that Venus will appear brightest at elongation θ given by

$$\cos \theta = \tfrac{2}{3} \{(3 + a^2)^{\frac{1}{2}} - a\},$$

where a is the heliocentric distance of Venus in astronomical units.

[*Lond.* 1926.]

14. Show that the "correction for defective illumination" of Mars (i.e. the apparent width in arc of the unilluminated crescent) is a maximum when Mars is at a geocentric distance $(b^2 - a^2)/b$, the orbits of the earth and Mars being treated as circles of radii a and b. Show that when the disc of the moon is seen at night by earthshine there is a narrow crescent not illuminated by either sunlight or earthshine and that the width of this crescent does not exceed $0''{\cdot}5$. (Lunar parallax $= 57'$; semi-diameter $= 15'$.) [*M.T.* 1926.]

15. Assuming the diameters of the earth and Venus to be negligible, show that ψ, the heliocentric elongation of Venus from the earth at the beginning or end of a transit, is given by the equation

$$b^2 r^2 \cos^2 \psi - 2br R^2 \cos \psi + R^2 (b^2 + r^2) - b^2 r^2 = 0,$$

where R is the sun's radius, and b, r the distances of Venus and the earth from the sun's centre. [*Ball.*]

CHAPTER VIII

ABERRATION

104. *The law of aberration.*

The phenomenon of aberration was discovered in 1728 by Bradley, later Astronomer Royal, as a result of a series of meridian observations of the second magnitude star γ Draconis. In 1675, the Danish astronomer Roemer had established that light travelled with a finite velocity and Bradley's observations were interpreted by recognising that a star's position in the sky could be displaced by an amount depending on the ratio of the earth's orbital velocity to the velocity of light, and on the position of the particular star concerned. The earth's average orbital speed is about $18\frac{1}{2}$ miles per second and the velocity of light is $186,324$ miles per second; thus the ratio just mentioned is small but not negligible.

In Fig. 71, let C represent the centre of a telescope's object-glass and E the eye-piece at the moment when a ray from a star X arrives at C. Let EF be parallel, at this moment, to the direction in which the earth is moving around the sun. Let τ be the time required by the ray to pass through the telescope; in this interval the earth has moved through a distance EE_1 or $V\tau$, where V is the earth's velocity. Denote by c the velocity of light. Then when the ray from the star reaches the eye-piece, the

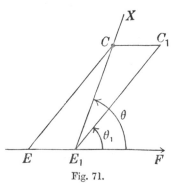

Fig. 71.

latter is at E_1 and we have $CE_1 = c\tau$. If the earth had no velocity, the direction in which the telescope would be pointed is along E_1C, which we describe as the true direction of the star. Actually, owing to the earth's motion, the telescope has to be pointed in the direction EC. Complete the parallelogram EE_1C_1C. Then E_1C_1 is parallel to EC and thus E_1C_1 is the apparent direction of the star at the moment of observation. Let θ, θ_1 denote $C\hat{E}_1F$,

$C_1 \hat{E}_1 F$ respectively. Then the angular displacement $\theta - \theta_1$ is said to be due to aberration. From Fig. 71 it is seen that aberration displaces the star's true direction towards the direction EF of the earth's motion and in the plane XE_1F.

In the triangle CE_1C_1,

$$\frac{\sin CE_1C_1}{\sin CC_1E_1} = \frac{CC_1}{CE_1}.$$

But $CC_1 = EE_1 = V\tau$ and $CE_1 = c\tau$. Hence we derive

$$\sin(\theta - \theta_1) = \frac{V}{c}\sin\theta_1.$$

Now V/c is small and consequently $\theta - \theta_1$ is small. We can then write with all needful accuracy ($\theta - \theta_1$ being expressed in seconds of arc)

$$\theta - \theta_1 = \frac{V}{c}\sin\theta_1 \operatorname{cosec} 1''$$

or $\qquad\qquad \theta - \theta_1 = \kappa\sin\theta_1 \qquad\qquad ......(1),$

in which $\qquad\qquad \kappa = \frac{V}{c}\operatorname{cosec} 1'' \qquad\qquad ......(2).$

κ is defined to be the constant of aberration; its value which can be calculated from the values of V and c is about $20''\cdot5$. The definition and value of κ will be considered later in greater detail. It is clear that as $\theta - \theta_1$ cannot exceed $20''\cdot5$ we can write (1) with sufficient accuracy in the form

$$\theta - \theta_1 = \kappa\sin\theta \qquad\qquad ......(3).$$

The law of aberration is contained in formula (1) or (3).

103. *Annual aberration in ecliptic longitude and latitude.*

For the present we shall neglect the eccentricity of the earth's orbit and assume simply that the orbit is a circle with the sun at the centre, and that the orbital velocity V is constant. If E (Fig. 72) is the position of the earth at any time, its direction of motion is along EF at right angles to SE. Let ET or ST_1 be the direction of the vernal equinox; then $T\hat{E}S$ is the geocentric longitude of the sun (denoted by $\odot$). The geocentric longitude of F is thus ($\odot - 90°$). Accordingly the point F of the ecliptic, towards which the earth's motion is directed, is $90°$ behind the position of the sun on the ecliptic. With reference to the earth as centre of the celestial sphere, the point F evidently makes a complete circuit of the ecliptic in a year; the aberration in this instance is called *annual aberration*, and the displacements to which it gives rise are repeated in yearly cycles.

In Fig. 73, F defines the direction of the earth's motion and $\Upsilon F = \odot - 90°$. Let X be the true position of a star (it defines the direction in which the star is viewed from the sun). Then as the aberrational displacement takes place in the plane XEF, the star will appear to be displaced along the great circle XF and

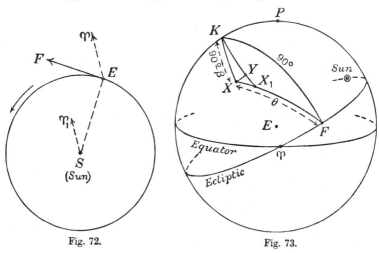

Fig. 72.   Fig. 73.

will be seen at X_1. With the notation of the previous section, $XF = \theta$, $X_1F = \theta_1$ and therefore XX_1 is given by (3), so that

$$XX_1 = \kappa \sin \theta \qquad \ldots\ldots(4).$$

Join KX and KX_1 by great circles, K being the pole of the ecliptic, and draw the small circle XY parallel to the ecliptic. If λ, β are the longitude and latitude of X and λ_1, β_1 are the corresponding co-ordinates of X_1, then $\lambda_1 - \lambda = X\hat{K}Y$ and, since $XY = X\hat{K}Y \sin KX$, we have $XY = (\lambda_1 - \lambda) \cos \beta$. Also $\beta - \beta_1 = X_1Y$. Write

$$\Delta\lambda = \lambda_1 - \lambda \quad \text{and} \quad \Delta\beta = \beta_1 - \beta.$$

Then $\qquad XY = \Delta\lambda \cos \beta \quad \text{and} \quad X_1Y = -\Delta\beta \qquad \ldots\ldots(5).$

In the infinitesimal plane triangle XX_1Y, let ϕ denote $Y\hat{X}X_1$.

Then $\qquad XY = XX_1 \cos \phi \quad \text{and} \quad X_1Y = XX_1 \sin \phi.$

Hence from (4) and (5)

$$\Delta\lambda = \kappa \sin\theta \cos\phi \sec\beta \qquad \ldots\ldots(6),$$
$$\Delta\beta = -\kappa \sin\theta \sin\phi \qquad \ldots\ldots(7).$$

Now in the spherical triangle KXF, $KX = 90° - \beta$, $KF = 90°$, $XF = \theta$, $K\hat{X}F = 90° + \phi$ and $X\hat{K}F$ is the difference of longitude of F and X, so that $X\hat{K}F = (\odot - 90°) - \lambda$.

By the sine-formula **B**,

$$\sin XF \sin KXF = \sin KF \sin XKF,$$

and consequently $\quad \sin\theta \cos\phi = -\cos(\odot - \lambda) \qquad \ldots\ldots(8).$

By formula **C**,

$$\sin XF \cos KXF = \cos KF \sin KX - \sin KF \cos KX \cos XKF,$$

and thus $\quad \sin\theta \sin\phi = \sin\beta \sin(\odot - \lambda) \qquad \ldots\ldots(9).$

From (6)...(9), we obtain the formulae

$$\Delta\lambda = -\kappa \sec\beta \cos(\odot - \lambda) \qquad \ldots\ldots(10),$$
$$\Delta\beta = -\kappa \sin\beta \sin(\odot - \lambda) \qquad \ldots\ldots(11).$$

These formulae—(10) and (11)—give the displacements in longitude and latitude due to annual aberration.

106. *The aberrational ellipse.*

Referring to Fig. 73 we see that the aberrational displacement from X to X_1 is equivalent to the two displacements from X to Y and from Y to X_1. Denote these by x and y respectively. Then

$$x = \Delta\lambda \cos\beta = -\kappa \cos(\odot - \lambda) \qquad \ldots\ldots(12),$$
$$y = -\Delta\beta \quad = \kappa \sin\beta \sin(\odot - \lambda) \qquad \ldots\ldots(13),$$

from which, by eliminating $(\odot - \lambda)$, we obtain

$$\frac{x^2}{\kappa^2} + \frac{y^2}{\kappa^2 \sin^2\beta} = 1 \qquad \ldots\ldots(14).$$

This is the equation of an ellipse, known as the aberrational ellipse. During the course of a year the star appears to describe this curve on the celestial sphere, the ellipse centre being the star's true position. The semi-major axis is κ, parallel to the ecliptic, and is therefore constant for all stars; the semi-minor axis is $\kappa \sin\beta$, perpendicular to the semi-major axis. The aberrational displacement is greatest when $x = \pm\kappa$, that is—from (12) —when $\odot - \lambda = 0°$ or $180°$. Assuming that we know the value

7 $\qquad\qquad\qquad\qquad\qquad\qquad\qquad$ SA

of λ for a particular star, we can thus obtain the two values of $\odot$ (the geocentric longitude of the sun) corresponding to maximum displacement and thus derive the date, by reference to the *Nautical Almanac*, when this occurs. Thus, if $\lambda = 0°$, the aberrational displacement is a maximum when $\odot = 0°$ or $180°$, that is, at the spring or autumn equinox.

For a star on the ecliptic ($\beta = 0°$), the aberrational ellipse degenerates into a straight line (an element of the ecliptic) and for a star at the pole of the ecliptic, it becomes a circle.

107. *Aberration in right ascension and declination.*

As before, let X be the true position of the star and X_1 its apparent position (Fig. 74). Let α, δ be the right ascension and

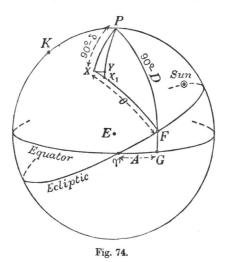

Fig. 74.

declination of X, and α_1, δ_1 the corresponding co-ordinates of X_1. Draw a small circle XY parallel to the equator. Then

$$X\hat{P}Y = \alpha_1 - \alpha \quad \text{and} \quad XY = X\hat{P}Y \sin PX.$$

Let $\qquad \Delta\alpha = \alpha_1 - \alpha \quad \text{and} \quad \Delta\delta = \delta_1 - \delta.$

Then $\qquad \Delta\alpha = XY \operatorname{cosec} PX = XY \sec \delta.$

From the figure, $X_1Y = -\Delta\delta$. Denote $Y\hat{X}X_1$ by ψ. Then

$$XY = XX_1 \cos\psi \quad \text{and} \quad X_1Y = XX_1 \sin\psi.$$

Hence by (3), since $XX_1 \equiv \theta - \theta_1$ in the notation of section 104,

$$\Delta\alpha = \kappa \sin\theta \cos\psi \sec\delta \qquad \ldots\ldots(15),$$
$$\Delta\delta = -\kappa \sin\theta \sin\psi \qquad \ldots\ldots(16).$$

Let A, D denote the right ascension and declination of F (the point on the ecliptic towards which the earth's motion is directed). In the spherical triangle PXF, we have: $PX = 90° - \delta$, $PF = 90° - D$, $X\hat{P}F = A - \alpha$, $XF = \theta$ and $P\hat{X}F = 90° + \psi$.

By formula **B**,

$$\sin XF \sin PXF = \sin PF \sin XPF,$$

from which $\qquad \sin\theta\cos\psi = \cos D \sin(A - \alpha) \qquad \ldots\ldots(17).$

By formula **C**,

$$\sin XF \cos PXF = \cos PF \sin PX - \sin PF \cos PX \cos XPF,$$

from which

$$-\sin\theta\sin\psi = \sin D \cos\delta - \cos D \sin\delta \cos(A - \alpha) \quad\ldots(18).$$

We thus derive from (15)...(18),

$$\Delta\alpha = \kappa \sec\delta \cos D \sin(A - \alpha) \qquad \ldots\ldots(19),$$
$$\Delta\delta = \kappa \sin D \cos\delta - \kappa \cos D \sin\delta \cos(A - \alpha) \quad \ldots\ldots(20).$$

Now consider the triangle FTG in which PFG is the meridian through F. We have: $TF = \odot - 90°$, $F\hat{T}G = \epsilon$ (the obliquity of the ecliptic), $TG = A$, $FG = D$ and $F\hat{G}T = 90°$. By formulae **A**, **B** and **C**, we derive

$$\left.\begin{array}{l} \sin\odot = \cos A \cos D \\ -\cos\odot \sin\epsilon = \sin D \\ -\cos\odot \cos\epsilon = \sin A \cos D \end{array}\right\} \qquad \ldots\ldots(21).$$

Now from (19) we have

$$\Delta\alpha = \kappa \sec\delta [\cos\alpha.\sin A \cos D - \sin\alpha.\cos A \cos D],$$

so that, using (21), we obtain

$$\Delta\alpha \equiv \alpha_1 - \alpha = -\kappa \sec\delta [\cos\alpha \cos\odot \cos\epsilon + \sin\alpha \sin\odot]$$
$$\ldots\ldots(22).$$

Similarly we obtain from (20) and (21),

$$\Delta\delta \equiv \delta_1 - \delta = -\kappa \cos\odot \cos\epsilon (\tan\epsilon \cos\delta - \sin\alpha \sin\delta)$$
$$-\kappa \cos\alpha \sin\delta \sin\odot \quad\ldots(23).$$

As κ is expressed in seconds of arc, the formula for $\Delta\alpha$ or $(\alpha_1 - \alpha)$ gives this quantity also in seconds of arc. It is of course usual to express right ascension in time-measure and we must then suppose the right of (22) to be divided by 15.

Let

$$C = - \kappa \cos \epsilon \cos \odot, \quad D = - \kappa \sin \odot$$
$$c = \tfrac{1}{15} \cos \alpha \sec \delta, \quad c' = \tan \epsilon \cos \delta - \sin \alpha \sin \delta \quad \ldots\ldots(24)$$
$$d = \tfrac{1}{15} \sin \alpha \sec \delta, \quad d' = \cos \alpha \sin \delta$$

Then we have from (22), (23) and (24),

$$\alpha_1 = \alpha + Cc + Dd \qquad \ldots\ldots(25),$$
$$\delta_1 = \delta + Cc' + Dd' \qquad \ldots\ldots(26).$$

In these equations, C and D depend on the sun's longitude and consequently on the time of the year to which they refer; the values of log C and log D (*Bessel's day numbers* or the *Besselian star numbers*) are tabulated in the almanacs for every day of the year. The quantities c, c', d, d' depend only on the co-ordinates of the star and the obliquity of the ecliptic, and can therefore be calculated once for all.

It is to be remarked that (25) and (26) give only the effect of aberration on the star's co-ordinates.

Another method of simplifying the computations is as follows.

Let
$$h \cos H = - \kappa \sin \odot; \quad h \sin H = - \kappa \cos \epsilon \cos \odot;$$
$$i = - \kappa \sin \epsilon \cos \odot.$$

Then when α_1 and α are expressed in time-measure, equations (22) and (23) become

$$\alpha_1 = \alpha + \tfrac{1}{15} h \sin (H + \alpha) \sec \delta \qquad \ldots\ldots(27),$$
and $\qquad \delta_1 = \delta + i \cos \delta + h \cos (H + \alpha) \sin \delta \qquad \ldots\ldots(28).$

The values of H, h and i (*the independent day numbers or star numbers*) and the quantities log h and log i are tabulated in the almanacs for every day of the year.

108. *The elliptic motion of the earth and aberration.*

So far, we have assumed that the earth moves with constant speed in a circular orbit. We consider now the problem in its general aspect. It has been shown in section 66 that the velocity along ET (the tangent at E) in an elliptic orbit (Fig. 75) is equivalent to a constant velocity h/p along EF perpendicular to the radius vector SE, together with a constant velocity eh/p

along Ef at right angles to the major axis AB. In the notation of Chapter v, we have $h^2 = n^2a^3p$ and $p = a\,(1 - e^2)$; hence, if we write $n = 2\pi/T$, where T is the period in the orbit, we have

$$\frac{h}{p} = \frac{2\pi a}{T\,(1 - e^2)^{\frac{1}{2}}} \qquad \dots\dots(29).$$

The total displacement, in any co-ordinate, of a star's position due to aberration can be regarded as made up of two displacements, the first due to the constant velocity h/p at right angles to the radius vector, and the second due to the constant velocity eh/p perpendicular to the major axis. We consider these in turn.

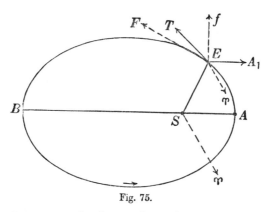

Fig. 75.

Since EF is perpendicular to SE, the geocentric longitude of F is the geocentric true longitude of the sun less 90° or $(\odot - 90°)$. We thus have the same geometrical conditions as are represented in Fig. 73. We consider the corresponding displacements in longitude and latitude. If these are denoted by $\Delta\lambda_1$ and $\Delta\beta_1$, they are given directly by (10) and (11), so that

$$\Delta\lambda_1 = -\,\kappa \sec\beta \cos\,(\odot - \lambda) \qquad \dots\dots(30),$$
$$\Delta\beta_1 = -\,\kappa \sin\beta \sin\,(\odot - \lambda) \qquad \dots\dots(31),$$

in which κ (in seconds of arc) is defined after the manner of formula (2) by

$$\kappa = \frac{h}{p}\cdot\frac{\operatorname{cosec} 1''}{c},$$

or, using (29), by
$$\kappa = \frac{2\pi a \operatorname{cosec} 1''}{cT\,(1 - e^2)^{\frac{1}{2}}} \qquad \dots\dots(32).$$

This is the precise definition of the *aberration constant* κ.

We now consider the aberrational displacements $\Delta\lambda_2$, $\Delta\beta_2$ arising from the constant velocity eh/p perpendicular to the major axis. In Fig. 75 let ST be the direction of the vernal equinox; then $T\hat{S}A$ is ϖ, the longitude of perihelion. Draw EA_1 parallel to SA; then the geocentric longitude of A_1 is ϖ. Also Ef is perpendicular to EA_1; therefore the geocentric longitude of f is $\varpi + 90°$. In Fig. 73, where TF is $\odot - 90°$, we derive the aberrational displacements $\Delta\lambda_1$, $\Delta\beta_1$ corresponding to the velocity h/p. It is now seen from Fig. 76, in which Tf is $\varpi + 90°$, that the formulae for $\Delta\lambda_2$, $\Delta\beta_2$ corresponding to the constant velocity eh/p can be written down

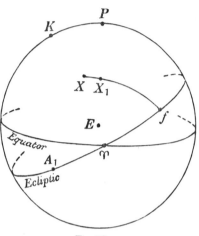

Fig. 76.

from (30) and (31) by substituting $\varpi + 90°$ for $\odot - 90°$, that is, $180° + \varpi$ for $\odot$, and by writing eh/p for h/p or $e\kappa$ for κ. We thus derive

$$\Delta\lambda_2 = + e\kappa \sec \beta \cos (\varpi - \lambda) \qquad \ldots\ldots(33),$$

$$\Delta\beta_2 = + e\kappa \sin \beta \sin (\varpi - \lambda) \qquad \ldots\ldots(34).$$

Hence the total aberrational displacements resulting from the motion of the earth in its elliptic orbit are given by

$$\Delta\lambda \equiv \Delta\lambda_1 + \Delta\lambda_2 = - \kappa \sec \beta \cos (\odot - \lambda) + e\kappa \sec \beta \cos (\varpi - \lambda) \qquad \ldots\ldots(35),$$

$$\Delta\beta \equiv \Delta\beta_1 + \Delta\beta_2 = - \kappa \sin \beta \sin (\odot - \lambda) + e\kappa \sin \beta \sin (\varpi - \lambda) \qquad \ldots\ldots(36).$$

It is to be noted that the expressions for $\Delta\lambda_2$ and $\Delta\beta_2$ in (33) and (34) are independent of the sun's longitude and are therefore the same for any given star throughout the year. As the value of the eccentricity e is about $1/60$, the value of $e\kappa$ is about $0''{\cdot}3$. Although $\varpi, e, \lambda, \beta$ are slowly varying quantities, the changes in $\Delta\lambda_2$, $\Delta\beta_2$ are so minute that for several hundred years at least the values of $\Delta\lambda_2$, $\Delta\beta_2$ can be regarded strictly as constants.

Since $\Delta\lambda = \lambda_1 - \lambda$, we have

$$\lambda_1 = \lambda + \Delta\lambda_1 + \Delta\lambda_2,$$

with a similar equation in β. Instead of applying the value of $\Delta\lambda_2$ in this equation, it is more convenient to regard $(\lambda + \Delta\lambda_2)$ as the true longitude of the star; similarly for the true latitude.

In the same way we can derive the aberrational effects, taking into account the ellipticity of the earth's orbit, on the equatorial co-ordinates α and δ. Again we suppose that the quantities $\Delta\alpha_2$, $\Delta\delta_2$, depending on the velocity eh/p perpendicular to the major axis, are incorporated in the true co-ordinates α and δ.

Thus the effective formulae for giving the aberrational displacements are, in ecliptic and equatorial co-ordinates,

$$\lambda_1 = \lambda - \kappa \sec \beta \cos (\odot - \lambda) \qquad \ldots\ldots(37),$$
$$\beta_1 = \beta - \kappa \sin \beta \sin (\odot - \lambda) \qquad \ldots\ldots(38),$$
and
$$\alpha_1 = \alpha + Cc + Dd \qquad \ldots\ldots(39),$$
$$\delta_1 = \delta + Cc' + Dd' \qquad \ldots\ldots(40),$$

where C, c, c', D, d and d' are defined by (24).

109. *The measurement of the constant of aberration.*

The accurate determination of κ is a practical problem that appears at first sight to be comparatively simple and straightforward. In order to reduce the effects of refraction which, even under the best conditions, may not be known with the certainty essential in this investigation, only stars culminating very near to the zenith are selected for observation. Let δ be the declination of a star culminating a little south of the zenith on a certain date. For simplicity we shall neglect the effects of precession and nutation, which will be discussed in a later chapter. The apparent declination δ_1 which will include the aberration displacement will be given by (23), which we write in the form

$$\delta_1 = \delta + \kappa x \qquad \ldots\ldots(41),$$

where x is the coefficient of κ on the right of (23).

In Fig. 77 let ZX' be the observed zenith distance (denoted by z) of a star on the meridian and let XX' represent the refraction r. Then $ZX = z + r$. Now $PX = 90° - \delta_1$ and $PZ = 90° - \phi$, where ϕ is the latitude, so that

$$ZX = \phi - \delta_1 = \phi - \delta - \kappa x.$$

Hence $\qquad z + r = \phi - \delta - \kappa x$

or $\qquad\qquad \phi = \delta + z + r + \kappa x \qquad \ldots\ldots(42).$

Suppose that another star culminates at Y', a little north of the zenith, and let Z denote its observed zenith distance and R the refraction. Then $ZY = Z + R$. Also if D is its true declina-

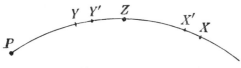

Fig. 77.

tion and D_1 its apparent declination (as affected by aberration), then by (23) $\qquad D_1 = D + \kappa X,$

where X is the value for this star of the coefficient of κ on the right of (23). Now $PY = 90° - D_1$ and $PZ = 90° - \phi$. Hence $ZY = D_1 - \phi$, and we have

$$Z + R = D + \kappa X - \phi$$

or $\qquad\qquad \phi = D - Z - R + \kappa X \qquad \ldots\ldots(43).$

Hence from (42) and (43),

$$2\phi = (\delta + D) + (z - Z) + (r - R) + \kappa (x + X) \quad\ldots(44).$$

In this equation we assume $(\delta + D)$ to be known and $(z - Z)$ is the difference of the measured zenith distances. As the stars culminate close to the zenith, the telescope can remain unaltered in position for the observations of the two stars, the interval between the transits—if the stars have been carefully chosen—being at most a few minutes. Although the meridian circle has been used in this problem it has been superseded by the photographic zenith telescope which records, on a photographic plate, the trails of the stars as they pass across the meridian. The measurement of the distance between the trails of the two stars we are considering gives the value of $(z - Z)$ very accurately. The difference $(r - R)$ of the refractions may be assumed known with the necessary accuracy. The quantity $(x + X)$ can be easily calculated.

About six months later, the observations are repeated. We have, as before, a subscript referring to the new observations,

$$2\phi = (\delta + D) + (z_1 - Z_1) + (r_1 - R_1) + \kappa (x_1 + X_1) \quad\ldots(45).$$

For simplicity we shall assume that the interval is such that the sun's longitude has increased by 180° exactly. By referring to (23) we see that the values of x_1 and X_1 will now be equal to $-x$ and $-X$ respectively, so that we can write (45) as

$$2\phi = (\delta + D) + (z_1 - Z_1) + (r_1 - R_1) - \kappa (x + X) \quad ...(46).$$

Subtracting (44) and (46) we obtain

$$2\kappa (x + X) + (z - Z) - (z_1 - Z_1) + (r - R) - (r_1 - R_1) = 0 \quad ......(47),$$

from which κ can be derived.

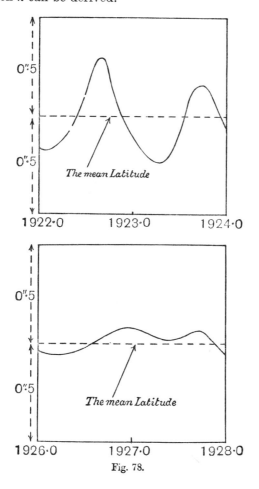

Fig. 78.

This procedure—of eliminating the latitude in (44) and (46) —is, however, unjustified when the object of the investigation is the *precise* determination of κ. It was discovered by Küstner that the earth's axis of rotation is not quite fixed relatively to the crust, and as the latitude ϕ is defined with reference to the axis of rotation, the value of ϕ undergoes minute changes of the order of a few tenths of a second of arc. Thus the values of ϕ in (44) and (46) must not be presumed to be identical. The determination of the aberrational constant is consequently bound up intricately with the problem of the *variation of latitude*. We shall not prolong the discussion further, but merely indicate some recent results. From a long series of observations with the Cookson photographic zenith telescope at Greenwich during the years 1919–27, the deduced value of the constant of aberration is 20″·445. Fig. 78 shows the changes in the latitude of Greenwich during the years 1922–3 and 1926–7.

110. *The theoretical value of the aberration constant.*

From (32), κ is given by

$$\kappa = \frac{2\pi a \cosec 1''}{cT\,(1-e^2)^{\frac{1}{2}}} \qquad \ldots\ldots(48),$$

in which a is the semi-major axis of the earth's orbit, c is the velocity of light, T is the sidereal year expressed in mean solar seconds and e is the eccentricity of the earth's orbit. The accepted values of these quantities are:

$$a = 149,500,000 \text{ kilometres,}$$
$$c = 299,791 \text{ kilometres per second,}$$
$$T = 31,558,149 \text{ seconds,}$$
$$e = 0·01674,$$

from which, by calculation, κ is found to be 20″·47. It is to be noted that the value of a which we have used for the above calculation is itself a quantity derived from astronomical observations and is, in consequence, subject to error. The same remark applies to the other quantities c, T and e, but for these the possible errors are believed to be too small to influence the result. The uncertain quantity is a. We shall refer again later (Chapter IX, p. 213) to the value of the aberration constant and the associated quantity a.

111. *Diurnal aberration.*

In addition to the yearly motion of the observer around the sun, there is also the daily motion due to the rotation of the earth about its axis. Let ρ denote the radius of the earth and ϕ the latitude of an observatory, which describes in the course of a sidereal day, due to rotation alone, the circumference of the small parallel of latitude whose radius is $\rho \cos \phi$. As there are 86,164 mean solar seconds in a sidereal day the speed concerned is $\dfrac{2\pi\rho \cos \phi}{86164}$ kilometres per second, where we suppose ρ to be expressed in kilometres. Taking ρ to be 6378 kilometres, the speed is easily found to be $0\cdot464 \cos \phi$ kilometres per second. Define k (in seconds of arc) by

$$k = \frac{0\cdot464}{c} \cos \phi \, \text{cosec} \, 1'' \qquad \ldots\ldots(49),$$

c being the velocity of light in kilometres per second. Then for an observatory in latitude ϕ, k is the *constant of diurnal aberration*. Inserting the value of c which we have stated previously, we find $k = 0''\cdot32 \cos \phi$.

Now the direction of motion is perpendicular to the plane of the observer's meridian and it is eastwards. In Fig. 79, let X be

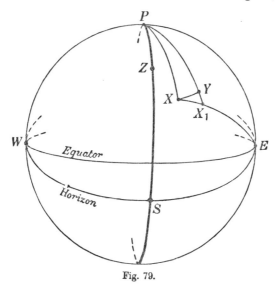

Fig. 79.

the true position of a star at a given moment on the celestial sphere. Owing to diurnal aberration it will be displaced to X_1 towards the east point E. Denote XE by θ and X_1E by θ_1. Then by the law of aberration [formula (3)] we have

$$XX_1 \equiv \theta - \theta_1 = k \sin \theta \qquad \ldots\ldots(50).$$

Draw a parallel of declination XY. Since hour angle is measured westward from the observer's meridian, the hour angle of X_1 is less than that of X, so that if H and H_1 are the respective hour angles, $H - H_1 = X\hat{P}Y$. Let $\Delta H = H_1 - H$. Now

$$XY = X\hat{P}Y \sin PX = X\hat{P}Y \cos \delta,$$

where δ is the true declination of the star.

Hence $\qquad\qquad XY = -\Delta H \cos \delta$.

Denote by δ_1 the declination of X_1. Then $X_1Y = \delta - \delta_1 = -\Delta\delta$. If $Y\hat{X}X_1 = \psi$, we have

$$XY = XX_1 \cos \psi; \quad X_1Y = XX_1 \sin \psi,$$

that is, using (50),

$$-\Delta H \cos \delta = k \sin \theta \cos \psi; \quad -\Delta\delta = k \sin \theta \sin \psi \ldots(51).$$

In the spherical triangle PXE, $PX = 90° - \delta$, $PE = 90°$, $XE = \theta$, $P\hat{X}E = 90° + \psi$ and $X\hat{P}E = H - 270°$. By formulae B and C we have

$$\sin XE \sin PXE = \sin XPE \sin PE,$$

and

$$\sin XE \cos PXE = \cos PE \sin PX - \sin PE \cos PX \cos XPE.$$

These become, on inserting the values of PX, etc.,

$$\sin \theta \cos \psi = \cos H,$$

and $\qquad\qquad \sin \theta \sin \psi = -\sin \delta \sin H$.

Hence from (51) and putting $k = 0''\!\cdot\!32 \cos \phi$, we obtain

$$\Delta H \equiv H_1 - H = -0''\!\cdot\!32 \cos \phi \cos H \sec \delta \quad\ldots\ldots(52),$$

and $\qquad \Delta\delta \equiv \delta_1 - \delta = 0''\!\cdot\!32 \cos \phi \sin H \sin \delta \quad\ldots\ldots(53).$

These equations—(52) and (53)—give the effect of diurnal aberration on the hour angle and declination of a star.

When the star is on the meridian it is seen from (53) that the effect on declination is zero.

The displacements due to diurnal aberration are so small that they are generally neglected. An exception occurs in the observation of stars with the meridian circle. It is seen from Fig. 79, or from formula (52), that diurnal aberration displaces the star eastwards of its true position; the star will therefore transit later than it would if the diurnal aberration were ineffectual. From (52) the transit is delayed by

$$\frac{0 \cdot 32}{15} \cos \phi \sec \delta \quad \text{(seconds of time)}$$

(the hour angle H at the time of transit is 0^h). For a fixed observatory, this expression is of the form $C \sec \delta$, which is precisely the form due to the effect of collimation error on the time of transit of a star. If t is the time of transit actually observed, the true time (considering only the effect of diurnal aberration) is

$$t - 0^s \cdot 021 \cos \phi \sec \delta \qquad \ldots\ldots(54).$$

The second term in (54) is the correction mentioned in section 46 and inserted in formula (9) of Chapter IV.

112. *Planetary aberration.*

Let t be the instant at which the sun or moon or a planet or comet is observed. Since light has a finite velocity, the moment of observation is—let us say—τ seconds after the instant at which the ray, which finally enters the observer's eye, has left the particular object concerned. Let us suppose the object to be a planet. The position observed at time t is affected by annual aberration and, after applying the general formulae for aberration derived in previous sections, we deduce the planet's position as it would be found if the earth were at rest. But this position is not the position of the planet at the time t of the observation; it is the position corresponding to the instant when the ray left the planet, that is to say, it is the position at time $t - \tau$. If we know the planet's distance from the earth at the time of observation, we can calculate τ from the known value of the velocity of light, and thus we have the true position at the definite time $(t - \tau)$. If we wish to obtain, for example, the right ascension and declination at time t, we require to add to the right ascension and declination at $t - \tau$ the quantities $\tau . \Delta \alpha$ and $\tau . \Delta \delta$ respectively, where $\Delta \alpha$, $\Delta \delta$ are the rates of change per second of the right ascension α and of the declination δ.

EXERCISES

1. Find the positions of stars which (i) are unaffected by aberration, (ii) are affected only in latitude.

2. The angular distance between two stars which have the same latitude β is θ and the mean of their longitudes is λ; show that the increase in θ due to aberration is

$$2\kappa \tan \tfrac{1}{2}\theta \sin(\lambda - \odot)(\cos^2 \beta - \sin^2 \tfrac{1}{2}\theta)^{\frac{1}{2}}. \qquad [M.T.]$$

3. Prove that the angular distance between two stars whose ecliptic coordinates are (λ, β) and (λ_1, β_1) is unaltered when the sun's longitude $\odot$ is given by

$$\cos \beta \sin(\odot - \lambda) + \cos \beta_1 \sin(\odot - \lambda_1) = 0. \qquad [Ball.]$$

4. Show that, in latitude ϕ, a star of declination δ will, owing to diurnal aberration, appear to move in an ellipse whose semi-axes are $m \cos \phi$ and $m \cos \phi \sin \delta$, where m is the ratio of the earth's circumference to the distance described by light in one day. [Coll. Exam.]

5. Prove that, when the aberration in declination has its greatest numerical value, the arcs on the celestial sphere joining the star to the sun and to the pole of the equator are at right angles. [Ball.]

6. If κ is the constant of aberration, c the velocity of light, G the constant of gravitation, M the mass of the sun, and l the semi-latus rectum of the earth's orbit, show that

$$GM = l\kappa^2 c^2. \qquad [M.T.\ 1922.]$$

7. Show that at any place and at any time there is a position of a star such that the aberrational effect is equal and opposite to that of refraction.

Show also that at midnight on the shortest day the zenith distance of this star is given by an equation of the form

$$\sin^2 z + \lambda \sin z = 1,$$

the correction for refraction being assumed proportional to the tangent of the zenith distance, and the earth's orbit to be circular. [M.T.\ 1900.]

8. If two planets revolve about the sun in circular orbits of radii a and b, show that the aberration of one as seen from the other is greater at opposition than at conjunction by an amount equal to

$$\frac{\sqrt{a} - \sqrt{b}}{\sqrt{a} + \sqrt{b}}. \qquad [M.T.]$$

CHAPTER IX

PARALLAX

113. *Introductory.*

The method of determining the distance of a heavenly body is, in principle, essentially similar to that adopted in land surveys. If a surveyor wishes to determine the distance of an object A from a point B, he first measures a base-line BC and with a suitable instrument, such as a theodolite, he observes the angles ABC and ACB. The solution of the plane triangle gives him the distance of the object A from B. As regards the astronomical survey of the solar system, we shall at first suppose the necessary base-line to be defined by the straight line joining two widely separated points on the earth's surface. It is thus necessary to consider more particularly the dimensions and figure of the earth. The maximum base-line obtainable in this way is not quite 8000 miles, corresponding to the earth's diameter; as this is wholly inadequate in the investigation of the distances of even the nearest stars, a different choice of base-line has to be made, as we shall see later, in dealing with stellar distances.

114. *The geoid.*

The surface defined by the ocean-level is called the *geoid*, and for astronomical purposes it is sufficient to consider it a spheroid

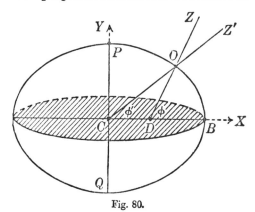

Fig. 80.

of revolution, its minor axis being coincident with the diameter joining the north and south poles. A terrestrial meridian, such as $POBQ$ in Fig. 80, is then an ellipse, the major axis CB being in the plane of the equator. If a denotes CB, the earth's equator is a circle of radius a. Denote the minor axis PC by b and the eccentricity of a meridian section by e. Then

$$b^2 = a^2 (1 - e^2) \qquad \ldots\ldots(1).$$

Referring the ellipse $POBQ$ to axes CX and CY, we have its equation in the well-known form

$$\frac{x^2}{a^2} + \frac{y^2}{b^2} = 1 \qquad \ldots\ldots(2).$$

115. *Astronomical and geocentric latitude.*

Consider an observer at O (Fig. 80). If we neglect any local gravitational irregularities, in the neighbourhood of O, which affect slightly the direction given by a plumb-line, the direction of the observer's zenith is defined to be perpendicular to the surface of the geoid at O; that is to say, it is in the direction DOZ, this line being the normal at O to the ellipse $POBQ$. Z is said to be the *astronomical zenith* of the observer; it is from this zenith that zenith distances are measured, for example, with the meridian circle. The angle which DOZ makes with the equatorial radius CB is the astronomical latitude of the observer; in Fig. 80, it is the angle ODB, denoted by ϕ. Join C and O and produce CO to Z'; then Z' is said to be the observer's geocentric zenith and the angle OCB, denoted by ϕ', is called the *geocentric latitude* of O. The angle ZOZ' between the directions of the astronomical and geocentric zeniths Z and Z' is called the *angle of the vertical*, usually denoted by v. We now investigate the relationship between v and ϕ.

If x, y are the rectangular co-ordinates of O with respect to the axes CX and CY, we have $y/x = \tan \phi'$. Also $\tan \phi$ is the slope of the normal DOZ, and it is given by a well-known formula

$$\tan \phi = \frac{y}{x} \frac{a^2}{b^2}.$$

Hence $$\frac{y}{x} = \tan \phi' = \frac{b^2}{a^2} \tan \phi \qquad \ldots\ldots(3)$$

On substituting in (2) the value of y, given by (3) in terms of x and ϕ, we obtain

$$\frac{x^2}{a^2} + x^2 \frac{b^2}{a^4} \tan^2 \phi = 1,$$

from which
$$x^2 = \frac{a^4 \cos^2 \phi}{a^2 \cos^2 \phi + b^2 \sin^2 \phi} \qquad \ldots\ldots(4),$$

or, replacing b^2 by $a^2 (1 - e^2)$,

$$x^2 = \frac{a^2 \cos^2 \phi}{1 - e^2 \sin^2 \phi} \qquad \ldots\ldots(5).$$

Similarly
$$y^2 = \frac{a^2 (1 - e^2)^2 \sin^2 \phi}{1 - e^2 \sin^2 \phi} \qquad \ldots\ldots(6).$$

Let ρ denote the radius vector CO; then

$$x = \rho \cos \phi'; \quad y = \rho \sin \phi'.$$

Hence, by (5) and (6),

$$x = \rho \cos \phi' = \frac{a \cos \phi}{(1 - e^2 \sin^2 \phi)^{\frac{1}{2}}} \qquad \ldots\ldots(7),$$

$$y = \rho \sin \phi' = \frac{a (1 - e^2) \sin \phi}{(1 - e^2 \sin^2 \phi)^{\frac{1}{2}}} \qquad \ldots\ldots(8),$$

from which
$$\rho^2 = \frac{a^2 [1 - (2e^2 - e^4) \sin^2 \phi]}{1 - e^2 \sin^2 \phi} \qquad \ldots\ldots(9),$$

an equation which gives the radius vector ρ in terms of the equatorial radius a, the eccentricity e and the astronomical latitude ϕ.

Multiply (7) by $\sin \phi$ and (8) by $\cos \phi$ and subtract; then, since $\phi - \phi' = v$, we obtain

$$\rho \sin v = \frac{ae^2 \sin \phi \cos \phi}{(1 - e^2 \sin^2 \phi)^{\frac{1}{2}}} \qquad \ldots\ldots(10).$$

Again, multiply (7) by $\cos \phi$ and (8) by $\sin \phi$ and add. Then

$$\rho \cos v = a (1 - e^2 \sin^2 \phi)^{\frac{1}{2}} \qquad \ldots\ldots(11).$$

Hence, from (10) and (11),

$$\tan v = \frac{e^2 \sin 2\phi}{2 (1 - e^2 \sin^2 \phi)}$$

$$= \frac{e^2 \sin 2\phi}{2 - e^2 + e^2 \cos 2\phi} \qquad \ldots\ldots(12).$$

Denoting $e^2/(2 - e^2)$ by m, formula (12) becomes

$$\tan v = \frac{m \sin 2\phi}{1 + m \cos 2\phi} \qquad \ldots\ldots(13).$$

Now e is about 0·08, so that m is about 0·003. We can, in consequence, obtain a rapidly converging series for v in the following manner. From (13) we obtain, i denoting the square root of -1,

$$\frac{1 + i \tan v}{1 - i \tan v} = \frac{1 + m (\cos 2\phi + i \sin 2\phi)}{1 + m (\cos 2\phi - i \sin 2\phi)},$$

or

$$e^{2iv} = \frac{1 + me^{2i\phi}}{1 + me^{-2i\phi}},$$

whence $2iv = \log (1 + me^{2i\phi}) - \log (1 + me^{-2i\phi})$.

Using the logarithmic expansion, we have

$$2iv = m (e^{2i\phi} - e^{-2i\phi}) - \frac{m^2}{2} (e^{4i\phi} - e^{-4i\phi}) + \ldots,$$

from which $v = m \sin 2\phi - \dfrac{m^2}{2} \sin 4\phi + \ldots.$

Expressing v in seconds of arc, we obtain

$$\phi - \phi' \equiv v = m \frac{\sin 2\phi}{\sin 1''} - \frac{1}{2} \frac{m^2 \sin 4\phi}{\sin 1''} + \frac{1}{3} \frac{m^3 \sin 6\phi}{\sin 1''} - \cdots$$

$$\ldots\ldots(14).$$

When ϕ is given and e (or m) is known, this formula enables the calculation of the angle of the vertical, and hence of the geocentric latitude ϕ', to be made simply and easily.

116. *Dimensions and compression of the geoid.*

In geodesy it is usual to deal with the *compression* or ellipticity (denoted by c) of the terrestrial meridian; it is related to the eccentricity by

$$c = \frac{a - b}{a} = 1 - (1 - e^2)^{\frac{1}{2}} \qquad \ldots\ldots(15).$$

The dimensions of the geoid, according to Hayford (1909), are:

$$a = 6378\cdot39 \text{ kms.} = 3963\cdot35 \text{ miles,}$$
$$b = 6356\cdot91 \text{ kms.} = 3950\cdot01 \text{ miles}$$

and $c = 1/297$,

whence $e = 0\cdot08199$ and $e^2 = 0\cdot006722$

and $m = 0\cdot003373$,

$$\frac{m}{\sin 1''} = 695\cdot65 \text{ and } \frac{m^2}{2 \sin 1''} = 1\cdot17.$$

Keeping only the first two terms in (14), which alone are of practical importance, we have

$$v \equiv \phi - \phi' = 695''{\cdot}65 \sin 2\phi - 1''{\cdot}17 \sin 4\phi \quad \ldots(16).$$

By means of this formula we can calculate the geocentric latitude ϕ' when the astronomical latitude ϕ is given. For example, the value of ϕ for Greenwich is $51° 28' 38''{\cdot}2$, and

$$\phi - \phi' = 677''{\cdot}93 + 0''{\cdot}51,$$

whence $\qquad \phi' = 51° 17' 19''{\cdot}8.$

117. *Geocentric parallax.*

In Fig. 81, let O be the observer, Z' his geocentric zenith and M the centre of the moon or sun or planet. Let z' and z denote the angles $Z'OM$ and $Z'CM$ and p the angle OMC. Let ρ denote

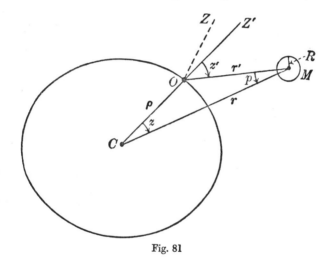

Fig. 81

the distance of O from the earth's centre C and r the geocentric distance CM of the body M. Then

$$z' = z + p \qquad \ldots\ldots(17),$$

$$\sin p = \frac{\rho}{r} \sin z' \qquad \ldots\ldots(18).$$

The angle p is called the *parallax* of the body M; it is the angle between the direction of the body as seen from O and the direction from the earth's centre regarded as the standard viewpoint. If p can be determined from observations, the distance r of the body can then be derived from (18), in which ρ may be presumed to be known and z' derived from zenith observations. The greater the value of r, the smaller is the angle of parallax p. As z' is greater than z, the effect of parallax is to increase the zenith distance of the body as observed at O, and this displacement takes place in the plane defined by CZ' and CM.

When z' is $90°$, the parallax p' is given by $\sin p' = \rho/r$. In this case, p' is called the *horizontal parallax* for the observer at O. If the observer is on the equator and M is on the horizon, its parallax P_0 will then be given by

$$\sin P_0 = \frac{a}{r} \qquad \ldots\ldots(19).$$

P_0 is called the *equatorial horizontal parallax*.

Owing to the effect of elliptic motion, the geocentric distance r of the moon or sun will vary during the orbital period, and if r_0 denotes the mean geocentric distance, we obtain from (19) the *mean equatorial horizontal parallax* P given by

$$\sin P = \frac{a}{r_0} \qquad \ldots\ldots(20).$$

Writing (18) in the form

$$\sin p = \frac{\rho}{a} \cdot \frac{a}{r_0} \cdot \frac{r_0}{r} \sin z',$$

we have $\qquad \sin p = \left(\frac{\rho}{a}\right)\left(\frac{r_0}{r}\right) \sin P \sin z' \qquad \ldots\ldots(21),$

in which it will be sufficient to regard ρ/a and r_0/r as known, the former by formula (9) and the latter from the circumstances of the orbital motion.

118. *The parallax of the moon.*

We shall now indicate how the distance of the moon can be accurately determined. Let us suppose that there are two observatories O and O_1 (Fig. 82) on the same terrestrial meridian, one in high northern latitude and the other in high southern latitude. This choice is made to ensure that the base-line determined by O and O_1 is as large as possible. Let us assume that the meridian zenith distances of the moon's centre are measured on the same day with meridian circles at O and O_1. These zenith distances are of course measured with reference to the astronomical zeniths Z and Z_1 at O and O_1 respectively (Fig. 82). After the corrections for refraction have been made, we shall denote these results by ζ and ζ_1. Thus $Z\hat{O}M = \zeta$ and $Z_1\hat{O}_1 M = \zeta_1$. If v and v_1 denote the angles of the vertical at O and O_1 and z' and z_1' the angles $Z'OM$ and $Z_1'O_1M$ (Z' and Z_1' are the geocentric zeniths at O and O_1), we have

$$z' = \zeta - v \quad \text{and} \quad z_1' = \zeta_1 - v_1.$$

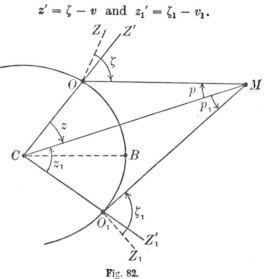

Fig. 82.

Denoting the angles of parallax for O and O_1 by p and p_1 and the angles OCM and O_1CM by z and z_1, we have

$$p = z' - z \quad \text{and} \quad p_1 = z_1' - z_1,$$

and therefore

$$p + p_1 = \zeta + \zeta_1 - (v + v_1) - (z + z_1) \quad \ldots\ldots(22).$$

But $z + z_1 = O\hat{C}O_1 = O\hat{C}B + O_1\hat{C}B$, where CB is the equatorial radius in the meridian of O and O_1. But OCB and O_1CB are respectively the geocentric latitudes of O and O_1, which we denote by ϕ' and ϕ_1'. Hence

$$z + z_1 = \phi' + \phi_1'.$$

Also if ϕ and ϕ_1 are the astronomical latitudes of O and O_1, $\phi = \phi' + v$ and $\phi_1 = \phi_1' + v_1$. Hence (22) becomes

$$p + p_1 = \zeta + \zeta_1 - \phi - \phi_1 \quad \ldots\ldots(23).$$

The quantities on the right are now all known and we shall write

$$p + p_1 = \theta \qquad \ldots\ldots(24).$$

Now

$$\sin p = \frac{\rho}{r} \sin (\zeta - v) \qquad \ldots\ldots(25),$$

and

$$\sin p_1 = \frac{\rho_1}{r} \sin (\zeta_1 - v_1) \qquad \ldots\ldots(26).$$

By means of (24), the equation (26) can be written

$$\sin \theta \cos p = \cos \theta \sin p + \frac{\rho_1}{r} \sin (\zeta_1 - v_1),$$

or, using (25),

$$\sin \theta \cos p = \frac{\rho}{r} \cos \theta \sin (\zeta - v) + \frac{\rho_1}{r} \sin (\zeta_1 - v_1) \quad (27).$$

Dividing (25) by (27), we eliminate r and obtain

$$\tan p = \frac{\rho \sin \theta \sin (\zeta - v)}{\rho \cos \theta \sin (\zeta - v) + \rho_1 \sin (\zeta_1 - v_1)} \quad \ldots(28).$$

All the quantities on the right of (28) are known and this equation thus determines p.

We now write (25) in the form

$$\sin p = \left(\frac{\rho}{a}\right) \left(\frac{r_0}{r}\right) \sin P \sin (\zeta - v) \qquad \ldots\ldots(29),$$

in which p, ρ/a, r_0/r, $\zeta - v$ are all presumed to be known. We thus determine from (29) the mean equatorial horizontal parallax P of the moon.

It has been assumed that the observations at O and O_1 refer to the moon's centre. In practice, however, it is the zenith distance of the moon's upper or lower limb, or of a small crater or mountain peak that is measured. The adjustments to the procedure outlined above are, however, slight and the details need not be more fully described here. Also it has been assumed that O and O_1 are on the same meridian and clearly it is unlikely that two fixed observatories can be found to satisfy this assumption exactly. The Royal Observatories at Greenwich and the Cape of Good Hope have collaborated in the past in the investigation of the moon's parallax, and as they differ in longitude by 1^{h} 13^{m} 55^{s}, the effect of the change in the moon's declination during the interval between the meridian transits at these two observatories has to be taken into account. The rate of change of declination is, however, known at any time on any day with the necessary accuracy and the zenith distance ζ observed, say, at Greenwich can be corrected to give the zenith distance which would be observed if the moon's declination remained unaltered between the transits at the two observatories.

The moon's mean equatorial horizontal parallax can also be determined from theoretical considerations combined with observations of the moon's motion. The value of this constant, according to Professor E. W. Brown, is $3422''\cdot7$. The mean distance of the moon from the earth is 384,400 kms. or 238,900 miles.

119. *Semi-diameter.*

In Fig. 81, let the linear radius of the moon be R miles and let r, r' denote the distances, in the same unit, of the moon's centre from C and O. Let S_0 denote the angle subtended by the moon's radius at C and s that subtended at O. Then we have $\sin S_0 = R/r$ and $\sin s = R/r'$. The first can be written

$$\sin S_0 = \frac{R}{a} \cdot \frac{a}{r},$$

where a is the earth's equatorial radius. Since $a/r = \sin P_0$, we have

$$\sin S_0 = \frac{R}{a} \sin P_0 \qquad \ldots\ldots(30).$$

If S denotes the angle subtended at C by the moon's radius when the moon is at its mean distance r_0, we have

$$\sin S = \frac{R}{a} \sin P \qquad \ldots\ldots(31);$$

S is called the mean semi-diameter of the moon.

Also $\qquad \sin s = \frac{R}{r'} = \frac{R}{r} \cdot \frac{\sin z'}{\sin z},$

or $\qquad \sin s = \sin S_0 \frac{\sin z'}{\sin z} \qquad \ldots\ldots(32).$

s can be measured, and as z' and z can be assumed to be known, equation (32) determines S_0. From (30) and (31) $\dfrac{\sin S}{\sin S_0} = \dfrac{\sin P}{\sin P_0}$ and assuming P_0 and P known we determine S. By (30) or (31) the moon's radius R is derived; it is found to be 1738 kms. or

1080 miles. Thus the moon's radius is about one-quarter that of the earth. The value of S is $16'32''\!\cdot\!6$. The semi-diameters of the sun and planets are defined in a similar manner.

120. Parallax in right ascension and declination.

We now investigate the effect of parallax on the right ascension and declination of the moon with respect to an observer at O. We shall develop rigorous formulae which are only necessary in the case of the moon, owing to the large value of P for this body. As regards the sun, P is $8''\!\cdot\!80$, a small angle compared with the corresponding value for the moon, and consequently the general formulae can be greatly modified and simplified in this case.

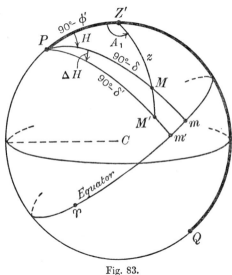

Fig. 83.

Consider, in Fig. 83, the celestial sphere centred at C (the earth's centre); P is the north pole, Z' is the observer's geocentric zenith and $PZ' = 90° - \phi'$. Let M be the position of the body on the celestial sphere as viewed from C; then $Z'M = z$. Produce the great circle arc $Z'M$ to M' so that $Z'M' = z'$, z' being the zenith distance of M with respect to the observer at O. Then, since $z' = z + p$, we have $MM' = p$, the angle of parallax corresponding to the observation made at O. Let

a, δ be the right ascension and declination of the point M on the celestial sphere and a', δ' the corresponding co-ordinates of M'. Then $PM = 90° - \delta$; $PM' = 90° - \delta'$. Let the hour angles of M and M' be H and H' respectively. We shall write $H' - H = \Delta H$. Now, from the figure, $H' - H = m'm$, and if T is the vernal equinox $a' = Tm'$ and $a = Tm$ so that, if we write $a' - a = \Delta a$, then $\Delta a = - m'm$ and therefore $\Delta H = - \Delta a$. Let A_1 denote the angle $PZ'M$.

We shall now investigate the problem of expressing the co-ordinates a' (or H') and δ' in terms of a (or H) and δ.

From the triangle $PZ'M$ we have, using formula **D**,

$$\sin \phi' \cos A_1 = \cos \phi' \cot z - \sin A_1 \cot H \ \ldots \ldots (33),$$

and from the triangle $PZ'M'$,

$$\sin \phi' \cos A_1 = \cos \phi' \cot z' - \sin A_1 \cot H' \ldots \ldots (34).$$

Subtracting (33) from (34), we obtain

$$\cos \phi' (\cot z - \cot z') = \sin A_1 (\cot H - \cot H'),$$

or, since $z' - z = p$, and $H' - H = \Delta H$,

$$\frac{\cos \phi' \sin p}{\sin z \sin z'} = \frac{\sin A_1 \sin \Delta H}{\sin H \sin H'} \ .$$

But by (18), $\sin p = \dfrac{\rho}{r} \sin z'$ and, by the sine-formula **B**,

$$\sin A_1 \sin z = \sin H \cos \delta.$$

Hence $$\frac{\rho}{r} \cos \phi' = \frac{\cos \delta \sin \Delta H}{\sin (H + \Delta H)},$$

from which we obtain

$$\tan \Delta H \equiv - \tan \Delta a = \frac{\rho}{r} \frac{\sin H \cos \phi'}{\cos \delta - \dfrac{\rho}{r} \cos H \cos \phi'}$$

$$\ldots \ldots (35).$$

This is a rigorous formula which enables us to calculate ΔH or Δa (the effect of parallax on right ascension) when ρ/r, ϕ' and the geocentric co-ordinates H and δ are known. We then obtain H' or a'. It is only in the case of the moon, however, when ρ/r is about $1/60$, that this rigorous formula is to be used.

We now find the general formula to give $\Delta\delta$, the effect of parallax on declination ($\Delta\delta = \delta' - \delta$).

By the cosine-formula **A**, from the triangles $PZ'M$ and $PZ'M'$,

$$\sin \delta = \sin \phi' \cos z + \cos \phi' \sin z \cos A_1,$$
$$\sin \delta' = \sin \phi' \cos z' + \cos \phi' \sin z' \cos A_1,$$

whence, eliminating $\cos A_1$, we obtain

$$\sin \delta \sin z' - \sin \delta' \sin z = \sin \phi' \sin p$$
$$= \frac{\rho}{r} \sin \phi' \sin z' \text{ by (18)},$$

which can be written

$$\sin \delta' \frac{\sin z}{\sin z'} = \sin \delta - \frac{\rho}{r} \sin \phi' \qquad \text{......(36)}.$$

Again, by **C**, we have

$$\cos \delta \cos H = \cos \phi' \cos z - \sin \phi' \sin z \cos A_1,$$
$$\cos \delta' \cos H' = \cos \phi' \cos z' - \sin \phi' \sin z' \cos A_1,$$

and, by eliminating $\cos A_1$ from these equations and using (18),

$$\cos \delta \cos H \sin z' - \cos \delta' \cos H' \sin z = \frac{\rho}{r} \cos \phi' \sin z',$$

whence $\qquad \cos \delta' \cos H' \dfrac{\sin z}{\sin z'} = \cos \delta \cos H - \dfrac{\rho}{r} \cos \phi' \text{......(37)}.$

Dividing (36) by (37), we have

$$\frac{\tan \delta'}{\cos H'} = \frac{\sin \delta - \frac{\rho}{r} \sin \phi'}{\cos \delta \cos H - \frac{\rho}{r} \cos \phi'}.$$

Writing $\delta' = \delta + \Delta\delta$ and $T = \tan \Delta\delta$, we obtain

$$\frac{\tan \delta + T}{1 - T \tan \delta} = \frac{\cos H' \left(\sin \delta - \frac{\rho}{r} \sin \phi' \right)}{\cos \delta \cos H - \frac{\rho}{r} \cos \phi'} \qquad \text{......(38)}.$$

$\cos H'$ occurs on the right of (38), and since $H' = H + \Delta H$ and ΔH is given by (35), we can then calculate T from (38). We thus obtain $\Delta\delta$ or $\delta' - \delta$, and hence δ' is deduced.

121. Parallax in zenith distance and azimuth.

We now consider briefly the effect of parallax on the zenith distance and azimuth of the moon or the sun when the body concerned is not on the meridian. In Fig. 84, let Z and Z' be the astronomical and geocentric zeniths of an observer at O, the centre of the celestial sphere being the earth's centre C. As before, let M and M' be the positions of the body on the celestial sphere as viewed from C and O respectively; $Z'M = z$, $Z'M' = z' = z + p$. Now observations of zenith distance and azimuth are made with reference to the astronomical zenith Z and so we can regard ZM' and PZM' as the observed quantities.

Denote these by ζ and A respectively. Now ZZ' is the angle of the vertical v, which we assume to be known, and $Z'\hat{Z}M' = 180° - A$. Formula **A** then enables us to express z' in terms of v, A and ζ. Also $\sin p = (\rho/r) \sin z'$, from which p and then $z \; (\equiv z' - p)$ may be derived if ρ/r is known. Thus z can be expressed in terms of v, ζ, A and ρ/r.

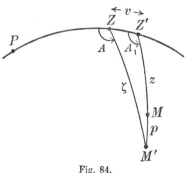

Fig. 84.

Let A_1 denote the geocentric azimuth $PZ'M$. By formula **D**, A_1 can be expressed in terms of v, ζ and A.

It is clear that if v, z, A_1, ρ/r are known, ζ and A can be derived by an analogous process.

The actual formulae giving $(z - \zeta)$ and $(A_1 - A)$, which can be derived from the procedure outlined above, are not of very great practical importance, and it is sufficient here to draw attention to the principles involved.

122. *The solar parallax.*

The method which we have described for the determination of the moon's parallax cannot be applied with sufficient accuracy to the direct measurement of the sun's parallax. The principle of the fundamental method of deriving the sun's distance is based on the measurement of the distance of an exterior planet, such as Mars or Eros, and on the application of Kepler's third law. If a (semi-major axis), T (orbital period) and m (mass) refer to the earth then, by formula (18), p. 102, M denoting the sun's mass,

$$\frac{4\pi^2 a^3}{T^2} = G(M + m),$$

and if a_1, T_1, m_1 refer to Mars or Eros,

$$\frac{4\pi^2 a_1{}^3}{T_1{}^2} = G\,(M + m_1).$$

We can safely neglect m and m_1 in comparison with M and consequently we have $a^3 : a_1{}^3 = T^2 : T_1{}^2.$

Now the periods of orbital revolution are known very accurately; consequently the ratio $a : a_1$ can be determined very accurately. This can be expressed more generally by the statement that the relative dimensions of the planetary orbits are known with high precision. If the distance (say, in miles) of a planet from the earth can be accurately measured at a particular time, the scale of the planetary system can then be deduced; in particular, the sun's distance from the earth can then be found in miles or in terms of the earth's equatorial radius as the unit.

Consider the earth's orbit and the highly elliptical orbit of the minor planet Eros (Fig. 85). If the earth and Eros are simultaneously at E and M respectively (the planet is then in opposition), it is clear that conditions are favourable for measuring the distance EM, for then the planets are at their minimum distance apart. The opposition shown by the configuration SE_1M_1 is evidently the most unsatisfactory. At the most favourable opposition, when Eros makes its closest approach to the earth, the distance between the two bodies is about 14 million miles, a very much smaller distance to be measured than the distance of the earth from

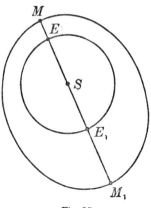

Fig. 85.

the sun, which is about 93 million miles. A favourable opportunity of measuring the distance of Eros occurred in 1900–1 and in 1930–1 the circumstances were again favourable.

In section 120, we obtained general formulae for the effect of parallax on the co-ordinates of a heavenly body. It is only necessary to use such rigorous formulae in connection with the moon for which, as we have seen, the horizontal parallax P is about 57'. For a planet such as Eros the value of the horizontal parallax may be no more than about 30''; it follows that for such an object ρ/r is a very small quantity (it is of the order of 0·0002), and therefore we can safely neglect squares and higher powers of ρ/r in the formulae mentioned. In this way, we can reduce (35), for example, to the form

$$\Delta\alpha = -\frac{\rho}{r}\sin H \cos \phi' \sec \delta.$$

We shall now derive this formula for $\Delta\alpha$ and the corresponding one for $\Delta\delta$ by another method. In Fig. 86 let M be the position of the planet on the celestial sphere as viewed from the earth's centre C and M' the position as seen by an observer at O whose geocentric zenith is Z'. Let p denote the displacement MM' due to parallax. Let α, δ and α', δ' be the equatorial co-ordinates of M and M' respectively, and H, H' the corresponding hour angles. Let ML be a small circle arc parallel

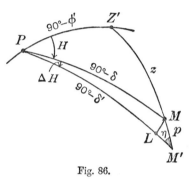

Fig. 86.

to the equator. Denote $P\hat{M}'Z'$ by η. Then as p is very small we have, from the sensibly plane triangle LMM',

$$LM = p\sin\eta \quad \text{and} \quad LM' = p\cos\eta.$$

Now from (18) we have, writing p for $\sin p$,

$$p = \frac{\rho}{r}\sin z' \qquad \qquad \text{......(39)}.$$

Put $\qquad \qquad \Delta H = H' - H \quad \text{and} \quad \Delta\delta = \delta' - \delta.$

We have then

$$LM = (H' - H)\cos\delta = \Delta H \cos\delta.$$

Hence $\qquad \qquad \Delta H \cos\delta = \frac{\rho}{r}\sin\eta\sin z'.$

But by **B**, $\qquad \sin\eta\sin z' = \cos\phi'\sin H'.$

Hence $\qquad \qquad \Delta H = \frac{\rho}{r}\sin H' \cos\phi'\sec\delta \qquad \text{......(40)}.$

We can write H for H' on the right of (40) without sensible loss of accuracy and, since $\alpha' - \alpha \equiv \Delta\alpha = -(H' - H) = -\Delta H$, we obtain

$$\alpha' - \alpha \equiv \Delta\alpha = -\frac{\rho}{r}\cos\phi'\sec\delta\sin H \qquad \text{......(41)}.$$

Again, $M'L = \delta - \delta' = - \Delta\delta$, and therefore

$$\Delta\delta = - p \cos \eta,$$

so that by (39), $\Delta\delta = - \dfrac{\rho}{r} \sin z' \cos \eta,$

and, by applying formula C, we obtain

$$\Delta\delta = - \frac{\rho}{r} (\sin \phi' \cos \delta' - \cos \phi' \sin \delta' \cos H').$$

Replacing δ', H' by δ and H on the right of this equation we have

$$\delta' - \delta \equiv \Delta\delta = - \frac{\rho}{r} (\sin \phi' \cos \delta - \cos \phi' \sin \delta \cos H)$$
$$\dots\dots(42),$$

which gives the displacement in declination due to parallax.

As in the case of the moon, the observation of the planet's zenith distance at two widely separated observatories on the same (or nearly the same) meridian will lead to the determination of ρ/r. Thus the distance r of the planet can be found at the time of observation and, the scale of the planetary system being now known, the mean distance of the earth from the sun can be deduced. The base-line in this instance is the straight line joining the two observatories.

123. *The solar parallax (diurnal method).*

The practical method now favoured is called the diurnal method. The principle is as follows. The planet is observed several hours before meridian passage and again several hours after meridian passage, at the *same* observatory. In the interval between observations the observer's position, with reference to the earth's centre and the planet, has been changed owing to the rotation of the earth, and the distance between the two positions constitutes essentially the base-line appropriate for the measurement of the planet's parallax.

Consider now a star whose position in the sky is near that of the planet. The parallax observations consist, in effect, of measuring the difference between the apparent right ascension α' of the planet and the right ascension α_0 of the star, or the corresponding difference in declination. For simplicity we shall consider the former only. This difference in right ascension can be measured by means of the meridian circle or filar micrometer or, more accurately, by making a photograph of the region of the sky in which the planet is situated. The distance between the image of the planet and the image of a neighbouring star can be accurately determined and, by the principles which we shall describe in Chapter XII, the difference $\alpha' - \alpha_0$ can be derived with great precision.

As the planet is in or near opposition when a parallax investigation is undertaken, the planet will be on or near the meridian at midnight. Our two sets of observations are then made a few hours before midnight and a few hours after midnight. We shall refer to these as the evening and morning observations. Consider first an evening observation when the difference m_1 in right ascension of the planet and of the star is measured. We have

$$m_1 = \alpha' - \alpha_0.$$

Now the displacement due to parallax is $\alpha' - \alpha$, that is, $(\alpha' - \alpha_0) + (\alpha_0 - \alpha)$. Hence by (41),

$$m_1 + \alpha_0 - \alpha = -\frac{\rho}{r} \cos \phi' \sec \delta \sin H,$$

or if we denote the values of α, r, δ, H at the time of the evening observation by α_1, r_1, δ_1, H_1 we have

$$m_1 + \alpha_0 - \alpha_1 = -\frac{\rho}{r_1} \cos \phi' \sec \delta_1 \sin H_1 \quad \ldots\ldots(43).$$

During the interval between the evening and morning observations the true right ascension, declination and distance of the planet from the earth's centre are altering; at the time of the morning observation let these quantities be α_2, δ_2, r_2 and let the planet's hour angle be H_2. Then we shall have an equation similar to (43); it is

$$m_2 + \alpha_0 - \alpha_2 = -\frac{\rho}{r_2} \cos \phi' \sec \delta_2 \sin H_2 \quad \ldots\ldots(44).$$

Subtracting (43) from (44) we eliminate the star's right ascension α_0 and obtain

$$m_2 - m_1 = (\alpha_2 - \alpha_1) - \rho \cos \phi' \left\{ \frac{\sec \delta_2 \sin H_2}{r_2} - \frac{\sec \delta_1 \sin H_1}{r_1} \right\};$$

$(\alpha_2 - \alpha_1)$ is the increase in the planet's geocentric right ascension between the observations due to the relative orbital motions of the earth and planet. From a series of observations, for example, meridian observations, this quantity may be deduced. Also we may write without loss of accuracy the mean values δ and r of δ_1, δ_2 and r_1, r_2 respectively. Hence

$$m_2 - m_1 = (\alpha_2 - \alpha_1) - \frac{\rho}{r} \cos \phi' \sec \delta (\sin H_2 - \sin H_1)$$
$$\ldots\ldots(45).$$

As $(m_2 - m_1)$ is given by the measures and $(\alpha_2 - \alpha_1)$, ρ, ϕ', δ, H_1, H_2 are all supposed known, we can determine r by (45). Thus the scale of the planetary system is found and the value of the solar parallax deduced.

It is to be noted that $\sin H_1$ is negative (planet east of the meridian); for example, if the observations are made when the planet is 6 hours east and again when 6 hours west of the meridian, $\sin H_1 = -1$ and $\sin H_2 = +1$, so that $(\sin H_2 - \sin H_1)$ is 2. This is the maximum value of this quantity. The determination of r will lead to the most favourable results if $(m_2 - m_1)$ is as large as possible and, omitting the effect of the variation of the planet's right ascension, this condition will be best attained if $\cos \phi'$ and $(\sin H_2 - \sin H_1)$ are both as large as possible. For the most reliable results a station near the equator is therefore desirable, and the evening and morning observations should be made so that H_2 and $(24^h - H_1)$ are as near 6 hours as circumstances allow.

We have assumed that the difference in right ascension of the planet and a single star is measured; in practice, this procedure is, in effect, applied to several stars.

The value of the solar parallax derived from observations made at several observatories at the opposition of Eros in 1930–1 is 8″·790, giving the mean distance of the earth from the sun to be 93,003,000 miles. This is known as the *astronomical unit of distance*.

124. *The solar parallax (other methods).*

(a) *Transit of Venus.* For a description of this method, which is now mainly of historical interest, see Ball's *Spherical Astronomy*, p. 312.

(b) *Gravitational methods.* These depend on comparing certain observed perturbations in the motion of the moon with the theoretical expressions derived from celestial mechanics, which involve the semi-major axis a of the earth's orbit.

(c) *The constant of aberration.* We have seen in Chapter VIII that the constant of aberration κ is given by

$$\kappa = \frac{2\pi a \operatorname{cosec} 1''}{cT (1 - e^2)^{\frac{1}{2}}}.$$

Denote the mean equatorial parallax of the sun by P and the earth's equatorial radius by ρ_0. Then $\sin P = \rho_0/a$ or, expressing P in seconds of arc,

$$P = \frac{\rho_0}{a} \operatorname{cosec} 1''.$$

Hence, combining these two formulae, we obtain

$$\kappa P = \frac{2\pi \rho_0 \operatorname{cosec}^2 1''}{cT (1 - e^2)^{\frac{1}{2}}} \qquad \ldots\ldots(46).$$

Here $\rho_0 = 6378\cdot4$ kms., $c = 299{,}774$ kms. per second, $T = 31{,}558{,}149$ seconds and $e = 0\cdot01674$. We thus obtain

$$\kappa P = 180\cdot21 \qquad \ldots\ldots(47),$$

κ and P being both expressed in seconds of arc. The constant of aberration can be determined from observations as we have shown in section 109, and therefore the solar parallax P can be derived from (47).

If we accept $20''\cdot45$ as the observed value of κ, we find $P = 8''\cdot812$, which is somewhat higher than the mean value of P obtained by the more direct method of section 123.

The value of the solar parallax adopted in the *Nautical Almanac* and other ephemerides is $8''\cdot80$.

(d) *Spectroscopic method.* The basis of this method depends on one of the achievements of spectroscopic astronomy, namely, the measurement* of the velocity of approach, or of recession, of any star to or from the earth. This velocity (called the *radial velocity*) is obtained directly in linear units, for example, as so many kilometres per second. Consider, for simplicity, a star whose position is in the ecliptic and assume for a moment that the earth's orbit is a circle and that its orbital velocity is constant. At some definite time during the year the earth's orbital velocity is directed towards the star. In general, the star will be moving relatively to the sun which we may consider to be at rest. Denote the component of the star's velocity, directed towards the sun,

---

* For an account of the spectroscopic principles and methods, see *Astronomical Physics* by F. J. M. Stratton (Methuen, 1925).

S A

by V. The spectroscopic method determines the star's velocity relative to and directed towards the earth. This is V plus the earth's orbital velocity. Six months later, the earth's velocity is directed away from the star and the spectroscopist then measures V minus the earth's orbital velocity. By subtraction V is eliminated and the earth's orbital velocity is determined; it is usually expressed in kilometres per second. The period of revolution in the earth's orbit being known, the circumference in kilometres is then deduced, from which the radius of the earth's orbit is finally found in kilometres. Thus the astronomical unit of distance is determined and from it the solar parallax. We now consider the general problem.

In Fig. 87, let λ and β be the longitude and latitude of a star X. The earth's orbital velocity, as we have seen on p. 185, can be resolved into two parts: (i), h/p directed to a point F on the ecliptic, 90° behind the sun in longitude (the longitude of F is thus $\odot - 90°$, where $\odot$ is the sun's true longitude), and (ii), eh/p directed to a point f on the ecliptic, 90° ahead of the point A, which is such that $\overrightarrow{EA}$ is parallel to the direction of perihelion as viewed from the sun. Consider first the velocity h/p and its component resolved in the direction of

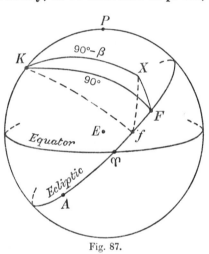

Fig. 87.

the star X. This component is $\dfrac{h}{p} \cos XF$. Now

$$F\hat{K}X = \lambda - (\odot - 90°), \quad KF = 90° \quad \text{and} \quad KX = 90° - \beta.$$

Hence by the cosine-formula **A**,

$$\cos XF = \cos \beta \sin (\odot - \lambda),$$

and consequently the component of the velocity h/p in the direction of the star is

$$\frac{h}{p} \cos \beta \sin (\odot - \lambda).$$

Consider now the component eh/p of the earth's velocity directed towards the point f. The component of eh/p along EX is $\dfrac{eh}{p} \cos Xf$. Now the longitude of perihelion is ϖ and therefore the longitude of f is $\varpi + 90°$. Hence the angle fKX is $\lambda - (\varpi + 90°)$. Using the cosine-formula we find that

$$\cos Xf = -\cos \beta \sin (\varpi - \lambda),$$

and the corresponding component of radial velocity is

$$-\frac{eh}{p} \cos \beta \sin (\varpi - \lambda).$$

For any given star this quantity is constant.

If the star is moving towards the sun (which we may regard as at rest) with velocity V, the velocity of approach of the star to the earth is

$$V + \frac{h}{p} \cos \beta \sin (\odot - \lambda) - \frac{eh}{p} \cos \beta \sin (\varpi - \lambda).$$

This is the radial velocity, which we shall denote by $-R$ (the convention is that a velocity of recession is taken to be positive).

Consider now two radial velocity determinations made several months apart. We shall then have

$$\left.\begin{aligned}
-R_1 &= V + \frac{h}{p} \cos \beta \{\sin (\odot_1 - \lambda) - e \sin (\varpi - \lambda)\} \\
-R_2 &= V + \frac{h}{p} \cos \beta \{\sin (\odot_2 - \lambda) - e \sin (\varpi - \lambda)\}
\end{aligned}\right\} \dots(48),$$

in which $\odot_1$ and $\odot_2$ are the sun's true longitudes at the two dates concerned. Then by subtraction

$$R_2 - R_1 = \frac{h}{p} \cos \beta \{\sin (\odot_1 - \lambda) - \sin (\odot_2 - \lambda)\} \dots(49).$$

$R_2 - R_1$ is numerically greatest when $\beta = 0$ and when $\odot_1 - \lambda = -(\odot_2 - \lambda) = 90°$ or $270°$. It follows that the best conditions for carrying out an investigation of this kind are obtained when (i) the star observed is in or near the ecliptic, (ii) the two sets of observations are made at an interval of six months, (iii) the dates are chosen so that the difference between the longitudes of the sun and star on either date is $90°$ or $270°$.

Now we have $\qquad \dfrac{h}{p} = \dfrac{2\pi a}{T (1 - e^2)^{\frac{1}{2}}},$

in which a is the semi-major axis of the earth's orbit. Hence from (49),

$$R_2 - R_1 = \frac{2\pi a}{T(1 - e^2)^{\frac{1}{2}}} \cos \beta \{\sin(\odot_1 - \lambda) - \sin(\odot_2 - \lambda)\}$$

$$\ldots\ldots(50).$$

In this equation, $R_2 - R_1$ is given by the spectroscopic observations in kilometres per second, T is the orbital period in seconds, e, $\odot_1$, $\odot_2$ and λ are known. Hence a is determined in kilometres and thus the solar parallax is found.

From the spectroscopic observations of nineteen stars made at the Cape Observatory, the solar parallax was recently (1927) found to be 8″·803.

In deriving the value of a in (50) from the radial velocity measures, we omitted one consideration. Unless the star is on the meridian at the time of observation, the linear velocity of the observatory due to the diurnal motion will have a component in the direction of the star, and this component will be included in the observed radial velocity. It is clear that this effect must first be removed before formula (50) can be applied. Consider the observer O on the earth's surface (Fig. 88). Due to the earth's rotation he is moving along OT (tangential to the parallel of latitude through O) with the linear velocity $\dfrac{2\pi\rho}{\tau} \cos \phi$ kms. per second, where ρ is

Fig. 88.

the radius of the earth in kilometres and τ is the length of the sidereal day in mean seconds (τ is 86164). Let X denote the point on the earth's surface at which the star is vertically overhead at the time of observation. Let E be the east point of the horizon. Then CE is parallel to OT. Hence the component of the observer's velocity along CX is

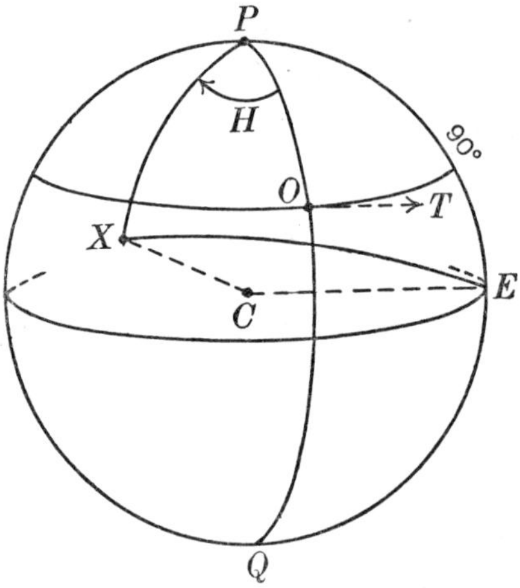

$$\frac{2\pi\rho}{\tau} \cos \phi \cos XE.$$

But if H is the hour angle of the star, $X\hat{P}O = H$ and $X\hat{P}E = H + 90°$. Hence by the cosine-formula **A**,

$$\cos XE = - \cos \delta \sin H.$$

Let R' denote the radial velocity of the star (positive, when it is a velocity of recession) due to the diurnal motion: then

$$R' = + \frac{2\pi\rho}{\tau} \cos \phi \cos \delta \sin H \qquad \ldots\ldots(51).$$

It is seen that R' is zero when H is 0^h or 12^h. The maximum value of R' is $0.47 \cos \phi$ kms. per second. The value of R' just found must be subtracted from the observed radial velocity so as to eliminate the effect of the diurnal motion.

In formula (50), R_1 and R_2 are the radial velocities so corrected.

125. Stellar parallax.

We now describe the fundamental method of determining stellar distances. Fig. 89 shows the earth's orbit, which we can regard as circular; S is the sun and a is the orbital radius. Let X denote the position of a star at a distance d from the sun, and we shall here assume simply that the star is stationary with respect to the sun. On a given date let the earth be at E and six months later at E_1. The distance EE_1 constitutes the base-line—it is about 186 million miles—by means of which the star's distance from the sun can eventually be determined.

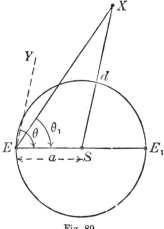

Fig. 89.

Let EY be parallel to SX and denote the angles SEY, SEX by θ, θ_1 respectively. From the triangle SXE, in which $S\hat{X}E = \theta - \theta_1$ we have

$$\sin (\theta - \theta_1) = \frac{a}{d} \sin \theta_1 \qquad \ldots\ldots(52).$$

Define Π by $\qquad\qquad \sin \Pi = \frac{a}{d} \qquad\qquad \ldots\ldots(53).$

Then Π is called the star's parallax (it is sometimes referred to as the *annual parallax*). It is the angle subtended at the star by the radius of the earth's orbit. For the nearest star, Π is $0''\cdot76$ and it is evident that we can write (52) with the help of (53) in the simple and sufficiently accurate form

$$\theta - \theta_1 = \Pi \sin \theta \qquad \qquad ......(54),$$

in which $(\theta - \theta_1)$ and Π are expressed in the same unit and θ is written for θ_1.

In Fig. 89, SX is the direction of the star as seen from the sun and EX is the direction as seen from the earth on the date in question; we refer to these as the heliocentric and geocentric directions respectively. If the star were at an infinite distance it would be seen from the earth in the direction EY, which has been drawn parallel to SX. The angle θ is the angle between the heliocentric direction EY and the direction ES of the sun from the earth. Fig. 89 shows that the geocentric direction EX is displaced from the heliocentric direction EY *towards* the direction ES of the sun, and that this displacement takes place in the plane ESX.

Consider now Fig. 90, which represents the celestial sphere centred at E. Let X denote the position of the star corresponding to its heliocentric direction. S is the sun's position on the

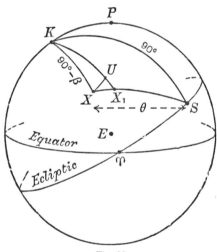

Fig. 90.

ecliptic at the date concerned. If X_1 is the star's position on the celestial sphere corresponding to its geocentric direction, then X_1 is in the plane of X, E and S and therefore X_1 is on the great circle arc XS. Moreover X_1 is between X and S. We shall refer to X and X_1 as the heliocentric and geocentric positions of the star respectively.

126. Effect of parallax on the star's longitude and latitude.

Let λ and β denote the star's heliocentric longitude and latitude (these quantities refer to the position X), and let λ_1 and β_1 denote the geocentric longitude and latitude (these refer to X_1). Draw the small circle arc UX parallel to the ecliptic. Denote $\lambda_1 - \lambda$ by $\Delta\lambda$ and $\beta_1 - \beta$ by $\Delta\beta$. Denote the angle UXX_1 by ϕ. Then in the infinitesimal plane triangle UXX_1,

$$UX \equiv \Delta\lambda \cos \beta = XX_1 \cos \phi,$$

and
$$UX_1 \equiv -\Delta\beta = XX_1 \sin \phi.$$

Now with the notation of Fig. 89, $XX_1 = \theta - \theta_1$ and $XS = \theta$. Hence using (54),

$$\Delta\lambda \cos \beta = \Pi \sin \theta \cos \phi \qquad \ldots\ldots(55),$$

and
$$\Delta\beta = -\Pi \sin \theta \sin \phi \qquad \ldots\ldots(56).$$

In the triangle KXS, $X\hat{K}S = \odot - \lambda$, where $\odot$ is the sun's true longitude ΥS; also $KS = 90°$, $KX = 90° - \beta$ and $K\hat{X}S = 90° + \phi$. Hence by the sine-formula **B**,

$$\sin \theta \cos \phi = \sin (\odot - \lambda),$$

and by **C**, $\qquad \sin \theta \sin \phi = + \sin \beta \cos (\odot - \lambda).$

The formulae (55) and (56) now become

$$\Delta\lambda \cos \beta = \Pi \sin (\odot - \lambda) \qquad \ldots\ldots(57),$$

$$\Delta\beta = -\Pi \sin \beta \cos (\odot - \lambda) \qquad \ldots\ldots(58).$$

These equations give the displacements of the star's position, due to parallax, in longitude and latitude.

127. The parallactic ellipse.

Let x denote the displacement UX parallel to the ecliptic and y the displacement UX_1 in latitude. Then by (57) and (58),

$$x = \Pi \sin (\odot - \lambda),$$
$$y = \Pi \sin \beta \cos (\odot - \lambda),$$

whence, eliminating ⊙, we derive

$$\frac{x^2}{\Pi^2} + \frac{y^2}{\Pi^2 \sin^2 \beta} = 1 \qquad \ldots\ldots(59),$$

the equation of an ellipse, known as the *parallactic ellipse*. This curve is the locus of the geocentric position X_1 throughout the year. The major axis is parallel to the ecliptic and the minor axis perpendicular to the ecliptic. Whatever the star observed, the semi-major axis is independent of the star's latitude and is equal to Π. For a star on the ecliptic the semi-minor axis vanishes and the parallactic displacement takes place wholly in the ecliptic.

128. *The effect of stellar parallax on right ascension and declination.*

In Fig. 91 let A, D be the right ascension and declination of the sun S. Let α, δ be the right ascension and declination of X and α_1, δ_1 the corresponding quantities for X_1. Denote $\alpha_1 - \alpha$

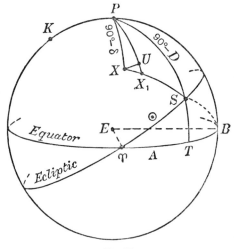

Fig. 91.

by $\Delta\alpha$ and $\delta_1 - \delta$ by $\Delta\delta$. Let UX be the parallel of declination through X. Denote $U\hat{X}X_1$ by ψ. Then in the small plane triangle UXX_1,

$$UX \equiv \Delta\alpha \cos \delta = XX_1 \cos \psi,$$
$$UX_1 \equiv -\Delta\delta = XX_1 \sin \psi.$$

As before $XS = \theta$ and $XX_1 = \theta - \theta_1$. Thus, using (54),

$$\Delta\alpha \cos \delta = \Pi \sin \theta \cos \psi \qquad \ldots\ldots(60),$$
$$\Delta\delta \qquad = -\Pi \sin \theta \sin \psi \qquad \ldots\ldots(61).$$

Now in the spherical triangle PXS, $PX = 90° - \delta$, $PS = 90° - D$, $X\hat{P}S = A - \alpha$, $XS = \theta$ and $P\hat{X}S = 90° + \psi$.

By formulae **B** and **C**,

$$\sin \theta \cos \psi = \cos D \sin (A - \alpha) \qquad \ldots(62),$$
$$- \sin \theta \sin \psi = \sin D \cos \delta - \cos D \sin \delta \cos (A - \alpha) \ldots(63).$$

Now from the triangle TST, in which $TS = \odot$, $S\hat{T}T = \epsilon$ (the obliquity of the ecliptic), $TT = A$, $T\hat{T}S = 90°$ and $TS = D$, we have, by **A**, **B** and **C** successively,

$$\left.\begin{array}{l} \cos \odot \qquad = \cos D \cos A \\ \sin \odot \sin \epsilon = \sin D \\ \sin \odot \cos \epsilon = \cos D \sin A \end{array}\right\} \qquad \ldots\ldots(64).$$

Hence from (60), (62) and (64) we have

$$\Delta\alpha \cos \delta = \Pi (\cos \alpha \cos \epsilon \sin \odot - \sin \alpha \cos \odot) \quad \ldots(65).$$

Similarly, from (61), (63) and (64) we obtain

$$\Delta\delta = \Pi (\cos \delta \sin \epsilon \sin \odot - \cos \alpha \sin \delta \cos \odot - \sin \alpha \sin \delta \cos \epsilon \sin \odot)$$
$$\ldots\ldots(66).$$

Formulae (65) and (66) may be modified as follows. Consider rectangular equatorial axes ET, EB and EP in Fig. 91, where TB is $90°$. Let X, Y and Z denote the co-ordinates of the sun, expressed in terms of the radius of the earth's orbit (which we have assumed to be circular) as the unit of distance. Then

$$X = \cos TS, \quad Y = \cos BS, \quad Z = \cos PS,$$
or $\qquad X = \cos \odot, \quad Y = \sin \odot \cos \epsilon, \quad Z = \sin \odot \sin \epsilon.$

Making the substitutions in (65) and (66) we have

$$\Delta\alpha \cos \delta = \Pi (Y \cos \alpha - X \sin \alpha) \qquad \ldots\ldots(65\,a),$$
and $\quad \Delta\delta \qquad = \Pi (Z \cos \delta - X \cos \alpha \sin \delta - Y \sin \alpha \sin \delta)$
$$\ldots\ldots(66\,a).$$

The values of X, Y and Z are tabulated in the almanacs for every day in the year and thus the coefficients of Π in $(65\,a)$ and $(66\,a)$ can be readily calculated.

The formulae (65) and (66) or $(65\,a)$ and $(66\,a)$ give the displacements, due to parallax, in right ascension and declination.

129. *The measurement of stellar parallax.*

We shall consider only the application of formula (65). We write it in the form

$$\Delta a \cos \delta \equiv (a_1 - a) \cos \delta = F\Pi \qquad \ldots\ldots(67),$$

where
$$F = \cos a \cos \epsilon \sin \odot - \sin a \cos \odot \qquad \ldots\ldots(68),$$

or
$$F = Y \cos a - X \sin a \qquad \ldots\ldots(68\,a).$$

F is called the *parallax factor* in right ascension and its value for any given star varies with the sun's true longitude. As we have seen, its value can be readily calculated.

Suppose there is a faint star B, presumably at a very great distance, near the star A whose distance is to be measured. Let a_0 be the right ascension of B. Suppose on a given date the difference between the geocentric right ascension a_1 of A and the right ascension a_0 of B is measured. Let $m_1 = a_1 - a_0$. Then $m_1 = (a_1 - a) + (a - a_0)$, where a is the heliocentric right ascension of A. If B be regarded as at an infinite distance, a_0 will be simply the heliocentric right ascension of B and the quantity $(a - a_0)$ will remain constant throughout the year. Using (67), we have

$$m_1 \equiv (a_1 - a) + (a - a_0) = F_1 \Pi \sec \delta + (a - a_0) \ldots(69),$$

where F_1 is the parallax factor at the date in question. Suppose the observation is repeated some months later. We shall then have
$$m_2 = F_2 \Pi \sec \delta + (a - a_0) \qquad \ldots\ldots(70),$$

and by eliminating $(a - a_0)$ between (69) and (70) we obtain

$$m_2 - m_1 = (F_2 - F_1) \Pi \sec \delta$$

and hence
$$\Pi = \frac{(m_2 - m_1) \cos \delta}{F_2 - F_1} \qquad \ldots\ldots(71).$$

If there is an error e in the measured quantity $(m_2 - m_1)$, the parallax Π will be determined with an error $e \cos \delta/(F_2 - F_1)$. The error in Π will then be least when $F_2 - F_1$ is greatest. This condition leads us to the consideration of the most favourable circumstances under which parallax determinations can be carried out.

Let
$$\cos a \cos \epsilon = g \cos h \qquad \ldots\ldots(72),$$

and
$$\sin a \quad = g \sin h \qquad \ldots\ldots(73).$$

Then we can write F from (68) in the form

$$F = g \sin (\odot - h) \qquad \ldots\ldots(74).$$

Clearly the greatest numerical value of F is g and this occurs when $\odot - h = 90°$ or $270°$. Now by (72) and (73),

$$g^2 = \cos^2 \alpha \cos^2 \epsilon + \sin^2 \alpha,$$

or

$$g = (1 - \sin^2 \epsilon \cos^2 \alpha)^{\frac{1}{2}},$$

a formula which gives the maximum numerical value of F. Also h is given by

$$\tan h = \tan \alpha \sec \epsilon,$$

so that, for the maximum value of F, the sun's longitude is given by

$$\odot = \tan^{-1}(\tan \alpha \sec \epsilon) + 90° \quad \text{(or } 270°\text{)}.$$

Thus the dates can be calculated when the parallax factor attains its greatest numerical value. If the interval between the two dates is six months, then $F_1 = -F_2$ and the parallax is given by

$$\Pi = \frac{(m_2 - m_1)\cos \delta}{2F_2}.$$

The modern practice for the measurement of stellar parallax consists in taking photographs, at or near the most favourable dates determined by consideration of the parallax factor, of the region of the sky in which the star whose distance is to be measured (called the "parallax star") is situated. Several faint stars are chosen as "comparison stars" and the measurement of a plate leads, in effect, to the determination of quantities $(\alpha_1 - \alpha_0)$, which we have denoted by m_1, for the several comparison stars concerned. In the same way, a plate taken about six months later will enable the quantities $(\alpha_2 - \alpha_0)$ or m_2 to be derived. The general formula (71) is then applied. We shall discuss the details of the photographic method more fully in Chapter XII. As a rule about a score of plates, each with three or four distinct exposures, taken at the appropriate epochs are necessary for a reliable determination of the parallax of a single star. The parallax of the nearest star, Proxima Centauri, is $0''\cdot76$ and with modern instruments parallaxes as small as $0''\cdot005$ can be measured with fair accuracy.

130. *The parsec and light-year.*

The distance corresponding to a parallax of $1''$ is called a *parsec*. If d is this distance, $\sin 1'' = a/d$, and as $a = 93\cdot0 \times 10^6$ miles and $\sin 1'' = 1/206265$, the parsec is thus equivalent to 206265 astronomical units or $93\cdot0 \times 10^6 \times 206265$ miles or $19\cdot18 \times 10^{12}$ miles.

A star whose parallax is $0''\cdot010$ is at a distance of 100 parsecs and generally, if the parallax is $n/1000$ seconds of arc, the corresponding distance is $1000/n$ parsecs. In popular books on astronomy, the *light-year* is generally adopted as the unit of stellar distance—it is the distance traversed by light in one year. From the known value of the velocity of light, it is easily found that 1 light-year $= 5\cdot88 \times 10^{12}$ miles. Thus 1 parsec $= 3\cdot26$ light-years.

EXERCISES

1. If a and b are the equatorial and polar radii of the earth (assumed spheroidal), show that the greatest value of the angle of the vertical is

$$\tan^{-1}\frac{a^2 - b^2}{2ab}.$$ [*Ball.*]

2. Show that, if two bodies have declinations and hour angles δ, H and $-\delta$, H, respectively, and equal horizontal parallaxes, their parallaxes in right ascension are equal.

3. Show that the parallax in declination of a planet observed from a place in latitude ϕ vanishes if $\tan \phi = \tan \delta \cos H,$

δ and H being the planet's declination and hour angle, respectively, and the earth being assumed spherical.

4. Show that, if the horizontal parallax P of a body is small so that $\sin^2 P$ may be neglected, the apparent daily path of the body as seen from a place in latitude ϕ is a small circle of radius $90° - \delta + P \sin \phi \cos \delta$, described about a point depressed $P \cos \phi \sin \delta$ below the pole.

5. Assuming the sun's horizontal parallax to be $8''\cdot80$, show that the time during which the sun is below the horizon at either pole is increased owing to parallax by an amount equal to 7 cosec ϵ minutes, ϵ being the obliquity of the ecliptic.   [*M.T.* 1903.]

6. Assuming the earth to be spherical, show that parallax increases the apparent semi-diameter of the moon in the ratio $\sin z' : \sin (z' - \psi)$, where z' is the apparent zenith distance of the moon's centre and ψ is the angle subtended at the moon by the observer and the earth's centre.   [*Ball.*]

7. The apparent position of a point S (a, δ) on the celestial sphere is displaced a small distance SS' along the great circle towards a point Q (a_0, δ_0) so that $SS' = k \sin SQ$, k being small. Show that the increments in R.A. and declination are $\Delta a = k \sin (a_0 - a) \cos \delta_0 \sec \delta,$

$\Delta \delta = k \{\cos \delta \sin \delta_0 - \sin \delta \cos \delta_0 \cos (a_0 - a)\}.$

Apply these to obtain formulae for the changes in the co-ordinates of a star due to parallax and aberration.   [*M.T.* 1929.]

8. Prove that the cosine of the angle between the directions in which a star is displaced on the celestial sphere by annual aberration and by annual parallax is

$$\sin 2 \, (\odot - \lambda) \cos^2 \beta \, [4 \sin^2 \beta + \cos^4 \beta \sin^2 2 \, (\odot - \lambda)]^{-\frac{1}{2}},$$

where β, λ, $\odot$ are the latitude and longitude of the star and the longitude of the sun, respectively.   [*Coll. Exam.*]

9. The satellite Phobos revolves in the plane of the equator of Mars in a period of $7^h \, 40^m$, the radius of its orbit being 2·79 times the radius of Mars. The rotation period of Mars is $24^h \, 40^m$. Show that allowing for parallax the time between rising and setting of Phobos at stations on the equator is $4^h \, 16^m$, and that Phobos is never seen from latitudes above 69°.   [*M.T.* 1925.]

10. Show that if a planet be observed from a place P on the earth's surface, and from O, the centre of the earth, its angular radii S and S' will be related by the formula

$$\sin S' = \frac{\sin (\delta' - \gamma)}{\sin (\delta - \gamma)} \sin S,$$

where γ is defined by the relation

$$\tan \gamma = \cos \tfrac{1}{2} \, (a' - a) \tan \phi' \sec \{\theta - \tfrac{1}{2} \, (a + a')\},$$

where ϕ' is the geocentric latitude of the place, θ the sidereal time of the observation, a' and δ' the planet's right ascension and declination as seen from P, and a and δ its corresponding co-ordinates as seen from O.   [*Ball.*]

11. If Z be the zenith, C_1 and C_2 two planets, O the middle point of $C_1 C_2$, D the true distance $C_1 C_2$, and D' the apparent distance, show that $D' - D = Q \cos ZR$, where R is a point on the arc $C_1 C_2$, such that $OR = \gamma$, where $Q \cos \gamma = (P_1 + P_2) \sin \tfrac{1}{2} D$ and $Q \sin \gamma = (P_1 - P_2) \cos \tfrac{1}{2} D$, P_1 and P_2 being the horizontal parallaxes of the two planets.   [*M.T.* 1928.]

12. A star is observed at longitude λ, latitude β. Owing to annual parallax its longitude is observed to change through $0''\!\cdot\!5$. What is the maximum change in its latitude, and at what times of the year do the maxima and minima of latitude and longitude occur? What is its distance?   [*M.T.* 1920.]

13. If a, δ be the right ascension and declination of a star with annual parallax Π; p and D the position angle and distance of an adjacent star with negligible parallax; and if ρ, θ, λ, μ be auxiliary quantities defined by the equations

$$\rho \cos \theta = \sin \delta', \qquad\qquad \lambda \cos \mu = \cos \delta' \sin (a' - a),$$
$$\rho \sin \theta = \cos \delta' \cos (a' - a), \qquad\qquad \lambda \sin \mu = \rho \cos (\theta + \delta),$$

where (a', δ') are the sun's co-ordinates, show that the parallax corrections necessary to be applied to the observed position angle and distance, in order to obtain their *heliocentric* values are $\Pi \lambda \cos (p + \mu) \operatorname{cosec} D$ and $\Pi \lambda \sin (p + \mu)$, respectively, if we assume the earth's orbit to be a circle.   [*Ball.*]

CHAPTER X

PRECESSION AND NUTATION

131. *Introductory.*

The phenomenon of precession was discovered by Hipparchus in the second century B.C. By comparing contemporary observations with observations made about a century and a half earlier, he was led to the conclusion that the longitudes of the stars appeared to be increasing at the rate of 36″ per annum (the modern value is about 50″) while, as far as he could detect, their latitudes showed no definite changes. There are two possible explanations; either all the stars examined had real and identical motions in longitude—an improbable hypothesis—or the fundamental reference point, the vernal equinox Υ from which longitudes are measured along the ecliptic, could no longer be regarded as a fixed point on the ecliptic. Now Υ is defined to be one of the two points of intersection of the ecliptic and the equator on the celestial sphere; the observations showed no changes in the latitudes of the stars and therefore it was legitimate to conclude that the ecliptic was a fixed plane. According to the second hypothesis (which was adopted by Hipparchus), it was necessary to assume that the equator and, in consequence, the vernal equinox moved in such a way that the longitudes of the stars increased uniformly by an amount in accordance with the observations.

In Fig. 92 let $L\Upsilon M$ denote the fixed ecliptic, ΥTR the celestial equator at time t and $TT_1 R$ the celestial equator one year later. In one year the vernal equinox has moved from Υ to Υ_1 and thus the longitude of a star S has increased from ΥD to $\Upsilon_1 D$, that is, by about 50″. The uniform backward movement of Υ along the ecliptic is called the *precession of the equinox.* Now Hipparchus satisfied himself that the obliquity ϵ of the ecliptic had suffered no appreciable change and it therefore followed that the motion of the equator must be such that the pole P moved from P to P_1 around K in a small circle, KP or KP_1 being the obliquity ϵ. As KP is perpendicular to the great circle joining K and Υ, and

KP_1 is perpendicular to the great circle $K\Upsilon_1$, the angle PKP_1 is equal to the arc $\Upsilon\Upsilon_1$. From the known movement of Υ along the ecliptic it is easily seen that the pole P describes the small circle of which K is the pole in a period of about 26,000 years.

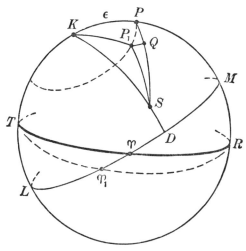

Fig. 92.

Newton first gave the correct dynamical explanation of precession. Consider the familiar phenomenon of an ordinary top (Fig. 93) spinning rapidly about its axis OC inclined at an angle i to the vertical OD, and with its point O on a rough horizontal table. If G is the centre of gravity of the top, its weight acting vertically downwards along GE has a moment about O which at first sight would appear to result in increasing the angle COD (that is, i) and in bringing the surface of the top almost immediately into contact with the table. But owing to the rapid rotation about OC, the motion is very

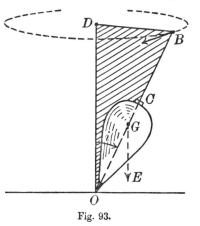

Fig. 93.

much different, for the axis OC is seen to move uniformly around the vertical, so that any point B on the axis describes a circle about a point D on the vertical OD, and the axis OC sweeps out a conical surface. Instead of falling towards the table, the axis OC is at any moment moving at right angles to the plane containing OC and OD. This motion of the axis is called *precession*. Consider now the earth spinning rapidly about its polar axis under the forces of attraction exerted by the sun and moon. If the earth were a sphere, the density of its matter at any point depending only on the distance from the centre, the direction of the solar and lunar attraction would pass through the centre and there would be no cause operating to change the direction of the rotational axis. But the earth is actually a spheroid, with the equator as principal plane, and as the sun and moon are not situated in the equatorial plane—except on two occasions during their respective orbital periods when their declinations are zero—the direction of the gravitational attractions of the sun and moon do not pass through the earth's centre and, accordingly, the attractive forces have moments about the centre. Consider the sun only. Then the moment of the solar attraction would appear to have the effect of making the earth's equatorial plane move towards the ecliptic, just as in the case of the top (Fig. 93), the moment of the top's weight about the point O would appear to cause the axis OC to approach the table. The actual dynamical result as regards the earth is similar to that in the example of the top; the earth's axis, at any instant, moves at right angles to the plane containing the axis and the sun and, accordingly, it has a conical motion about the direction given by the pole of the ecliptic; in other words, the pole of the equator describes uniformly a small circle around the pole of the ecliptic K (Fig. 92) and the vernal equinox Υ moves backward along the ecliptic in the direction $\Upsilon\Upsilon_1$. The principal action of the moon on the direction of the earth's axis is of a similar nature. The combined retrograde movement of the equinox Υ along the ecliptic in the direction $\Upsilon\Upsilon_1$ is called the *luni-solar precession*, about two-thirds of the effect being due to the moon and the remainder to the sun. Certain effects of a periodic character are excluded in the definition of luni-solar precession; reference will be made to these later (section 134).

132. *The effect of precession on the right ascension and declination of a star.*

Let θ denote the luni-solar precession in one year. Then as we have seen, the latitude of a star is unaffected, while the longitude increases at the rate of θ or $50''$ per annum. In Fig. 92 let TT_1 or $P\hat{K}P_1$ represent θ. If (α, δ) and (α_1, δ_1) are the co-ordinates of the star S, referred to the equators TTR and TT_1R respectively, we shall have: $PS = 90° - \delta$, $P_1S = 90° - \delta_1$, $K\hat{P}S = 90° + \alpha$, $K\hat{P}_1S = 90° + \alpha_1$, $KS = 90° - \beta$, $P\hat{K}S = 90° - \lambda$, $P_1\hat{K}S = 90° - \lambda_1$, where β is the star's latitude and λ, λ_1 the longitudes referred to T and T_1 respectively. Also $KP = KP_1 = \epsilon$.

We shall first investigate the value of $\delta_1 - \delta$. Through P_1 draw the small circle arc P_1Q perpendicular to PS. We regard θ as a small quantity and so QS is sensibly equal to P_1S; hence $QS = 90° - \delta_1$ and $PQ = \delta_1 - \delta$ or, as we shall write it, $\Delta\delta$. Now P_1P is perpendicular to KP and as $K\hat{P}Q = 90° + \alpha$ we have that $P_1\hat{P}Q = \alpha$. Hence

$$\Delta\delta \equiv PQ = PP_1 \cos\alpha \qquad \ldots\ldots(1).$$

Now $P\hat{K}P_1 \equiv TT_1 = \theta$ and

$$PP_1 = P\hat{K}P_1 \sin\epsilon = \theta \sin\epsilon \qquad \ldots\ldots(2).$$

Hence, from (1) and (2),

$$\delta_1 - \delta \equiv \Delta\delta = \theta \sin\epsilon \cos\alpha \qquad \ldots\ldots(3).$$

This equation gives the yearly change in declination due to the luni-solar precession θ.

Consider now the change in right ascension. We shall denote $\alpha_1 - \alpha$ by $\Delta\alpha$. From the triangle KPS, by the cosine-formula **A**, we have

$$\sin\beta = \cos\epsilon \sin\delta - \sin\epsilon \cos\delta \sin\alpha \qquad \ldots\ldots(4)$$

and from the triangle KP_1S,

$$\sin\beta = \cos\epsilon \sin\delta_1 - \sin\epsilon \cos\delta_1 \sin\alpha_1 \qquad \ldots\ldots(5).$$

In (5) we write $\delta_1 = \delta + \Delta\delta$, $\alpha_1 = \alpha + \Delta\alpha$ and expand; keeping only first order terms in $\Delta\alpha$ and $\Delta\delta$, we shall have

$$\sin\delta_1 = \sin\delta + \Delta\delta \cos\delta,$$
$$\cos\delta_1 = \cos\delta - \Delta\delta \sin\delta,$$
$$\sin\alpha_1 = \sin\alpha + \Delta\alpha \cos\alpha,$$

so that (5) becomes

$$\sin \beta = \cos \epsilon \, (\sin \delta + \Delta\delta \cos \delta)$$
$$- \sin \epsilon \, (\cos \delta - \Delta\delta \sin \delta) \, (\sin \alpha + \Delta\alpha \cos \alpha) \quad \ldots\ldots(6).$$

Subtracting (4) from (6) and neglecting the term in $\Delta\alpha . \Delta\delta$ we obtain

$$\sin \epsilon \cos \alpha \cos \delta \, \Delta\alpha = (\cos \epsilon \cos \delta + \sin \epsilon \sin \alpha \sin \delta) \, \Delta\delta.$$

But from (3), $\qquad \Delta\delta = \theta \sin \epsilon \cos \alpha.$

Hence $\qquad \alpha_1 - \alpha \equiv \Delta\alpha = \theta \, (\cos \epsilon + \sin \epsilon \sin \alpha \tan \delta) \quad \ldots\ldots(7).$

This last equation enables us to compute the yearly change in right ascension due to the luni-solar precession θ.

133. *The effect of precession on right ascension and declination (alternative method).*

Owing to precession, the spherical triangle defined by the star and the poles of the equator and ecliptic, that is KPS, is changed after an interval of one year to the triangle KP_1S (Fig. 92). We can thus suppose that the latter triangle is obtained by applying infinitesimal changes $\Delta\lambda$, $\Delta\alpha$ and $\Delta\delta$ to the values of λ, α and δ, which occur in the elements of the triangle KPS. The elements ϵ and $90° - \beta$ remain constant. From the triangle KPS, by formula **A**,

$$\sin \delta = \cos \epsilon \sin \beta + \sin \epsilon \cos \beta \sin \lambda$$

whence, by differentiation,

$$\cos \delta \, \Delta\delta = \sin \epsilon \cos \beta \cos \lambda \, \Delta\lambda \qquad \ldots\ldots(8).$$

But by formula **B**,

$$\cos \alpha \cos \delta = \cos \beta \cos \lambda \qquad \ldots\ldots(9).$$

Hence, from (8) and (9)

$$\Delta\delta = \Delta\lambda \sin \epsilon \cos \alpha \qquad \ldots\ldots(10).$$

If $\Delta\lambda$ is the change in λ due to luni-solar precession in one year, $\Delta\lambda = \theta$ and we obtain the formula (3) which gives the annual change in declination.

From (9), by differentiation,

$$\sin \alpha \cos \delta \, \Delta\alpha + \cos \alpha \sin \delta \, \Delta\delta = \cos \beta \sin \lambda \, \Delta\lambda$$

and by means of (10) this becomes

$$\sin \alpha \cos \delta \, \Delta\alpha = \Delta\lambda \, [\cos \beta \sin \lambda - \sin \epsilon \cos^2 \alpha \sin \delta] \ldots\ldots(11).$$

But, by formula C,

$$\cos \beta \sin \lambda = \sin \delta \sin \epsilon + \cos \delta \cos \epsilon \sin \alpha.$$

Hence (11) becomes, after a little simplification,

$$\Delta \alpha = \Delta \lambda \left[\cos \epsilon + \sin \epsilon \sin \alpha \tan \delta \right]$$

which, on writing θ for $\Delta \lambda$, gives formula (7) of the previous section.

134. *Nutation.*

In (11) of the previous section, $\Delta \lambda$ represents the change of the equinox in longitude in one year, and perhaps it is assumed that $\Delta \lambda$ is constant, to be identified with a value of about 50″ per annum which leads to the revolution of the equinox around the ecliptic in the period of about 26,000 years. Now, $\Delta \lambda$ would be constant provided that the declinations of the sun and the moon remain fixed and had non-zero values for, then the gravitational attraction of each on the spheroidal earth would be unaltered and the direction of each attraction would not pass through the earth's centre; the result would be that each body would exert a couple on the earth.

First consider the changing position of the sun in the ecliptic. The couple now tending to tilt the earth's axis is of the variable form $b \left(1 - \cos 2\odot \right)$, where b is a constant partly depending on the obliquity. By integration, the change of longitude in t years is of the form

$$a_1 t + l \sin 2\odot \qquad \qquad \ldots\ldots(12)$$

(small terms being omitted), in which $\odot$ is the sun's longitude and a_1, l are constants whose values are obtained from dynamical theory. The term $a_1 t$ indicates a progressive and uniform motion of Υ along the ecliptic and terms of this character are generally referred to as *secular terms*, and in the particular problem under discussion as *precessional terms*. Thus $a_1 t$ represents the part of the precession in t years due to the sun.

The term $l \sin 2\odot$ is periodic in character; it is zero when the sun's longitude is $0°, 90°, \ldots 360°$. If Υ_2 represents the equinox as defined by the term $a_1 t$ and Υ_3 the equinox as defined by (12), it is clear that the distance between Υ_2 and Υ_3 oscillates between the values $+ l$ and $- l$, and that the period of the changes in the value of $\Upsilon_2 \Upsilon_3$ is six months. The term $l \sin 2\odot$ is called a *periodic term*.

In a similar way the corresponding nature of the moon's orbit will lead to principal terms of the form

$$a_2 t + m \sin 2\mathbb{C} \qquad \text{......(13)},$$

where $\mathbb{C}$ denotes the moon's longitude.

Thus combining the effects of the sun and moon the motion of the equinox can be described, so far as the preceding argument is concerned, by the formula (principal terms only)

$$at + l \sin 2\odot + m \sin 2\mathbb{C} \qquad \text{......(14)},$$

where a is written for $a_1 + a_2$.

We have assumed in the foregoing that the plane of the moon's orbit is coincident with the ecliptic. Actually the moon's orbital plane is inclined at an angle (denoted by i) of $5° 11'$ to the ecliptic and we have now to examine the effects of this inclination on the motion of the equinox Υ.

In Fig. 94 let the dotted circle intersecting the ecliptic at N_1 and N represent the intersection of the moon's orbital plane at

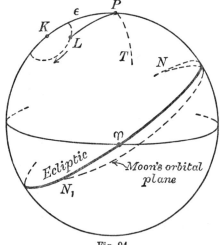

Fig. 94.

a given time with the celestial sphere. The pole of the moon's orbital plane will be at L where the arc KL is the inclination i or $5° 11'$. If the pole L is regarded for the moment as a fixed point on the celestial sphere, the effect of the moon's attraction on the spheroidal earth would be to cause the pole P to move around L in a small circle PT of radius LP. Firstly, this would lead to the backward movement of Υ along the ecliptic; secondly, the angular distance between K and P would not remain constant or, in other words, the obliquity of the ecliptic would be altered. But from considerations of dynamical theory, as well as from observation, it is known that the moon's orbital plane is not fixed with respect to the ecliptic, but moves in such a way that the pole L describes a small circle around K (in the direction of the arrow) in a period of 18·6 years. We infer that so far as the moon is concerned, the backward movement of the equinox Υ along the ecliptic will consist of two parts; one part due to the action of the moon on the supposition that the pole of its orbit coincides with K, and the other depending on the position of the moon's orbital plane with respect to the ecliptic, that is to say, on the position of the pole L which can be defined in terms of the longitude of the node N_1 or N and the inclination i. The second part will evidently be periodic in character and the principal terms of the mathematical expression are of the form

$$b \sin \Omega + c \sin 2\Omega \qquad \ldots\ldots(15),$$

where Ω is the longitude of the moon's ascending node N and b and c are constants whose values are obtained from dynamical theory.

Combining (14) and (15) we have the following expression (principal terms only) giving the motion of the equinox on the ecliptic:

$$at + b \sin \Omega + c \sin 2\Omega + l \sin 2\odot + m \sin 2\mathbb{C} \quad \ldots\ldots(16).$$

The first term, at, gives the *luni-solar precession in longitude* and the remaining terms—all periodic in character—give the *nutation in longitude*. The nutation in longitude is denoted by $\Delta\psi$.

In the statements made in this section only the more important terms for the nutation in longitude have been mentioned; we shall refer more fully to these later.

135. *Nutation in the obliquity.*

We have already indicated in the preceding section that, inasmuch as the pole of the moon's orbital plane does not coincide with the pole of the ecliptic, the obliquity will not remain constant. As the motion of L around K is periodic, the changes in the obliquity are also periodic. If $\Delta\epsilon$ denotes the change in the obliquity, the principal terms are found to be of the form (we include one small term due to the sun):

$$\Delta\epsilon = b_1 \cos\Omega + c_1 \cos 2\Omega + l_1 \cos 2\odot + m_1 \cos 2\mathbb{C} \quad \ldots\ldots(17).$$

These terms, being periodic, give the *nutation in the obliquity.*

We now investigate the changes in the right ascension and declination of a star due only to the change $\Delta\epsilon$ in the obliquity, the ecliptic being regarded as fixed. In Fig. 95, let P be the pole

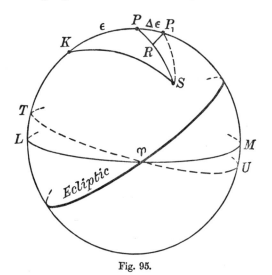

Fig. 95.

of the equator $L\Upsilon M$, say, at 1900·0 and P_1 the pole of the equator $T\Upsilon U$ at time $1900 + t$. Then the arc PP_1 on the great circle KP is $\Delta\epsilon$. Let (α, δ) and (α_1, δ_1) be the co-ordinates of a star S referred to the equators $L\Upsilon M$ and $T\Upsilon U$ respectively. Then $PS = 90° - \delta$, $P_1 S = 90° - \delta_1$, $KP = \epsilon$, $KP_1 = \epsilon + \Delta\epsilon$, $K\hat{P}S = 90° + \alpha, K\hat{P}_1 S = 90° + \alpha_1, P\hat{K}S = 90° - \lambda$ and $KS = 90° - \beta$. From P_1 draw a small circle arc $P_1 R$ perpendicular to PS. As $\Delta\epsilon$ is a small quantity, $P_1 S$ and RS are sensibly equal and therefore $PR = \delta_1 - \delta$ which we shall denote by $\Delta\delta$. Also $P_1\hat{P}R = 90° - \alpha$. Then

$$PR = PP_1 \cos P_1 PR,$$

or $\qquad\qquad \delta_1 - \delta \equiv \Delta\delta = \Delta\epsilon \sin\alpha \qquad\qquad \ldots\ldots(18).$

From the triangle KPS, by the sine-formula **B**,

$$\cos \alpha \cos \delta = \cos \lambda \cos \beta \qquad \ldots \ldots (19)$$

and from the triangle KP_1S, similarly,

$$\cos \alpha_1 \cos \delta_1 = \cos \lambda \cos \beta \qquad \ldots \ldots (20).$$

Write $\alpha_1 - \alpha = \Delta \alpha$. Then with sufficient accuracy, keeping only the first powers of $\Delta \alpha$ and $\Delta \delta$, we have by simple expansion

$$\cos \alpha_1 = \cos \alpha - \Delta \alpha \sin \alpha,$$
$$\cos \delta_1 = \cos \delta - \Delta \delta \sin \delta.$$

Hence from (20), neglecting the product term in $\Delta \alpha . \Delta \delta$,

$$\cos \lambda \cos \beta = \cos \alpha \cos \delta - \Delta \delta \cos \alpha \sin \delta - \Delta \alpha \sin \alpha \cos \delta,$$

and using (19) we have

$$\Delta \alpha \sin \alpha \cos \delta = - \Delta \delta \cos \alpha \sin \delta \qquad \ldots \ldots (21).$$

But $\Delta \delta$ is given by (18); hence

$$\Delta \alpha = - \Delta \epsilon \cos \alpha \tan \delta \qquad \ldots \ldots (22).$$

Formulae (18) and (22) give the changes in the declination and right ascension of a star due only to the nutation $\Delta \epsilon$ in the obliquity.

They may also be derived simply as follows. From the triangle KPS, we have by formula **A**,

$$\sin \delta = \sin \beta \cos \epsilon + \cos \beta \sin \epsilon \sin \lambda.$$

Regarding λ and β as constant, as the nutation in obliquity does not affect these co-ordinates, we obtain by differentiation

$$\cos \delta \Delta \delta = - \Delta \epsilon (\sin \beta \sin \epsilon - \cos \beta \cos \epsilon \sin \lambda).$$

But by **C**,

$$- \cos \delta \sin \alpha = \sin \beta \sin \epsilon - \cos \beta \cos \epsilon \sin \lambda.$$

Hence $\qquad \Delta \delta = \Delta \epsilon \sin \alpha,$

which is equation (18).

Also by differentiating (19) we obtain

$$\sin \alpha \cos \delta \Delta \alpha + \cos \alpha \sin \delta \Delta \delta = 0,$$

which is the same as (21) and the resulting formula (22) is derived as before.

136. *Planetary precession.*

Up to this point it has been assumed that the ecliptic is a fixed plane. Let us first examine what is meant by a "fixed plane". Consider the celestial sphere with the sun as centre. Now if we are given any two points on the sphere, not diametrically opposite, one and only one great circle can be drawn through them and accordingly we may specify the great circle in terms of the two points. Let us suppose that the two points on the sphere are given by two stars which, for simplicity, will be assumed at an infinite distance. The assumptions made will dispose of any necessity to consider aberration, parallax and proper motion. Then we can say that the plane containing the sun and the two stars is a fixed plane relative to the sun. In this sense, the ecliptic is not quite a fixed plane.

Consider the ecliptic ATB (Fig. 96) at 1900·0, which we shall assume to be coincident with a fixed plane as just defined. We shall thus consider the ecliptic ATB to be a fixed plane of reference, and we shall refer to it as the "fixed ecliptic". We

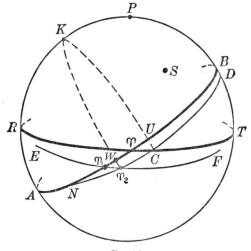

Fig. 96.

shall suppose that one year later the ecliptic is NCD, intersecting the fixed ecliptic in N. This change in the position of the ecliptic is produced by the mutual gravitational attraction of the planets, giving rise to secular variations in the inclination, and in the longitude of the node N of the earth's orbital plane, with reference to the fixed plane of reference. Due to this cause alone, the equinox moves in one year from Υ to C; this motion of the equinox is called *planetary precession*, which we shall denote by l. The value of l is about $0''\cdot13$ per annum. It is evident from Fig. 96 that the effect of planetary precession is to decrease the right ascensions of all stars by the same yearly amount, namely l, without altering their declinations. It is also clear from Fig. 96 that a minute change in the obliquity results which, due to this cause, decreases at the rate of about $0''\cdot5$ per annum. Actually it is found that the displacement in t years of the equinox Υ— along the equator in the direction ΥC—is given with sufficient accuracy by a formula of the type $ct + dt^2$ and that the change in the obliquity is given by a similar expression. The coefficients in each case are very small.

Consider now the combined effect of the luni-solar and planetary precessions. In Fig. 96 $A\Upsilon B$ and $R\Upsilon T$ are the ecliptic and equator for 1900·0. One year later the fundamental planes are NCD and $E\Upsilon_1\Upsilon_2$. $\Upsilon\Upsilon_1$ is the luni-solar precession in one year. As all the changes in the fundamental planes are small we shall have, with sufficient accuracy, that $\Upsilon_1\Upsilon_2 = \Upsilon C$. $\Upsilon_1\Upsilon_2$ is the planetary precession in one year. Join K to C and K to Υ_2 by great circle arcs cutting the fixed ecliptic in U and W respectively. Then ΥU or $\Upsilon_1 W$ is the planetary precession in longitude. Its value is clearly $l \cos \epsilon$ per annum.

We now define *general precession*. In one year (we are still excluding nutation), the equinox moves from Υ to Υ_2. Referred to the fixed ecliptic $A\Upsilon B$ and the equinox Υ, the longitude of Υ_2 can be written $360° - \chi$, where χ is clearly a small quantity. The quantity χ is called the general precession (in longitude). When we are dealing with the positions of the fundamental planes separated by a short interval—a few years, say—the general precession can be represented geometrically with high accuracy. We shall consider the interval of one year for which Fig. 96 has already been used. Then the general precession in longitude is ΥW, that is $\Upsilon\Upsilon_1 - \Upsilon_1 W$, or the luni-solar precession minus the planetary precession in longitude.

As we have seen, planetary precession produces a slight change in the obliquity.

We now give the numerical expressions for the general

precession χ and the obliquity at a time t years after 1900; the first is $\chi = 50''\cdot2564 + 0''\cdot000222t$ per annum,

and the second is

$$\epsilon = 23°\ 27'\ 8''\cdot26 - 0''\cdot4684t.$$

137. *The mean equator and the mean co-ordinates of a star.*

In Fig. 96, $A \Upsilon B$ and $R \Upsilon T$ are respectively the ecliptic and equator at the beginning of 1900 (this is usually written 1900·0); these are regarded as fixed planes of reference. After one year the luni-solar and planetary precessions produce changes in the ecliptic and equator so that for 1901·0 these planes are now $N \Upsilon_2 D$ and $E \Upsilon_2 F$. $N \Upsilon_2 D$ and $E \Upsilon_2 F$ are defined to be the *mean ecliptic* and *mean equator* for 1901·0. Also the equinox Υ_2 is the *mean equinox* for 1901·0.

In this chapter we shall omit the general consideration of the proper motion of a star and we shall now define provisionally the *mean position* of a star. The mean position of a star at any time is its position on the celestial sphere centred at the sun, and referred to the mean equator and equinox at that time. In practice we consider only the mean position with reference to the equinox at the beginning of a year. It is to be noted that in our definition no cognisance is taken of nutation, aberration, annual parallax and (provisionally) of the star's proper motion.

We now investigate the problem: to derive the mean position of a star for 1901·0, given the mean position for 1900·0. Let (α, δ) be the co-ordinates referred to the mean equator and equinox for 1900·0, and (α_1, δ_1) referred to the mean equator and equinox for 1901·0. Due to luni-solar precession, the star's longitude has increased in one year by θ. Formulae (7) and (3) are immediately applicable to give the changes in right ascension and declination due to this cause. We have, rewriting these formulae,

$$\alpha_1 - \alpha = \theta\ (\cos \epsilon + \sin \epsilon \sin \alpha \tan \delta)   \quad ......(23),$$
$$\delta_1 - \delta = \theta \sin \epsilon \cos \alpha   \quad\quad\quad ......(24).$$

But planetary precession decreases the right ascension in one year by l ($\Upsilon_1\Upsilon_2$ in Fig. 96); hence the complete expression for $\alpha_1 - \alpha$ is $\alpha_1 - \alpha = (\theta \cos \epsilon - l) + \theta \sin \epsilon \sin \alpha \tan \delta$ ...(25).

The declination is unaltered by planetary precession, and hence (24) is the complete equation for the change in declination.

Put $m = \theta \cos \epsilon - l; \quad n = \theta \sin \epsilon$ ......(26).

Then $\alpha_1 - \alpha = m + n \sin \alpha \tan \delta$ ......(27),

$\delta_1 - \delta = n \cos \alpha$ ......(28).

The quantities m and n are functions of the luni-solar precession and of the obliquity ϵ, both of which vary slowly with the time; consequently m and n vary slowly with the time.

The following table* gives the values of m and n at several epochs (as n occurs in both formulae (27) and (28), it is tabulated both in time and angular measures).

Table: values of m *and* n.

	m	n	n
1800	3ˢ·07048	1ˢ·33703	20″·0554
1850	3ˢ·07141	1ˢ·33674	20″·0511
1900	3ˢ·07234	1ˢ·33646	20″·0468
1950	3ˢ·07327	1ˢ·33617	20″·0426

Taking the year as the unit and writing $\dfrac{d\alpha}{dt}$ and $\dfrac{d\delta}{dt}$ for the rates of change of the co-ordinates α and δ due to precession, we rewrite (27) and (28):

$$\frac{d\alpha}{dt} = m + n \sin \alpha \tan \delta \qquad ......(29),$$

$$\frac{d\delta}{dt} = n \cos \alpha \qquad ......(30).$$

Example. To calculate the mean right ascension of α Orionis for the epoch 1902·0, given that for 1900·0 the mean co-ordinates are: $\alpha = 5^h\ 49^m\ 45^s\cdot481, \quad \delta = + 7° 23' 18''\cdot41.$

We first compute the annual precession given by formula (29). From the table above, for 1900·0,

$$m = 3^s\cdot0723, \quad n = 1^s\cdot3365.$$
$$\log n = 0\cdot12597$$
$$\log \sin \alpha = 9\cdot99956$$
$$\log \tan \delta = 9\cdot11284$$
$$\overline{9\cdot23837}$$

Hence $n \sin \alpha \tan \delta = 0^s\cdot1731$

Also $m = 3^s\cdot0723$

Therefore $\dfrac{d\alpha}{dt} = 3^s\cdot2454$

---

* Newcomb: *Spherical Astronomy*, p. 406.

The annual precession in right ascension for this star is thus
$+ 3^{s}\!\cdot\!2454$, and in two years the change in the right ascension
due to precession is $+ 6^{s}\!\cdot\!491$. Hence the mean right ascension
referred to the mean equinox of 1902·0 is obtained by adding
$6^{s}\!\cdot\!491$ to the mean right ascension for 1900·0, and the result is:

$$\text{Mean right ascension (1902·0)} = 5^{h}\ 49^{m}\ 51^{s}\!\cdot\!972.$$

In a similar way, the mean declination for 1902·0 is found to be
$+ 7° 23' 20''\!\cdot\!20$.

133. *The secular variation.*

The procedure adopted in the example just given is correct
only when the interval of time concerned is a small number of
years. It is to be noticed that on the right-hand sides of (29) and
(30) the values of α and δ are varying quantities as well as m
and n, and this fact must be taken into account when the interval
of time exceeds, say, 5 to 10 years. The rate of change of $\dfrac{d\alpha}{dt}$ per
century is defined to be the *secular variation* in right ascension.
Now α is a function of t (in years, measured say from 1900·0)
and if α_0, $\dfrac{d\alpha_0}{dt}$ and $\dfrac{d^2\alpha_0}{dt^2}$ denote the values of the right ascension,
of $\dfrac{d\alpha}{dt}$ and of $\dfrac{d^2\alpha}{dt^2}$ at $t = 0$, that is for 1900·0, then α can be written

$$\alpha = \alpha_0 + t\,\frac{d\alpha_0}{dt} + \frac{t^2}{2}\frac{d^2\alpha_0}{dt^2} \qquad \text{......(31)}.$$

If s denotes the secular variation in right ascension, $\dfrac{d^2\alpha_0}{dt^2}$ is the
rate of change (per annum) of $\dfrac{d\alpha_0}{dt}$, that is, $\dfrac{d^2\alpha_0}{dt^2} = \dfrac{s}{100}$. Hence
from (31),

$$\alpha - \alpha_0 = t\left(\frac{d\alpha_0}{dt} + \frac{st}{200}\right) \qquad \text{......(32)}.$$

Now, in general,

$$\frac{s}{100} \equiv \frac{d^2\alpha}{dt^2} = \frac{d}{dt}\,(m + n\sin\alpha\tan\delta)$$

$$= \frac{dm}{dt} + \sin\alpha\tan\delta\,\frac{dn}{dt} + n\cos\alpha\tan\delta\,\frac{d\alpha}{dt}$$

$$+ n\sin\alpha\sec^2\delta\,\frac{d\delta}{dt} \qquad \text{......(33)}.$$

Now from the values of m it is found that, for 1900·0, the annual rate of change of m is given by

$$\frac{dm}{dt} = + 0^{s}\cdot 0000186,$$

and, similarly, $\qquad \frac{dn}{dt} = - 0''\cdot 0000853.$

Also $\frac{d\alpha}{dt}$ and $\frac{d\delta}{dt}$ in (33) are given by (29) and (30). Hence the value of s can be computed.

The formula in declination corresponding to (32) is:

$$\delta - \delta_0 = t\left(\frac{d\delta_0}{dt} + \frac{s_1 t}{200}\right) \qquad \ldots\ldots(34),$$

where s_1 is the secular variation in declination.

In the principal catalogues, quantities called the *annual variation* in right ascension and in declination are tabulated for each star. The annual variation in right ascension, for example, is the value of the annual precession $\frac{d\alpha}{dt}$ (for the appropriate epoch, say 1900·0) together with the effect of proper motion, which we have hitherto excluded from our discussion. The secular variations s and s_1 are also given.

Example. To compute the mean co-ordinates of ϵ Octantis for 1950·0, given that the mean co-ordinates (α_0, δ_0) for 1900·0 are

$$22^{h}\ 8^{m}\ 49^{s}\cdot 30 \quad \text{and} \quad - 80°\ 56'\ 14''\cdot 7.$$

From Boss's *Preliminary General Catalogue*, the annual variations in right ascension and declination for 1900·0 are respectively

$$+ 6^{s}\cdot 9931 \quad \text{and} \quad + 17''\cdot 699,$$

and the secular variations s, s_1 are

$$s = - 0^{s}\cdot 5926 \quad \text{and} \quad s_1 = + 0''\cdot 468.$$

Then, by (32), α being the mean right ascension for 1950·0,

$$\alpha - \alpha_0 = 50\ (6^{s}\cdot 9931 - \tfrac{50}{200} \times 0^{s}\cdot 5926)$$
$$= 50\ (6^{s}\cdot 9931 - 0^{s}\cdot 1481)$$
$$= 5^{m}\ 42^{s}\cdot 25.$$

Hence the mean right ascension for 1950·0 is

$$(22^{h}\ 8^{m}\ 49^{s}\cdot 30 + 5^{m}\ 42^{s}\cdot 25) \quad \text{or} \quad 22^{h}\ 14^{m}\ 31^{s}\cdot 55.$$

Also by (34),
$$\delta - \delta_0 = 50\,(17''{\cdot}699 + \tfrac{50}{200} \times 0''{\cdot}468)$$
$$= +\,14'\,50''{\cdot}8.$$

Hence the mean declination for 1950·0 is
$$(-\,80°\,56'\,14''{\cdot}7 + 14'\,50''{\cdot}8)\quad \text{or}\quad -\,80°\,41'\,23''{\cdot}9.$$

139. *The true equator and the true co-ordinates of a star.*

To fix our ideas we shall consider a particular time and date, say, U.T. 15ʰ, 1931 March 9, The *true equator*, true ecliptic and the *true equinox* are the actual equator, ecliptic and equinox respectively, at this instant, and their respective positions with reference to a fixed ecliptic, say, at 1900·0 are dependent on the amount of precession and nutation computed for the interval between 1900·0 and the instant in question. We define the *true place* of a star at this instant as its position on the celestial sphere, centred at the sun, referred to the true equinox and equator of the date.

Let us denote the mean co-ordinates of a star for 1900·0 by (α_0, δ_0), the mean co-ordinates for 1931·0 by (α, δ) and the true co-ordinates for U.T. 15ʰ, 1931 March 9 by (α_1, δ_1). Also let τ denote the fraction of the year between the beginning of the year and the date in question. Then, considering the change in the right ascension of the star due to precession and nutation, we have

$\alpha_1 - \alpha_0 =$ precession in R.A. for $(31 + \tau)$ years
$+$ the effect of nutation on the R.A.

We write $\alpha_1 - \alpha_0 = (\alpha - \alpha_0) + (\alpha_1 - \alpha)$.

Now $\alpha_1 - \alpha =$ the difference between the true R.A. of the star, referred to the true equinox of date, and the mean R.A. referred to the mean equinox of the beginning of the year (1931·0); it is made up of

(i) precession in R.A. for the interval τ,

(ii) the effect of nutation in R.A. for the date in question.

Also $\alpha - \alpha_0$ is the difference of the mean co-ordinates for 1931·0 and 1900·0. We thus have the general rule:

To derive the true co-ordinates of the star at $(1931 + \tau)$, the mean co-ordinates for 1900·0 (say) being given, firstly compute the mean co-ordinates for 1931·0 (by the procedure illustrated on p. 241) and, secondly, compute (i) the precession for the interval τ, and (ii) the effect of nutation. Denote by $\Delta\alpha_1$, $\Delta\alpha_2$ the effects on the right ascension due to (i) and (ii) respectively.

Then by formula (27)—which gives the *annual* change in R.A. due to precession—we have for the interval τ,

$$\Delta\alpha_1 = \tau\,(m + n \sin\alpha \tan\delta).$$

Consider now nutation. As we have seen, nutation changes the longitude of a star and also the obliquity of the ecliptic. For the date concerned, let $\Delta\psi$, $\Delta\epsilon$ denote the nutation in longitude and in the obliquity respectively. $\Delta\psi$ consists of the periodic terms given in (16) and $\Delta\epsilon$ is given by (17). The change in the right ascension due to $\Delta\psi$ is given from (7) by

$$\Delta\psi\,(\cos\epsilon + \sin\epsilon \sin\alpha \tan\delta),$$

and the change in R.A. due to $\Delta\epsilon$ is given from (22) by

$$-\Delta\epsilon \cos\alpha \tan\delta.$$

Hence the total change $\Delta\alpha_2$ in right ascension due to nutation is given by

$$\Delta\alpha_2 = \Delta\psi\,(\cos\epsilon + \sin\epsilon \sin\alpha \tan\delta) - \Delta\epsilon \cos\alpha \tan\delta.$$

But $\alpha_1 - \alpha$ (the difference between the true R.A. at date and the mean R.A. for the beginning of the year) is $\Delta\alpha_1 + \Delta\alpha_2$; thus we can write

$$\alpha_1 - \alpha = (m\tau + \Delta\psi \cos\epsilon) + \sin\alpha \tan\delta\,(n\tau + \Delta\psi \sin\epsilon)$$
$$- \Delta\epsilon \cos\alpha \tan\delta \quad\ldots\ldots(35).$$

But from (26), m and n are defined in terms of the luni-solar precession θ and the planetary precession l by

$$m = \theta \cos\epsilon - l, \quad n = \theta \sin\epsilon \qquad \ldots\ldots(36).$$

Substituting in (35) the expressions for $\cos\epsilon$ and $\sin\epsilon$ given by (36), we obtain

$$\alpha_1 - \alpha = \left(\tau + \frac{\Delta\psi}{\theta}\right)(m + n \sin\alpha \tan\delta) + \frac{l\Delta\psi}{\theta} - \Delta\epsilon \cos\alpha \tan\delta$$
$$\ldots\ldots(37).$$

In (37) we suppose that m and n are expressed in seconds of time and l, $\Delta\psi$, θ, $\Delta\epsilon$ in seconds of arc. Write

$$A = \left(\tau + \frac{\Delta\psi}{\theta}\right); \quad B = -\Delta\epsilon; \quad E = \frac{l\Delta\psi}{15\theta} \left.\vphantom{\frac{\Delta\psi}{\theta}}\right\} \ldots(38).$$
$$a = m + n \sin\alpha \tan\delta; \quad b = \tfrac{1}{15} \cos\alpha \tan\delta$$

Then
$$\alpha_1 - \alpha = Aa + Bb + E \qquad \ldots\ldots(39),$$

in which the right-hand side is expressed uniformly in seconds of time.

In (39), the quantities A, B and E are computed from the known interval τ, the nutations $\Delta\psi$, $\Delta\epsilon$ corresponding to the date concerned, and the luni-solar and planetary precessions θ and l; they are independent of the position of the star. They are known as *Bessel's day numbers* (or *star numbers*) and are tabulated in the almanacs for every day in the year.

The quantities a and b depend on the values of m and n and on the co-ordinates of the star; they change slowly with the time, but over considerable intervals they may be regarded as constants associated with the particular star concerned.

In a similar way, we have the analogous formula in declination:
$$\delta_1 - \delta = \left(\tau + \frac{\Delta\psi}{\theta}\right) n \cos\alpha + \Delta\epsilon \sin\alpha \qquad \ldots\ldots(40),$$

in which n and $\Delta\epsilon$ are supposed expressed in seconds of arc.

Put
$$a' = n \cos\alpha; \quad b' = -\sin\alpha \qquad \ldots\ldots(41).$$

Then, using the values of A and B in (38), we have
$$\delta_1 - \delta = Aa' + Bb' \qquad \ldots\ldots(42).$$

Equations (39) and (42) enable us to calculate the *true* co-ordinates (α_1, δ_1) at any date when the mean co-ordinates (α, δ) at the beginning of the year are given.

The same equations enable the complementary problem to be solved, namely, the calculation of the mean co-ordinates for the beginning of the year from the true co-ordinates of date.

It is to be remembered that, in this discussion, we have omitted the effects of the star's proper motion on the co-ordinates.

The calculations may also be carried out in another manner. From (38) and (39), we can write
$$\alpha_1 - \alpha = mA + E + \tfrac{1}{15}nA \sin\alpha \tan\delta - \tfrac{1}{15}\Delta\epsilon \cos\alpha \tan\delta \qquad \ldots\ldots(43),$$

in which $mA + E$ is expressed in seconds of time and n and $\Delta\epsilon$ in seconds of arc. The last two terms of (43) have the divisor 15 to reduce the terms to seconds of time. Also from (38) and (40)
$$\delta_1 - \delta = nA \cos\alpha + \Delta\epsilon \sin\alpha \qquad \ldots\ldots(44).$$

Let f, g, G be defined by

$$f = mA + E; \quad g \sin G = - \Delta\epsilon; \quad g \cos G = nA \quad \ldots\ldots(45).$$

Then
$$\alpha_1 - \alpha = f + \tfrac{1}{15} g \sin (G + \alpha) \tan \delta \quad \ldots\ldots(46),$$
$$\delta_1 - \delta = \quad g \cos (G + \alpha) \quad \ldots\ldots(47).$$

In these formulae, g is given in seconds of arc and f in seconds of time.

The quantities f, g and G are known as *independent day numbers* (or *star numbers*) and are tabulated for every day in the almanacs.

140. *The apparent place of a star.*

We have seen that the *true place* of a star at any instant is its position on the celestial sphere, centred at the *sun*, with reference to the true equinox and equator at that instant. The true place is thus independent of the effects of aberration and annual parallax. The *apparent place* at any instant of a star or other heavenly body is defined to be its position on the celestial sphere, whose centre is the *earth's centre*, with reference to the true equinox and true equator at that instant. We thus have:

Apparent place = true place plus the corrections due to aberration and annual parallax.

The formulae (25) and (26) of Chapter VIII give the corrections to the right ascension and declination of a star, due to aberration, in the form $Cc + Dd$ and $Cc' + Dd'$ respectively. As the corrections due to annual parallax (Chapter IX, p. 221) are practically negligible (except for a few stars), they can in general be ignored. We then write:

Apparent R.A. = True R.A. + $Cc + Dd$ $\quad \ldots\ldots(48),$

Apparent Dec. = True Dec. + $Cc' + Dd'$ $\ldots\ldots(49).$

141. *Reduction from mean place to apparent place (or vice versa).*

We shall suppose that (α_0, δ_0) are the mean co-ordinates of a star for some such epoch as 1900·0. It is required to determine the apparent co-ordinates (α', δ'), for example, at U.T. 15^h, 1931 March 9.

The first step is to compute the mean co-ordinates (α, δ) for

9

the mean equinox of 1931·0 by the principles of section 138.
Using (39), (42) and (48), (49) we shall have

$$\alpha' - \alpha = Aa + Bb + Cc + Dd + E \qquad \ldots\ldots(50),$$
$$\delta' - \delta = Aa' + Bb' + Cc' + Dd' \qquad \ldots\ldots(51).$$

These formulae give the differences between the apparent co-
ordinates at the date concerned and the mean co-ordinates for
the beginning of the year. The quantities in capital letters are
Bessel's day numbers and the quantities in small letters (*star-
constants*) are functions of the star's co-ordinates. The star-
constants a, b, etc., can be rapidly obtained from R. Schorr's
Präzessions-Tafeln (Bergedorf, 1927).

We may also carry out the computations by means of (46) and
(47) of section 139 and (27) and (28) of Chapter VIII in terms of
the *independent day numbers* as follows:

$$\alpha' - \alpha = f + \tfrac{1}{15}g \sin (G + \alpha) \tan \delta + \tfrac{1}{15}h \sin (H + \alpha) \sec \delta$$
$$\ldots\ldots(52),$$
$$\delta' - \delta = \qquad g \cos (G + \alpha) \qquad + h \cos (H + \alpha) \sin \delta + i \cos \delta$$
$$\ldots\ldots(53).$$

The formulae can also be used in deriving the mean co-ordinates
for the beginning of the year when the apparent co-ordinates at
the date are known.

142. *Cataloguing the stars.*

Let us suppose that a star is observed with the meridian circle
at U.T. 15^{h} on 1931 March 9. After the instrumental and
refraction corrections have been applied, we obtain the apparent
right ascension α' and the apparent declination δ' corresponding
to the true equinox of date. By applying the formulae (50) and
(51)—or (52) and (53)—we obtain the star's mean co-ordinates
α and δ referred to the beginning of the year 1931. For the
purposes of a catalogue it is convenient to refer the positions to
the mean equinox of a common epoch, say, 1900·0. The mean
co-ordinates (α_0, δ_0) referred to this common epoch are then
obtained according to the procedure in section 138.

143. *The numerical values of the nutation in longitude and in the
obliquity of the ecliptic.*

In (16) we have indicated the nature of the principal terms of
the nutation in longitude, $\Delta\psi$—they are the periodic terms in

(16)—and in (17) the principal terms of the nutation in the obliquity, $\Delta\epsilon$. The coefficients b and b_1 are not quite constant, varying slowly with the time. If T denotes the time interval expressed in centuries and reckoned from 1900·0 the expressions (16)—nutation only—and (17) take the following numerical forms:

$$\Delta\psi = -\,(17''\cdot234 + 0''\cdot017T)\sin\Omega + 0''\cdot209\sin 2\Omega$$
$$-\,1''\cdot272\sin 2L - 0''\cdot204\sin 2\mathbb{C} \quad \ldots\ldots(54),$$
$$\Delta\epsilon = (9''\cdot210 + 0''\cdot0009T)\cos\Omega - 0''\cdot090\cos 2\Omega$$
$$+\,0''\cdot551\cos 2L + 0''\cdot088\cos 2\mathbb{C} \quad \ldots\ldots(55),$$

in which we have written the sun's mean longitude L in place of the sun's true longitude $\odot$.

In calculating Bessel's day numbers and the independent day numbers, several other terms which are not given in (54) and (55) are used. The complete expressions for $\Delta\psi$, $\Delta\epsilon$ are given in the almanacs.

The coefficient of $\cos\Omega$ in (55) for any epoch is called the *constant of nutation.*

EXERCISES

1. If P, K are the poles of the equator and ecliptic, and X is a star such that $P\hat{X}K$ is 90°, show that X has no precession in right ascension.

2. Make a rough estimate of the present declination and right ascension of the point of the celestial sphere which was First Point of Aries in 120 B.C., the date when the precession was discovered. [*M.T.* 1921.]

3. The equatorial co-ordinates of a star on the ecliptic are (a, δ) and its longitude is λ. If Δa, $\Delta\delta$ and $\Delta\lambda$ are the annual precession in right ascension, declination and longitude respectively, prove that

$$\Delta a \cot a \cos^2\delta = \Delta\delta \cot\delta = \Delta\lambda \cot\lambda. \qquad [\textit{Coll. Exam.}]$$

4. At a certain date two stars came simultaneously to the horizon of a place in latitude $\cot^{-1}(\sqrt3\sin\epsilon)$ at 0^{h} sidereal time. At a later date, when the precession had amounted to 60°, show that the same stars came simultaneously to the horizon of another place, whose latitude was $\epsilon + \cot^{-1}(2\tan\epsilon)$, at sidereal time 6^{h}, ϵ being the obliquity of the ecliptic. [*M.T.* 1929.]

5. Show that a southern star of declination $-\delta$ and right ascension a must at some epoch have been visible from a station in north latitude ϕ, provided that

$$\sin\delta\cos\epsilon + \cos\delta\sin\epsilon\sin a < \cos(\phi - \epsilon),$$

where ϵ is the obliquity of the ecliptic. Obtain formulae for calculating the date at which it ceased (or will cease) to be visible.   [*M.T.* 1926.]

6. Show that, owing to precession, the position angle θ of a non-polar double star alters at a rate given by $\dfrac{d\theta}{dt} = +1\cdot4 \sin a \sec \delta \sin \epsilon$ degrees per century.

The approximate position of the double star 22 Cygni is $(19^{\text{h}}\,22^{\text{m}}; +27° 30')$. The observed position angles (increasing) for the mean epochs 1830·0 and 1905·0 are respectively 256°·0 and 311°·8. What is the real change in position angle?   [*Lond.* 1926.]

7. Owing to precession the interval between two consecutive transits of a star across a given meridian differs from a mean sidereal day. If $\beta + \epsilon > 90°$, show that this difference vanishes if the star's longitude λ is given by

$$\sin \lambda = \cot \beta \cot \epsilon,$$

where β is the star's latitude and ϵ is the obliquity of the ecliptic.

8. The co-ordinates of a star for 1900·0 are: $16^{\text{h}}\,56^{\text{m}}\,12^{\text{s}};\ \delta = +82° 12'$. Show that the annual precession in R.A. for 1900·0 is $-6^{\text{s}}\cdot30$.

CHAPTER XI

THE PROPER MOTIONS OF THE STARS

141. *Definition of proper motion.*

In 1718 it was discovered by Halley that the positions of certain bright stars had altered appreciably, since the time of Hipparchus, in relation to the general stellar background. Suppose for the moment that all stars with the exception of Arcturus are at an infinite distance from the sun, so that they form a definite system of fixed reference points. The comparison of the observations of Arcturus, made in the days of Hipparchus and of Halley, showed conclusively that this star had moved in the interval of about twenty centuries through the considerable angle of about one degree, with reference to the stars in its immediate neighbourhood on the celestial sphere. This suggested that Arcturus had a definite space-velocity relative to the sun, and that its distance was finite.

Consider Fig. 97. Let S be the sun and suppose that a star with a space-velocity relative to the sun moves in the course of a year from A to B. The star's track in space can be as-

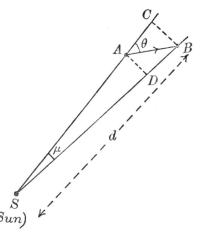

Fig. 97.

sumed to be a straight line; certainly no observations during so long an interval as two centuries have suggested that the path of a single star deviates by a measurable amount from a straight line. Let μ denote the angle ASB. Then μ is the angle through which the star is seen to move in one year; it is called the *proper motion* of the star, and is generally measured in seconds of arc per annum.

145. *Relation between proper motion, tangential velocity and parallax.*

In Fig. 97 let BC and DA be drawn perpendicularly to SA. Let d denote the distance SB (in kilometres) of the star from the sun and θ the angle between its direction of motion AB and the direction SA. Let V denote the star's linear velocity from A to B, expressed in kilometres per second, so that, if n is the number of seconds in one year, we have

$$AB = nV \quad \text{kms.} \qquad \ldots\ldots(1).$$

Let v denote the component of the linear velocity at right angles to the direction SA (the line of sight); v is called the *tangential velocity* or *cross-velocity*, and it is expressed in kilometres per second. For the star with the largest proper motion the value of μ is about $10''$ per annum, and in every instance we can assume that $A\hat{S}B$ is a very small angle so that, in Fig. 97, AD and CB may be taken to be equal; also AD or CB is the distance described in one year perpendicularly to SA, so that

$$AD = CB = nv \quad \text{kms.} \qquad \ldots\ldots(2).$$

Now $CB = d \sin \mu$, or since μ is small, we can write

$$CB = d\mu \sin 1'' \qquad \ldots\ldots(3).$$

Let Π denote the star's annual parallax and a the radius of the earth's orbit in kilometres; then the distance d and the parallax Π are related (Chapter IX) by the formula

$$\sin \Pi = a/d$$

or, since Π is small, by

$$d = \frac{a}{\Pi \sin 1''} \qquad \ldots\ldots(4),$$

where Π is expressed in seconds of arc. Hence from (2), (3) and (4) we have

$$v = \frac{\mu a}{n\Pi} \qquad \ldots\ldots(5).$$

We now insert the values of a and n in (5); $a = 149 \cdot 5 \times 10^6$ kms. and $n = 31 \cdot 56 \times 10^6$ (the number of seconds in a year). Hence we derive

$$v = 4 \cdot 74 \frac{\mu}{\Pi} \qquad \ldots\ldots(6),$$

a relation which gives the cross-velocity v in kilometres per second when the values of μ and Π are known.

For example, the annual proper motion μ of Capella is $0''\cdot439$ (the methods of measuring proper motion will be described later) and its parallax Π is $0''\cdot075$. Inserting these values in (6), we find that the tangential velocity v is $27\cdot7$ kms. per second.

146. Radial velocity.

In one year (Fig. 97), the distance of the star has increased from SA to SB or, with sufficient accuracy, from SA to SC, that is, by the distance AC. The rate at which the star's distance from the sun is changing owing to its linear velocity in space is called the *radial velocity* with respect to the sun. If ρ denotes the radial velocity in kilometres per second, we have

$$AC = n\rho = AB \cos \theta.$$

But by (1), $AB = nV$; hence

$$\rho = V \cos \theta \qquad \ldots\ldots(7).$$

The values of the radial velocity ρ can be determined for the brighter stars by spectroscopic methods, directly in terms of kilometres per second. As in section 124 (d), we accept such results here without explaining the spectroscopic principles involved.

When a star is increasing its distance from the sun (as in Fig. 97), the radial velocity is defined to be positive; when its distance is diminishing, the radial velocity is negative.

Now $BC = AB \sin \theta$; hence, from (1) and (2),

$$v = V \sin \theta \qquad \ldots\ldots(8).$$

If v is determined from (6) and ρ is known from spectroscopic observations, we then have the two equations (7) and (8), from which the star's linear space-velocity V and the angle θ can be determined. For Capella, as we have seen, $v = 27\cdot7$ kms. per second and ρ is found to be $+ 30\cdot2$ kms. per second. Hence

$$V \sin \theta = 27\cdot7 \quad \text{and} \quad V \cos \theta = 30\cdot2.$$

By an easy calculation, it is found that $V = 41\cdot0$ kms. per second (this is the space-velocity of Capella relative to the sun) and θ is $42\frac{1}{2}°$.

147. The measurement of proper motion.

Referring again to Fig. 97, we notice that the component of the star's linear space-velocity which gives rise to proper motion is the tangential velocity in the direction AD or CB; the radial velocity by itself does not affect the direction in which the star is seen. Now A, D and S are coplanar; hence on the celestial sphere with the sun as centre, the star will appear to move along a great circle. In Fig. 98 let S be the position of the star, say, at 1930·0 and T its position one year later. The great circle arc ST is thus the annual proper motion μ. Let (α, δ) and (α_1, δ_1) be the co-ordinates of S and T respectively, referred to the *same* equinox and equator— say the mean equinox and

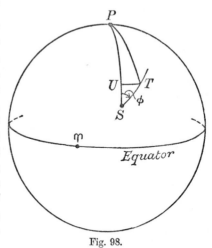

Fig. 98.

equator of 1930·0. Then the differences $(\alpha_1 - \alpha)$ and $(\delta_1 - \delta)$ are due to the annual proper motion. We shall write:

$$\alpha_1 - \alpha = \mu_a; \quad \delta_1 - \delta = \mu_\delta \qquad \ldots\ldots(9).$$

Then μ_a and μ_δ are the components of the proper motion in right ascension and declination respectively. We shall call μ the *total proper motion*, which is always expressed in seconds of arc per annum.

Draw the small circle arc UT parallel to the equator. Then $UT = U\hat{P}T \sin PT$; also $UT = ST \sin \phi$, where ϕ is the position angle PST. ϕ is measured from the meridian joining the star to the north pole P from 0° to 360°, in the direction shown by the arrow in Fig. 98. But $U\hat{P}T = \mu_a$, $PT = 90° - \delta_1$ and $ST = \mu$. Hence

$$\mu_a \cos \delta_1 = \mu \sin \phi$$

or, writing δ for δ_1 since μ is small, we have

$$\mu_a = \mu \sin \phi \sec \delta \qquad \ldots\ldots(10).$$

As μ is supposed to be expressed in seconds of arc per annum, μ_a is also given by (10) in terms of the same units. Similarly, $SU = \delta_1 - \delta = \mu_\delta$, so that

$$\mu_\delta = \mu \cos \phi \qquad \ldots\ldots(11),$$

μ_δ being expressed in seconds of arc per annum. If μ_α, μ_δ are known, the formulae (10) and (11) are sufficient to enable the values of μ and ϕ to be calculated.

The values of μ_α and μ_δ for the brighter stars (down to about the ninth magnitude) are obtained from meridian observations separated by a long interval of time. We shall illustrate the procedure by means of an example. In the *A.G. Catalogue* the mean co-ordinates of Arcturus for 1875·0 are

$$14^{\text{h}}\ 9^{\text{m}}\ 57^{\text{s}}{\cdot}63;\ +19°\ 50'\ 2''{\cdot}6.$$

These mean co-ordinates are derived from several meridian observations of the star, which, for simplicity, we shall suppose to have been made round about the beginning of the year 1875. After the instrumental and refraction corrections have been applied, each meridian observation yields the apparent co-ordinates referred to the true equinox of date; after the application of the formulae for precession, nutation and aberration, the mean co-ordinates for the epoch 1875·0 are obtained. This is done for each of the observations; when the average is taken for the several observations, the results, already quoted, are the mean co-ordinates of the star for the beginning of 1875 referred to the mean equator of 1875·0.

A similar set of observations round about the beginning of 1925 give the mean co-ordinates of the star at the beginning of 1925 referred to the mean equator of 1925·0. With our assumptions the interval between the two sets of observations is 50 years. Now if the star had no proper motion, the mean co-ordinates for 1925·0 ought to be the mean co-ordinates for 1875·0 *plus* the amount of precession in the interval. To find the actual amount of proper motion in right ascension and declination, we first derive the star's position at the beginning of 1875 referred to the mean equator of 1925·0. This involves the application of precession between 1875 and 1925—call this interval t. From the *A.G. Catalogue* we extract the following for Arcturus:

	R.A.	Dec.
Annual Precession	$+2^{\text{s}}{\cdot}8132$;	$-16''{\cdot}915$
[Chapter x: formulae (29), (30)]		
Secular variation (s and s_1)	$+0^{\text{s}}{\cdot}0003$;	$+0''{\cdot}228$

Using formulae (32) and (34) of Chapter x we have, for $t = 50$,

$$\frac{st}{200} = + 0^s \cdot 0001; \quad \frac{s_1 t}{200} = + 0'' \cdot 057;$$

which, added to the annual precessions above, give respectively $+ 2^s \cdot 8133$ and $- 16'' \cdot 858$. Hence the total changes in the co-ordinates due to precession are respectively

$$+ 140^s \cdot 66; \quad - 842'' \cdot 9,$$

or $\quad\quad\quad\quad + 2^m 20^s \cdot 66; \quad - 14' 2'' \cdot 9.$

Applying these to the mean co-ordinates for 1875·0, we obtain the mean co-ordinates of Arcturus, *as observed at the beginning of* 1875, referred to the mean equator of 1925·0; the results are

$$14^h 12^m 18^s \cdot 29; \quad + 19° 35' 59'' \cdot 7 \quad\quad ......(12).$$

We now consider the second set of observations made at the beginning of 1925. The mean co-ordinates of Arcturus, as observed in 1925 and referred to the mean equator of 1925·0, are given as follows:

$$14^h 12^m 14^s \cdot 39; \quad + 19° 34' 19'' \cdot 9 \quad\quad ......(13).$$

The co-ordinates given by (12) and (13) are all referred to the *same* co-ordinate system, namely, the mean equator and equinox of 1925·0. The differences between the two right ascensions and the two declinations in (12) and (13) are thus due to the proper motion of the star in 50 years. The R.A. has decreased by $3^s \cdot 90$ in 50 years and the declination has decreased by $1' 39'' \cdot 8$. We then have $\quad\quad \mu_a = - 0^s \cdot 078; \quad \mu_\delta = - 2'' \cdot 00.$

These are the components of the proper motion of Arcturus, with reference to the mean equator and equinox of 1925·0.

For methods of combining a large number of such determinations and for a discussion of the systematic errors of the various catalogues, the reader is referred to Newcomb's *Spherical Astronomy.*

For stars fainter than the ninth magnitude the determination of proper motions is undertaken with the photographic telescope; the method is described later in Chapter XII.

148. *The components of proper motion at different epochs referred to the same equatorial system.*

Let S (Fig. 99) denote the mean position of a star in 1850, say, referred to the mean equator and equinox of 1850·0. To avoid confusion, we shall at first express μ in circular measure. Owing to proper motion the star will move along the great circle SAC at the rate of μ radians per annum. We shall assume that μ is constant—any changes in μ are, even in the most likely instances, so small as to be wholly negligible for very many years to come. After an interval of t years let the star's position on the celestial sphere be A.

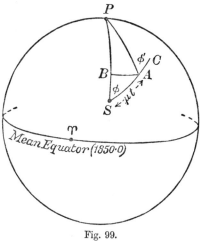

Fig. 99.

Let us write μ_a, μ_δ for the components of proper motion when the star is at S, that is, in 1850. Then by (10) and (11), μ_a and μ_δ being expressed in radians per annum, we have

$$\mu_a = \mu \sin \phi \sec \delta; \quad \mu_\delta = \mu \cos \phi \quad \ldots\ldots(14),$$

the co-ordinates of S being (α, δ) and ϕ the angle PSA.

Let us write μ_a', μ_δ' for the components of proper motion when the star is at A, that is, in $1850 + t$. Then if (α', δ') are the co-ordinates of A (referred to the mean equator of 1850·0) and ϕ' is the angle PAC, we have similarly

$$\mu_a' = \mu \sin \phi' \sec \delta'; \quad \mu_\delta' = \mu \cos \phi' \quad \ldots\ldots(15).$$

In these formulae we neglect the influence of radial velocity—for the latter's effect see Exercise 7 below. It is thus clear from (14) and (15) that the components of proper motion alter with regard to the epoch for which they are defined, even although the fundamental plane of reference (the mean equator of 1850·0) is the same and the total proper motion μ is constant.

Through A draw the small circle arc AB parallel to the equator. In the triangle PSA, $PS = 90° - \delta$, $PA = 90° - \delta'$, $S\hat{P}A = \alpha' - \alpha$, $P\hat{S}A = \phi$, $P\hat{A}S = 180° - \phi'$ and $SA = \mu t$. By the sine-formula B we have

$$\sin \phi' \cos \delta' = \sin \phi \cos \delta.$$

Writing $\phi + \Delta\phi$ for ϕ' and $\delta + \Delta\delta$ for δ' we obtain (regarding $\Delta\phi$, $\Delta\delta$ as small angles, for SA ($\equiv \mu t$) is small even for considerable intervals of time)

$$(\sin\phi + \Delta\phi\cos\phi)(\cos\delta - \Delta\delta\sin\delta) = \sin\phi\cos\delta,$$

from which, neglecting infinitesimals of the second order, we have

$$\Delta\phi\cos\phi\cos\delta = \Delta\delta\sin\phi\sin\delta \qquad \ldots\ldots(16).$$

This equation gives the variation of ϕ with δ.

In the infinitesimal triangle SAB, $SB = \delta' - \delta = \Delta\delta$, and therefore

$$\Delta\delta = \mu t\cos\phi \qquad \ldots\ldots(17).$$

Hence from (16) and (17),

$$\Delta\phi = \mu t\sin\phi\tan\delta \qquad \ldots\ldots(18).$$

From (15) we have

$$\mu_a' = \frac{\mu(\sin\phi + \Delta\phi\cos\phi)}{\cos\delta - \Delta\delta\sin\delta}$$

$$= \mu\sec\delta\,(\sin\phi + \Delta\phi\cos\phi)(1 + \Delta\delta\tan\delta),$$

or

$$\mu_a' = \mu\sin\phi\sec\delta + \mu\Delta\phi\cos\phi\sec\delta + \mu\Delta\delta\sin\phi\sec\delta\tan\delta,$$

again neglecting small quantities of the second order. Hence using (14), (17) and (18) we obtain

$$\mu_a' - \mu_a = 2\mu^2 t\sin\phi\cos\phi\sec\delta\tan\delta,$$

and by (14)
$$\mu_a' - \mu_a = 2t\,\mu_a\,\mu_\delta\tan\delta \qquad \ldots\ldots(19).$$

This equation has been derived on the supposition that μ, μ_a, μ_a' and μ_δ are all expressed in circular measure. Express μ_a, μ_a' in seconds of time and μ_δ in seconds of arc; then (19) is written

$$\mu_a' - \mu_a = 2t\,\mu_a\,\mu_\delta\tan\delta\sin 1'' \qquad \ldots\ldots(20).$$

In a similar way, we have from (15), the proper motions being expressed in circular measure,

$$\mu_\delta' = \mu(\cos\phi - \Delta\phi\sin\phi),$$

or
$$\mu_\delta' - \mu_\delta = -\mu\Delta\phi\sin\phi$$

$$= -\mu^2 t\sin^2\phi\tan\delta \qquad \text{[by (18)]}$$

$$= -t\mu_a^2\sin\delta\cos\delta \qquad \text{[by (14)]}.$$

Expressing μ_a, $(\mu_\delta' - \mu_\delta)$ in seconds of time and arc respectively, we obtain
$$\mu_\delta' - \mu_\delta = -t\,(15\mu_a)^2\sin\delta\cos\delta\sin 1'' \qquad \ldots\ldots(21).$$

Equations (20) and (21) thus give the components of proper motion at the epoch $(1850 + t)$ when the components for epoch 1850·0 are known; and *vice versa*. It is only, however, when the components of proper motion are large that the quantities on the right of (20) and (21) need be taken into account.

Example. The mean co-ordinates of the star Groombridge 1830 for 1900·0 are $11^h\ 47^m\ 13^s\cdot0,\ +\ 38°\ 26'\ 10''$, and the components of the proper motion for 1900 are

$$\mu_a = +\ 0^s\cdot3405,\quad \mu_\delta = -\ 5''\cdot801.$$

To find the components of proper motion μ_a', μ_δ' for this star in the year 2000, referred to the mean equator of 1900·0.

Apply (20) in which $t = 100$, $\tan \delta = 0\cdot793$ and $\sin 1'' = 1/206265$. Then

$$\mu_a' - \mu_a = -\ 0^s\cdot0016,$$

so that μ_a' is $+\ 0^s\cdot3389$.

Similarly from (21), $\mu_\delta' = -\ 5''\cdot801 - 0''\cdot006$ or $-\ 5''\cdot807$.

149. *The components of proper motion referred to the mean equators of two different epochs.*

In Fig. 100 let S be the position of a star in 1900·0, referred to the mean equator and equinox of that year. Owing to luni-solar precession the pole P will move to Q after t years, describing a

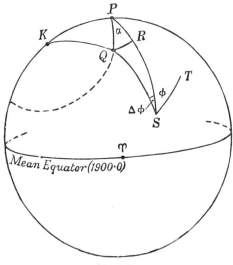

Fig. 100.

small circle arc around K, the pole of the ecliptic. The angle PKQ is the luni-solar precession θt in t years, the value of θ as we have seen in Chapter x being about $50\frac{1}{4}''$. As in the previous section we shall at first express θ and the proper motions in circular measure. We regard $P\hat{K}Q$ as a small angle, so that the great circle arc joining P and Q will be indistinguishable from the small circle arc along which the precessional motion of the pole takes place. Let (α, δ) be the co-ordinates of S referred to the mean equator and equinox of 1900·0 and (α_1, δ_1) the co-ordinates of S referred to the mean equator and equinox of $1900 + t$. Then $PS = 90° - \delta$ and $QS = 90° - \delta_1$. Also $K\hat{P}S = 90° + \alpha$, so that $Q\hat{P}S = \alpha$.

Let the proper motion μ of the star be directed along the great circle ST. Then if μ_a, μ_δ are the components of proper motion of S *in 1900* referred to the mean equator of 1900·0 we have, by (10) and (11),
$$\mu_a = \mu \sin \phi \sec \delta; \quad \mu_\delta = \mu \cos \phi \quad \text{......(22)},$$
where ϕ is the angle PST.

Let μ_a'', μ_δ'' denote the components of proper motion of the star, *also in 1900*, referred to the mean equator and pole Q for $1900 + t$. Then
$$\mu_a'' = \mu \sin \phi_1 \sec \delta_1; \quad \mu_\delta'' = \mu \cos \phi_1 \quad \text{......(23)},$$
where ϕ_1 is the angle QST. We shall write $\phi_1 = \phi + \Delta\phi$, so that $P\hat{S}Q = \Delta\phi$. Then by formula **B** we have, from the triangle PQS,
$$\sin \Delta\phi = \sin PQ \sin \alpha \sec \delta_1,$$
or, since PQ and $\Delta\phi$ are supposed to be small and $PQ = \theta t \sin \epsilon$,
$$\Delta\phi = \theta t \sin \alpha \sec \delta \sin \epsilon \quad \text{......(24)},$$
in which we have written δ for δ_1 without sensible loss of accuracy.

Draw the small circle arc QR perpendicular to QS. Then we have, very nearly, $PR = \delta_1 - \delta$. Writing $\Delta\delta$ for $\delta_1 - \delta$, we obtain from the infinitesimal triangle PQR,
$$\delta_1 - \delta \equiv \Delta\delta = \theta t \cos \alpha \sin \epsilon \quad \text{......(25)}.$$
From the first of (23), we have
$$\mu_a'' = \frac{\mu \sin (\phi + \Delta\phi)}{\cos (\delta + \Delta\delta)}$$
$$= \frac{\mu (\sin \phi + \Delta\phi \cos \phi)}{\cos \delta (1 - \Delta\delta \tan \delta)}$$
$$= \mu \sec \delta (\sin \phi + \Delta\phi \cos \phi) (1 + \Delta\delta \tan \delta),$$

or, neglecting small quantities of the second order,

$$\mu_a{}'' = \mu \sec \delta \sin \phi + \mu \Delta\delta \sec \delta \sin \phi \tan \delta + \mu \Delta\phi \sec \delta \cos \phi,$$

or, using (22), (24) and (25),

$$\mu_a{}'' - \mu_a = \mu \theta t \sin \epsilon \sec \delta \; (\cos \alpha \tan \delta \sin \phi + \sin \alpha \sec \delta \cos \phi).$$

But $\qquad \mu_a = \mu \sin \phi \sec \delta$ and $\mu_\delta = \mu \cos \phi.$

Hence $\quad \mu_a{}'' - \mu_a = \theta t \sin \epsilon \; (\mu_a \cos \alpha \tan \delta + \mu_\delta \sin \alpha \sec^2\delta)$ (26).

Similarly, $\mu_\delta{}'' = \mu \; (\cos \phi - \Delta\phi \sin \phi)$

$$= \mu \cos \phi - \mu \theta t \sin \epsilon \sin \alpha \sec \delta \sin \phi,$$

so that $\mu_\delta{}'' - \mu_\delta = - \theta t \mu_a \sin \alpha \sin \epsilon.$ $\qquad$ ......(27).

In the derivation of the formulae (26) and (27), μ_a, μ_δ and θ are all supposed to be expressed in circular measure. Express μ_a, $\mu_a{}''$ in seconds of time and μ_δ, $\mu_\delta{}''$ and θ in seconds of arc, and these equations become

$$\mu_a{}'' - \mu_a = \theta t \sin \epsilon \; (\mu_a \cos \alpha \tan \delta + \tfrac{1}{15}\mu_\delta \sin \alpha \sec^2 \delta) \sin 1''$$
$$......(28),$$

$$\mu_\delta{}'' - \mu_\delta = - 15\theta t \mu_a \sin \alpha \sin \epsilon \sin 1'' \qquad ......(29).$$

Example. To find the value of $\mu_a{}''$ for the motion of the star Groombridge 1830 in 1900 referred to the mean equator of 1930·0, given that $\mu_a = + 0^{s}\!\cdot\!3405$, $\mu_\delta = - 5''\!\cdot\!801$ (referred to the mean equator for 1900·0). Here $t = 30$, $\alpha = 11^{h} 47^{m} 13^{s}$, $\delta = 38° 26' 10''$, $\cos \alpha = - 0\!\cdot\!998$, $\sin \alpha = 0\!\cdot\!056$, $\tan \delta = 0\!\cdot\!793$, $\sec^2 \delta = 1\!\cdot\!63$, $\theta = 50\tfrac{1}{4}$, $\sin \epsilon = 0\!\cdot\!398$. Hence, from (28),

$$\mu_a{}'' - \mu_a = - 0^{s}\!\cdot\!0009,$$
so that $\qquad\qquad \mu_a{}'' \qquad = + 0^{s}\!\cdot\!3396.$

The computation for the value of $\mu_\delta{}''$ can be performed in a similar manner by means of (29).

The principal formulae of this and the preceding section enable us to derive the components of proper motion of a star at any epoch referred to the fundamental co-ordinate system at that or a different epoch. These formulae need only be applied when the components of proper motion are very large.

150. *Mean and apparent co-ordinates.*

In the chapter on precession and nutation we omitted the consideration of proper motion, and we now revise the fundamental definitions, taking proper motion into account. Suppose we start with the mean co-ordinates (α_0, δ_0) of a star for 1900·0. The mean co-ordinates (α, δ) for $1900 + t$ are then given by applying the precession for t years with the secular variation, together with the proper motion for t years. The annual precession plus the annual proper motion is called the *annual variation*, and this quantity is generally tabulated in the catalogues for each star. The mean R.A. and declination for $1900 + t$ are then obtained by means of formulae (32) and (34) of Chapter X, in which $\dfrac{d\alpha_0}{dt}$ and $\dfrac{d\delta_0}{dt}$ are now taken to mean the annual variation in right ascension and declination respectively.

The *apparent place* of a star, say, on 1931 March 9, is its position on this date on the celestial sphere (with the earth as centre) and referred to the actual equator and equinox of date. The apparent place on the date is obtained from the mean place at the beginning of the year by applying precession, nutation, aberration (by means of the Besselian or independent day numbers) and the proper motion for the fraction of the year in question.

151. *The solar motion and parallactic motion.*

We have seen that the proper motion of a star is its rate of change of direction as viewed from the sun. It has been convenient to describe the cause of the star's proper motion as the star's linear space-velocity relative to the sun, but the question might well be asked: "Is it not possible that all the stars are 'fixed in space' and that their proper motions are the results of the sun's motion in some definite direction?" In this section we examine the problem thus suggested.

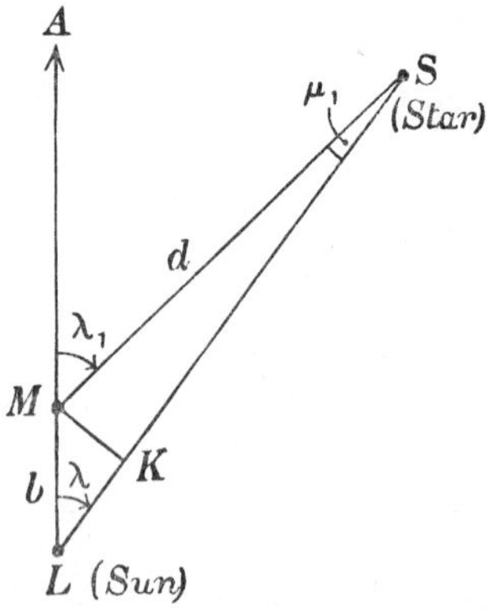

In Fig. 101 consider a star S at rest with respect to certain hypothetical axes in space. Suppose the sun moves with respect to these axes along the straight line LA, traversing the distance LM in one year. If U is its velocity (in kilometres per second) and n the number of seconds in one year,

Fig. 101.

then the distance LM—which we denote by b—is given in kilometres by
$$b = nU \qquad \qquad \dots\dots(30).$$

Let λ denote the angle ALS—it is the angle between the direction of the solar motion and the direction in which the star is seen from L. Let λ_1 denote the angle AMS. Further let $\lambda_1 - \lambda = \mu_1$, so that $M\hat{S}L = \mu_1$. Then μ_1 is the change in the direction of the star, as seen from the sun, in one year. This appears in the observations as proper motion.

If d is the distance MS in kilometres, we have
$$\frac{\sin \mu_1}{\sin \lambda} = \frac{b}{d},$$

or, since μ_1 is a very small angle, we can write
$$\mu_1 = \frac{b}{d} \sin \lambda \operatorname{cosec} 1'' \qquad \dots\dots(31),$$

in which μ_1 is expressed in seconds of arc.

We can transform (31) by introducing the annual parallax Π of the star by means of
$$\Pi = \frac{a}{d} \operatorname{cosec} 1'' \qquad \dots\dots(32),$$

where a is the radius of the earth's orbit around the sun in kilometres. From (30), (31) and (32) we derive
$$\mu_1 = \frac{n}{a} \Pi U \sin \lambda.$$

But $n = 31{\cdot}56 \times 10^6$ and $a = 149{\cdot}7 \times 10^6$ kms., hence
$$\mu_1 = \frac{\Pi U \sin \lambda}{4{\cdot}74} \qquad \dots\dots(33).$$

It is to be noted that formula (33) provides a method of deriving the annual parallax Π of a star, provided that the solar speed U and its direction are known and provided also that the assumption regarding the "fixity" of the stars still holds.

The proper motion μ_1 which arises from the solar motion is called the *parallactic motion* of the star. It varies directly as the parallax Π and, consequently, by (32) it varies inversely as the distance of the star from the sun.

The point on the celestial sphere towards which the solar motion is directed is called the *solar apex.*

In Fig. 102 let A be the solar apex and S a star. We shall show later how the position of A is determined; meanwhile we shall assume that its co-ordinates are known. The great circle arc AS defines the angle between the direction in which the sun is moving and the direction of the star. Thus AS is λ. Also, referring to Fig. 101, we see that the change in direction, or μ_1, takes place in the plane ALS, so that the star appears to move on the celestial sphere along the great circle AST (Fig. 102). As λ_1 is greater than λ, the star appears to move away from A, that is, in the direction ST.

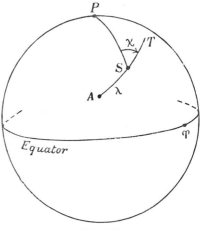

Fig. 102.

Denote by χ the position angle PST; χ is measured from $0°$ to $360°$ in the direction of the arrow. Then since the parallactic motion is μ_1 along ST we have, applying (10) and (11) and denoting the components of the parallactic motion in R.A. and declination by P_a, P_δ respectively,

$$P_a = \mu_1 \sin \chi \sec \delta; \quad P_\delta = \mu_1 \cos \chi \quad \ldots\ldots(34).$$

Inserting the expression for μ_1 given by (33), we obtain

$$P_a = \frac{\Pi U}{4\cdot74} \sin \lambda \sin \chi \sec \delta \quad \ldots\ldots(35),$$

and

$$P_\delta = \frac{\Pi U}{4\cdot74} \sin \lambda \cos \chi \quad \ldots\ldots(36).$$

These formulae are of importance when it is desired to find the components of the parallactic motion of a star, or of a group of stars, when the quantities on the right of (35) and (36) are known.

152. *Secular parallax.*

It is evident that if the distance LM (Fig. 101) through which the sun moves in one year is known, then this line forms a convenient base-line for the measurement of stellar distances.

By analogy with the annual parallax Π the *secular parallax* H, as it is called, is defined by

$$\sin H = \frac{b}{d},$$

or, since H is small, by

$$H = \frac{b}{d} \operatorname{cosec} 1''.$$

Also, from (30), $b = nU$, so that

$$H = \frac{nU}{d} \operatorname{cosec} 1'' \qquad \ldots\ldots(37),$$

in which H is expressed in seconds of arc. Hence, from (32) and (37),

$$H = \frac{\Pi U}{4\cdot74} \qquad \ldots\ldots(38).$$

The measurement of U will be discussed later; meanwhile we state its value, which is 19·5 kms. per second. Thus we obtain the numerical relation between the annual parallax Π and the secular parallax H in the form:

$$\Pi = 0\cdot243\,H \qquad \ldots\ldots(39).$$

By means of (38) the formulae (35) and (36) take on a simpler form: they become

$$P_a = H \sin \lambda \sin \chi \sec \delta \qquad \ldots\ldots(40),$$

$$P_\delta = H \sin \lambda \cos \chi \qquad \ldots\ldots(41).$$

153. *The solar motion and radial velocity.*

In Fig. 101 let MK be drawn perpendicularly to LS. Now LM is very small in comparison with LS and we may consider MS and SK equal. Thus in one year, as the sun moves from L to M, the distance of the star decreases from LS to MS, that is, by the distance LK. This is equivalent to the effect of a negative radial velocity ρ_1 (in kilometres per second). Hence $LK = n\rho_1$, where n is the number of seconds in one year. Also $LK = b \cos \lambda$ and, since $b = nU$ by (30), we obtain

$$\rho_1 = - U \cos \lambda \qquad \ldots\ldots(42),$$

in which the minus sign has been inserted, since the radial velocity is negative. Formula (42) gives the radial velocity of the star arising from the solar motion.

154. *The solar motion in the general case.*

In section 151 we started from the hypothesis that the stars are "fixed in space" and that only the sun was in motion with respect to fixed axes. This assumption can be examined by means of the observed proper motions. Referring to (35) and (36) or (40) and (41), we see that there are two points on the celestial sphere—the solar apex and the diametrically opposite point, the *ant-apex*—where the parallactic motion vanishes, for then λ is $0°$ or $180°$, so that both P_a and P_δ vanish. Thus there ought to be two regions of the sky in which the observed proper motions of all the stars concerned should be zero. This is contrary to the facts. More generally, if the hypothesis is correct, the total proper motion of any star ought to be in a great circle passing through a definite point A (Fig. 102)—the solar apex—so that by considering only two stars in different parts of the sky it should be possible to determine the apex as one of the two points of intersection of the respective great circles. Actually it is found from the proper motions that the points of intersection derived from one pair of stars differ widely from the points of intersection derived from any other pair of stars. The conclusion is that the stars have individual motions of their own in addition to the parallactic motion due to the sun's velocity.

Under these conditions the solar motion must be given a more precise interpretation. We have no longer fixed points in space to which the motion of the sun can be referred, for all the stars as well as the sun are in motion. If we are considering the proper motions of 10,000 stars scattered over the sky, the solar motion is defined with reference to all these stars regarded as a group or, more precisely, as the velocity relative to the mean centre of the group.

Consider Fig. 103, in which S is a star and A is the solar apex whose position we shall assume to be known. Let the proper motion μ of the star be along the great circle SU, and let $P\hat{S}U$ be denoted by ϕ. Both μ and ϕ can be calculated from the observed values of μ_a and μ_δ by means of (10) and (11). The parallactic motion μ_1 is along the great circle AST, from S towards T; the angle PST is χ. Let (a, δ) be the co-ordinates of S and (A, D) the co-ordinates of the apex A. Then, in the triangle PAS,

$$PS = 90° - \delta, \quad PA = 90° - D, \quad A\hat{P}S = a - A, \quad P\hat{S}A = 180° - \chi,$$

$AS = \lambda$. The values of λ and χ can be easily found as follows
By the cosine-formula **A**,

$$\cos \lambda = \sin D \sin \delta + \cos D \cos \delta \cos (\alpha - A) \quad ...(43),$$

from which λ can be computed, and by formula **C**,

$$- \sin \lambda \cos \chi = \sin D \cos \delta - \cos D \sin \delta \cos (\alpha - A) ...(44),$$

from which χ can be obtained.

λ is the angular distance of the star from the apex. The
direction ST is the direction of the ant-apex from S and the
angle χ between the meridian SP and the great circle ST is called

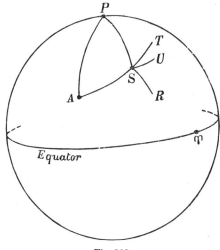

Fig. 103.

the *position angle of the ant-apex*. As the ant-apex is 180° from
A, the angular distance of the star from the ant-apex is $180° - \lambda$.
These angles, $(180° - \lambda)$ and χ, can be found with sufficient
accuracy by means of specially prepared charts,* thus saving
the rather tedious computations by (43) and (44).

Now resolve the observed proper motion μ, which is in the
direction SU (Fig. 103), along the great circle ST and along the

* W. M. Smart: Charts published by the Royal Astronomical Society and described in *Monthly Notices*, vol. LXXXIII, p. 465; J. A. Pearce and S. N. Hill: *Publications of the Dominion Astrophysical Observatory, Victoria, B.C.*, vol. IV, No. 4; J. M.
Baldwin: Chart published by the Royal Astronomical Society and described in
Monthly Notices, vol. LXXXIX, p. 453.

great circle SR drawn at right angles to AST. Let v and τ denote these components. Then, since $T\hat{S}U = \phi - \chi$,

$$v = \mu \cos(\phi - \chi),$$

and $$\tau = \mu \sin(\phi - \chi).$$

Since, by (10) and (11), $\mu_a = \mu \sin \phi \sec \delta$ and $\mu_\delta = \mu \cos \phi$, we can write the formulae for v and τ as follows:

$$v = \mu_a \cos \delta \sin \chi + \mu_\delta \cos \chi \qquad \ldots \ldots (45),$$

$$\tau = \mu_a \cos \delta \cos \chi - \mu_\delta \sin \chi \qquad \ldots \ldots (46).$$

Assuming the co-ordinates of the solar apex to be known, we can obtain the value of χ by formula (44) or from charts, and thus we can calculate the values of v and τ, the values of μ_a and μ_δ being supposed known.

Let us write $$v = \mu_1 + v_1 \qquad \ldots \ldots (47),$$

where μ_1 is the parallactic motion along ST. Then v_1 is the portion of the observed proper motion in the direction ST due to the star's individual motion, μ_1 being due to the solar motion. The component τ is evidently due entirely to the star's individual motion.

155. *Statistical parallaxes.*

The determination of the annual parallax of even a single star involves a large amount of observational and computational work. In many investigations it is of great importance to know the mean parallax of a particular class of stars, say stars between the tenth and eleventh magnitudes in a particular region of the sky, and this mean parallax can be determined from the v and τ components of proper motion. We shall consider the v components of N stars close together in the sky, the stars belonging, for example, to some definite magnitude class.

We shall define v_1 to be positive when it is in the direction ST, and negative when it is in the opposite direction SA (Fig. 103). If the N stars have haphazard individual motions, the expectation will be that there will be as many stars with positive values of v_1 as with negative values, so that the algebraic sum of all the v_1's (denoted by Σv_1) will either vanish or be very small compared with $\Sigma \mu_1$. Adding the N equations of the type (47), we have $$\Sigma v = \Sigma \mu_1 + \Sigma v_1,$$

and, neglecting Σv_1 as the result of the previous remark, we are left with

$$\Sigma \mu_1 = \Sigma v \qquad \ldots\ldots(48).$$

For a given star, v is supposed to be known—it is calculated by means of (45) from known data. Hence for all the stars the sum Σv will be easily found. Now for a star whose annual parallax is Π, μ_1 is given by (33), namely,

$$\mu_1 = \frac{\Pi U \sin \lambda}{4\cdot 74}.$$

Hence
$$\Sigma \mu_1 = \frac{U \sin \lambda}{4\cdot 74} \Sigma \Pi.$$

If Π_0 denotes the mean parallax of the N stars, $N\Pi_0 = \Sigma \Pi$, and therefore we obtain, by means of (48),

$$\Pi_0 = \frac{4\cdot 74}{N U \sin \lambda} \Sigma v \qquad \ldots\ldots(49).$$

As all the quantities on the right of (49) are supposed known, the mean parallax Π_0 is readily derived.

This is one illustration of statistical methods applied in astronomy. In the table in Appendix E (p. 424) we give Kapteyn's values* of the mean secular parallax H of stars of different magnitude classes (Harvard Scale) and in different galactic latitudes (galactic latitudes are defined with reference to the Milky Way regarded as a great circle). The corresponding values of the mean annual parallax are obtained by multiplying the entries in the table by 0·243 [see formula (39)].

153. *Determination of the solar apex from proper motions.*

If we are given that the sun is moving in a certain direction with velocity U, then relatively to the sun each star will appear to have an additional velocity of U in the opposite direction. In Fig. 104 let O (the sun) be the centre of the celestial sphere and let Ox, Oy, Oz be rectangular axes: Ox is drawn through the vernal equinox Υ, Oy through the point B on the equator with right ascension $90°$, and Oz through the pole P. Then if $-X$, $-Y$, $-Z$ (in kilometres per second) denote the components of the solar motion with respect to the axes, then relatively to the sun a star at S will have a velocity whose components are

---

* *Groningen Publications*, No. 29, Table 26.

$+ X, + Y, + Z$. We neglect for the moment the star's individual linear velocity, so that we are simply investigating the effect of parallactic motion. Let the star be at a distance of d kms. from the sun, and for simplicity we shall take the radius of the sphere to be d.

The velocity of the star S has components $+ X$ parallel to OT, $+ Y$ parallel to OB and $+ Z$ parallel to OP. It is our first object to find the components of the star's linear velocity along

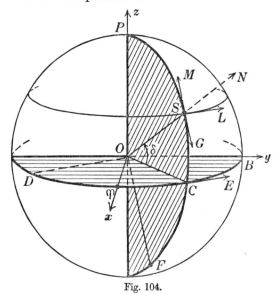

Fig. 104.

the rectangular axes SL, SM and SN, where SL is tangential at S to the parallel of declination (in the sense of increasing R.A.), SM is tangential to the meridian PSC (in the sense of increasing declination) and SN is the radius OS produced.

Now the velocity X parallel to OT is equivalent to a velocity $X \cos \alpha$ parallel to OC (the angle TOC is α) and a velocity $X \sin \alpha$ parallel to OD, where DOC is $90°$. Similarly the velocity Y parallel to OB is equivalent to a velocity $Y \sin \alpha$ parallel to OC and a velocity $- Y \cos \alpha$ parallel to OD. We thus can replace X and Y by

$$X \cos \alpha + Y \sin \alpha \quad \text{parallel to } OC \quad \text{......(50)},$$

and $\qquad X \sin \alpha - Y \cos \alpha \quad \text{parallel to } OD \quad \text{......(51)}.$

Let CE be drawn tangential to the equator at C. Then CE is perpendicular to OC and therefore CE is parallel to OD. Since CE is drawn in the direction opposite to that of OD, we can write for (51),

$$-X \sin \alpha + Y \cos \alpha \quad \text{parallel to } CE \quad \ldots\ldots(52).$$

Also SL is parallel to CE, so that (52) can be replaced by

$$-X \sin \alpha + Y \cos \alpha \quad \text{parallel to } SL \quad \ldots\ldots(53).$$

Let OF be drawn in the plane of the meridian PSC perpendicularly to OS. Then OF is parallel to the tangent MSG at S.

Consider (50). We resolve this velocity parallel to OS and OF. Now $C\hat{O}S = \delta$. Hence $(X \cos \alpha + Y \sin \alpha)$ parallel to OC is equivalent to

$$(X \cos \alpha + Y \sin \alpha) \sin \delta \quad \text{parallel to } OF,$$

or $\quad -(X \cos \alpha + Y \sin \alpha) \sin \delta \quad$ parallel to SM $\ldots\ldots(54)$,

and $\quad (X \cos \alpha + Y \sin \alpha) \cos \delta \quad$ along OS or SN $\ldots(55)$.

Similarly the velocity Z parallel to OP is equivalent to $Z \cos P\hat{O}S$ along OS and $Z \sin P\hat{O}S$ along SM. But $P\hat{O}S = 90° - \delta$, so that Z is equivalent to

$$Z \cos \delta \quad \text{along } SM \qquad \ldots\ldots(56),$$

and $\qquad Z \sin \delta \quad$ along SN $\qquad \ldots\ldots(57)$.

Collecting the components of velocity along SL, SM and SN in turn and denoting the sums by ξ, η and ζ respectively, we have:

from (53), $\qquad \xi = -X \sin \alpha + Y \cos \alpha \qquad \ldots\ldots(58)$;

from (54) and (56),

$$\eta = -X \cos \alpha \sin \delta - Y \sin \alpha \sin \delta + Z \cos \delta \ldots\ldots(59);$$

from (55) and (57),

$$\zeta = X \cos \alpha \cos \delta + Y \sin \alpha \cos \delta + Z \sin \delta \quad \ldots\ldots(60).$$

As X, Y, Z are expressed in kilometres per second, so also are ξ, η, ζ. Now consider the velocity ξ along SL. Due to this velocity alone, the star will move along SL through a distance $n\xi$ kms. in one year, where n is the number of seconds in one year, and therefore, since OS is d kms., the angle through which S moves along the parallel of declination is $n\xi/d$ in circular measure or $\dfrac{n\xi}{d}$ cosec $1''$ in seconds of arc. Hence the right ascension

of the star increases by $\dfrac{n\xi}{d}$ sec δ cosec $1''$ seconds of arc per annum.

But this is the component of the parallactic motion in right ascension, P_a, which we assume to be expressed in seconds of arc. Hence

$$\frac{n\xi}{d} = P_a \cos\delta \sin 1''.$$

Introducing the annual parallax Π by (4) we obtain, after inserting the values of n and a and writing for ξ the expression given by (58),

$$- X \sin\alpha + Y \cos\alpha = \frac{4\cdot74}{\Pi} P_a \cos\delta \quad \ldots\ldots(61).$$

This is an accurate equation for the star S and clearly it holds for any other star. The reader is reminded that the parallactic motion is alone considered at the moment.

In a similar manner, by considering the component η along SM which gives rise to the component of parallactic motion in declination, P_δ, we derive from (59),

$$- X \cos\alpha \sin\delta - Y \sin\alpha \sin\delta + Z \cos\delta = \frac{4\cdot74}{\Pi} P_\delta$$
$$\ldots\ldots(62).$$

Now consider the observed components of proper motion, μ_a and μ_δ. The parallactic motion contributes to μ_a and so does the star's individual motion. Similarly for μ_δ. We can then write

$$\mu_a = P_a + \mu_a'; \quad \mu_\delta = P_\delta + \mu_\delta' \quad \ldots\ldots(63),$$

where μ_a', μ_δ' are the contributions of the star's individual motion to the proper motion in R.A. and declination respectively. All the quantities in (63) are assumed to be expressed in seconds of arc.

By means of (63), equation (61) becomes

$$- X \sin\alpha + Y \cos\alpha = \frac{4\cdot74}{\Pi} \mu_a \cos\delta - \frac{4\cdot74}{\Pi} \mu_a' \cos\delta$$
$$\ldots\ldots(64).$$

In this equation, X, Y are quantities to be found, α and δ are known and μ_a is given from proper motion observations. In general Π is unknown and μ_a' is also unknown.

Consider N stars in a small area of the sky, so that we can regard each star as having the same values of α and δ. Each star contributes an equation of the type (64), and adding the N equations we derive

$$- X \sin \alpha + Y \cos \alpha = \frac{4 \cdot 74}{N} \cos \delta \, \Sigma \left(\frac{\mu_a}{\Pi} \right) - \frac{4 \cdot 74}{N} \cos \delta \, \Sigma \left(\frac{\mu_a'}{\Pi} \right)$$

$$\dots \dots (65).$$

If Π_0 corresponds to the mean distance d_0 of the stars, we can write (65) as

$$- X \sin \alpha + Y \cos \alpha = \frac{4 \cdot 74}{N \Pi_0} \cos \delta \, \Sigma \mu_a - \frac{4 \cdot 74}{N \Pi_0} \cos \delta \, \Sigma \mu_a'$$

$$\dots \dots (66),$$

in which the assumption is now made that the stars have each the same parallax Π_0.

Now consider the last term of (66). μ_a' is the component of proper motion due to the star's individual velocity, and it is as likely to be positive as negative. We can either assume that $\Sigma \mu_a'$ vanishes or that, being small, it is of the nature of an accidental error in the measurement of the proper motions. Neglecting the last term we have

$$- X \sin \alpha + Y \cos \alpha = \frac{4 \cdot 74}{N \Pi_0} \cos \delta \, \Sigma \mu_a \quad \dots \dots (67),$$

in which Π_0 is unknown and $\Sigma \mu_a$ is obtained from the observed proper motions. Write $\frac{1}{N} \Sigma \mu_a = \bar{\mu}_a$, so that $\bar{\mu}_a$ is the average observed value of μ_a for the N stars.

Suppose that we deal with a large number of regions in the same way, and for simplicity assume that there are N stars in each region and that the mean distance of the stars in each region is the same, so that Π_0 is constant. Then we have for each region an equation of the type

$$- X \sin \alpha + Y \cos \alpha = K \bar{\mu}_a \cos \delta \quad \dots \dots (68),$$

where $K \equiv 4 \cdot 74 / \Pi_0$ and is regarded as an unknown constant.

In a similar manner and with the same assumptions we derive from (62),

$$- X \cos \alpha \sin \delta - Y \sin \alpha \sin \delta + Z \cos \delta = K \bar{\mu}_\delta \dots (69).$$

If there are M regions, there are M equations of the type (68) and M of the type (69). The solution of these equations is effected by the method of least squares. The procedure is as follows. Multiply (68) throughout by $-\sin\alpha$ (the coefficient of X) for each of the M equations and add together all the resulting equations. Denoting summation by Σ we obtain

$$X\Sigma\sin^2\alpha - Y\Sigma\sin\alpha\cos\alpha = -K\Sigma\bar{\mu}_a\sin\alpha\cos\delta \dots(70).$$

Similarly by multiplying by $\cos\alpha$ (the coefficient of Y) and summing we derive

$$-X\Sigma\sin\alpha\cos\alpha + Y\Sigma\cos^2\alpha = K\Sigma\bar{\mu}_a\cos\alpha\cos\delta \dots(71).$$

Treating (69) in a similar manner, we obtain the three equations:

$$X\Sigma\cos^2\alpha\sin^2\delta + Y\Sigma\sin\alpha\cos\alpha\sin^2\delta - Z\Sigma\cos\alpha\sin\delta\cos\delta$$
$$= -K\Sigma\bar{\mu}_\delta\cos\alpha\sin\delta \quad\dots\dots(72),$$

$$X\Sigma\sin\alpha\cos\alpha\sin^2\delta + Y\Sigma\sin^2\alpha\sin^2\delta - Z\Sigma\sin\alpha\sin\delta\cos\delta$$
$$= -K\Sigma\bar{\mu}_\delta\sin\alpha\sin\delta \quad\dots\dots(73),$$

$$-X\Sigma\cos\alpha\sin\delta\cos\delta - Y\Sigma\sin\alpha\sin\delta\cos\delta + Z\Sigma\cos^2\delta$$
$$= K\Sigma\bar{\mu}_\delta\cos\delta \quad\dots\dots(74).$$

In the five equations, (70) to (74), the coefficients of X, Y and Z can all be calculated from the known values of α and δ, and also the coefficients of K on the right-hand sides of these equations, since the values of $\bar{\mu}_a$, $\bar{\mu}_\delta$ are obtainable from the observed proper motions. Adding (70) and (72), adding (71) and (73) and rewriting (74), we have the three equations in symbolical form

$$\begin{aligned}
aX + bY + cZ &= KS_1 \\
bX + BY + dZ &= KS_2 \\
cX + dY + CZ &= KS_3
\end{aligned} \right\} \quad\dots\dots(75),$$

in which the coefficients of X, Y and Z are all calculable functions of α and δ and S_1, S_2, S_3 are also known. The solution of these equations (75) determines

$$\frac{X}{K}, \quad \frac{Y}{K}, \quad \frac{Z}{K},$$

but *not* X, Y and Z, since K is unknown.

We can now derive the co-ordinates of the solar ant-apex towards which the sun is moving with a velocity whose components are $+X$, $+Y$, $+Z$. (We assumed originally that the components of the solar motion are $-X$, $-Y$, $-Z$, and as this motion is directed towards the apex the velocity whose components are $+X, +Y, +Z$ will be directed towards the opposite point on the celestial sphere, that is, the ant-apex.)

In Fig. 105 let J be the ant-apex and take rectangular axes as shown. Let PJC be the meridian through J and let A, D be the equatorial co-ordinates of J. Then $\widehat{TOC} = A$ and $\widehat{JOC} = D$. The components X and Y combine to give a velocity $(X^2 + Y^2)^{\frac{1}{2}}$

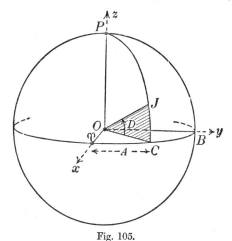

Fig. 105.

along OC and A is given by $\tan A = Y/X$. Also $(X^2 + Y^2)^{\frac{1}{2}}$ along OC combines with Z along OP to give the reverse of the solar motion in the direction OJ, so that

$$\tan D = Z/(X^2 + Y^2)^{\frac{1}{2}}.$$

We rewrite these equations in $\tan A$ and $\tan D$ as follows:

$$\tan A = \frac{Y}{K} \Big/ \frac{X}{K}; \quad \tan D = \frac{Z}{K} \Big/ \left[\left(\frac{X}{K}\right)^2 + \left(\frac{Y}{K}\right)^2\right]^{\frac{1}{2}} \dots(76).$$

The ratios $\dfrac{X}{K}, \dfrac{Y}{K}, \dfrac{Z}{K}$ are known from the solution of (75); hence A and D are determined. The co-ordinates of the solar apex are then $180° + A$ and $-D$.

Numerous investigations of the solar apex from the proper motions of stars have been made since the time of Sir William Herschel. Recent work seems to indicate that so far as the brighter stars are concerned, the co-ordinates of the solar apex are:

R.A. 18^{h};   Dec. $+34°$.

For fainter stars, the declination of the apex is generally found to be a few degrees further north.

The method of determining the solar apex which we have just described is essentially due to Sir George Airy and is generally known as Airy's method.

157. *Determination of the solar motion from radial velocities.*

We now consider formula (60), which gives the component velocity ζ in the line of sight OS (Fig. 104) due to the solar velocity. Thus ζ is simply $-\rho_1$ of section 153, and by (42) we can write
$$\zeta = U \cos \lambda \qquad \dots\dots(77),$$

where U is the solar velocity. Let ρ now denote the observed radial velocity.* Then the solar motion contributes $U \cos \lambda$ to ρ and the star's individual space-velocity contributes ρ_2, say. Hence
$$\rho = U \cos \lambda + \rho_2 \qquad \dots\dots(78).$$

Hence, from (60), (77) and (78), we clearly have
$$X \cos \alpha \cos \delta + Y \sin \alpha \cos \delta + Z \sin \delta = \rho - \rho_2 \quad \dots\dots(79),$$

in which ρ is given by spectroscopic observations and ρ_2 is unknown. If the radial velocities of N stars, scattered over the sky, are observed, each star contributes an equation of the form (79) and the N equations are solved by the method of least squares. Multiplying each equation in turn by the coefficient of X and summing we obtain
$$X \, \Sigma \cos^2 \alpha \cos^2 \delta + Y \, \Sigma \sin \alpha \cos \alpha \cos^2 \delta + Z \, \Sigma \cos \alpha \sin \delta \cos \delta$$
$$= \Sigma \rho \cos \alpha \cos \delta - \Sigma \rho_2 \cos \alpha \cos \delta \quad \dots\dots(80).$$

Now the radial component ρ_2 of the star's individual linear velocity is as likely to be positive as negative, and the sum $\Sigma \rho_2 \cos \alpha \cos \delta$ will tend to vanish. Neglecting this term in (80) we write
$$X \, \Sigma \cos^2 \alpha \cos^2 \delta + Y \, \Sigma \sin \alpha \cos \alpha \cos^2 \delta + Z \, \Sigma \cos \alpha \sin \delta \cos \delta$$
$$= \Sigma \rho \cos \alpha \cos \delta \quad \dots\dots(81).$$

---

* The *observed radial velocity* with which we are concerned here is the radial velocity relative to the sun. But the observation of a star's radial velocity made on the earth includes an effect due to the earth's orbital motion round the sun and an effect due to the earth's rotation (pp. 213–217). We suppose that these effects are removed from the actual observations of radial velocity.

Similarly,

$$X \Sigma \sin \alpha \cos \alpha \cos^2 \delta + Y \Sigma \sin^2 \alpha \cos^2 \delta + Z \Sigma \sin \alpha \sin \delta \cos \delta$$
$$= \Sigma \rho \sin \alpha \cos \delta \quad \ldots\ldots(82),$$

$$X \Sigma \cos \alpha \sin \delta \cos \delta + Y \Sigma \sin \alpha \sin \delta \cos \delta + Z \Sigma \sin^2 \delta$$
$$= \Sigma \rho \sin \delta \quad \ldots\ldots(83).$$

The coefficients of X, Y, Z in these equations can be computed and also the quantities on the right-hand sides. Thus we have three linear equations from which X, Y and Z are derived. It is to be remarked that X, Y and Z are given from this solution in kilometres per second.

The co-ordinates of the solar ant-apex A and D are then given by (76) as in the preceding section.

The solar velocity U is given by

$$U^2 = X^2 + Y^2 + Z^2 \quad \ldots\ldots(84).$$

The most recent determination* of the solar motion gives the following results:

Right Ascension of the Solar Apex   18ʰ or 270°,

Declination of the Solar Apex   ...   + 29°,

The solar velocity ...   .   19·6 kms. per second.

The declination is rather lower than the value derived from proper motions; this may be partly due to the fact that the same stars are not as a rule observed in the two different kinds of investigation, and partly to the special assumption regarding the parallaxes of the stars made in Airy's method.

EXERCISES

1. The parallax of Lalande 21258 is $0''\cdot177$ and the total annual proper motion is $4''\cdot52$. Show that the star's velocity at right angles to the line of sight is 121 kms. per second.   [*Lond.* 1928.]

2. If the parallax of a star is $0''\cdot037$ and its total annual proper motion is $0''\cdot52$, show that its tangential linear velocity is 2·24 times the orbital velocity of the earth, the latter's orbit being assumed circular.   [*Lond.* 1925.]

3. The parallax of 61 Cygni is $0''\cdot30$, and its motion perpendicular to the line of sight is $5''\cdot2$ per year. Compare its tangential linear velocity with that of the earth in its motion round the sun.

* W. W. Campbell and J. H. Moore: *Publications of the Lick Observatory*, vol. XVI (1928).

4. The total annual proper motion of a star (R.A. 3^h, declination $+ 10°$) is $0''·1$. Assuming that this proper motion is entirely parallactic, determine the star's parallax, given that the solar motion is 19·5 kms. per second towards the point $(18^h, + 34°)$. [*Lond.* 1927.]

5. A star $(a, δ)$ has a tangential linear velocity, relative to the sun, of V kms. per second in the direction of a point on the celestial sphere whose right ascension is A and declination is D. Prove that

$$\mu_a = 0^s·0141 V \Pi \cos D \sin (A - a) \sec δ,$$

where Π is the star's parallax. [*Lond.* 1926.]

6. The R.A. and declination of Sirius are $6^h 41^m$ and $- 16° 35'$ respectively, and the proper motions in R.A. and declination are $- 0^s·0374$ and $- 1''·209$; the observed radial velocity is $- 7·5$ kms. per second, and the parallax is $0''·38$.

Show that the velocity of Sirius relative to the sun is inclined to the line of sight at an angle of about $114\frac{1}{2}°$, and calculate the magnitude of this velocity in kms. per second. [*Lond.* 1929.]

7. If μ_a, $\mu_δ$ are the components of annual proper motion of a star $(a, δ)$ at time t with reference to a particular equinox and equator, and if μ_a', $\mu_δ'$ are the components of proper motion at time t' with reference to the *same* equinox and equator, prove that, $t' - t$ being expressed in years,

$$\mu_a' - \mu_a = \{2\mu_a \mu_δ \tan δ - 2V\Pi \mu_a/4·74\} (t' - t) \sin 1'',$$
$$\mu_δ' - \mu_δ = - \{225\mu_a{}^2 \sin δ \cos δ + 2V\Pi \mu_δ/4·74\} (t' - t) \sin 1'',$$

Π being the star's parallax in seconds of arc and V its line of sight velocity in kms. per second.

[μ_a is expressed in seconds of time and $\mu_δ$ in seconds of arc.] [*Lond.* 1930.]

8. μ_a, $\mu_δ$ are the components of proper motion, in equatorial co-ordinates, of a star $S (a, δ)$. If (l, b) are the galactic co-ordinates of S and μ_l, μ_b the corresponding components of proper motion, show that

$$\mu_l = \mu_a \cos η \cos δ \sec b + \mu_δ \sin η \sec b,$$
$$\mu_b = - \mu_a \sin η \cos δ + \mu_δ \cos η,$$

where $η$ is the angle PSK, P and K being the north poles of the equator and galactic plane respectively.

[N.B. $η$ is to be reckoned positive if, on going round the triangle SPK in the order named, the area of the triangle is on the left.] [*Lond.* 1926.]

9. The R.A. of the node of the galactic plane on the equator is $Ω$, and i is its inclination to the equator.

$(ξ, η, ζ)$ are the components of velocity (in kms. per sec.) of a star $(a, δ)$ referred to galactic rectangular axes, the direction of the $ξ$-axis being given by the node N.

If $\mu_a, \mu_δ$ are the components of annual proper motion in seconds of time and seconds of arc respectively, and Π is the parallax of the star, show that

$$ξ \sin (a - Ω) - η \cos i \cos (a - Ω) + ζ \sin i \cos (a - Ω) = - \frac{K}{\Pi} \mu_a \cos δ,$$

where K is a certain numerical constant, and derive the corresponding equations in μ_δ and in the radial velocity. What is the value of K in the system of units employed?   [*Lond.* 1926.]

10. If (u, v, w) are the components of the linear velocity V, referred to equatorial axes for a fixed date, of an open cluster, prove for any cluster star (a, δ) with components of annual proper motion μ_a, μ_δ (both in seconds of arc) that

$u \left(\mu_\delta \sin a - \mu_a \cos a \sin \delta \cos \delta \right)$

$$- v \left(\mu_\delta \cos a + \mu_a \sin a \sin \delta \cos \delta \right) + w \mu_a \cos^2 \delta = 0.$$

Hence show how the co-ordinates (A, D) of the convergent point can be derived.   [*Lond.* 1930.]

11. Show that if θ be the angle between the convergent point of a moving cluster and one of its component stars, whose radial velocity is v, the velocity of the cluster relative to the sun is $v \sec \theta$, and the distance of the star is $v \tan \theta \div$ observed proper motion. If $v = 45 \cdot 6$ km./sec., $\theta = 60°$, and the P.M. is $1'' \cdot 8$, find the parallax of the star.   [*Lond.* 1922.]

CHAPTER XII

ASTRONOMICAL PHOTOGRAPHY

158. *The photographic refractor.*

The two principal features of the photographic refractor with which we need concern ourselves in this chapter are, first, the object-glass and, second, the photographic plate (Fig. 106). We

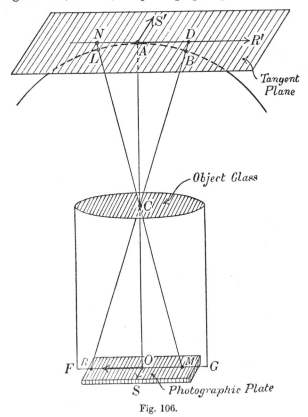

Fig. 106.

shall assume that the object-glass (which is specially designed for photographic purposes) is correctly adjusted in the telescope tube and that CO is the optical axis, C being the centre of the object-glass. The focal plane of the object-glass is FG, at right angles to CO. The photographic plate (which we shall assume to be a square) is held in position by a carrier (not shown in the figure) in such a way that the sensitive side of the plate is exactly in the focal plane. We shall also suppose for simplicity that the optical axis CO passes through the geometrical centre of the plate. Consider a star seen in the direction CA (OC produced). All the rays from the star, falling on the object-glass, will be brought to a focus at O and a small circular image of the star will be formed on the photographic plate at this point. Consider another star seen in the direction CB. The rays falling on the object-glass from this star will be brought to a focus at R (on BC produced). Let the plane ACB or RCO be parallel to an edge of the plate. Suppose that the image of the star B is just on the plate at R. Then if l denotes the length of the side of the plate, $l = 2OR$ and, since $OR = OC \tan RCO = OC \tan ACB$,

$$l = 2OC \tan \beta \qquad \ldots\ldots(1),$$

where β denotes the angular distance between the stars A and B. OC is of course the focal length of the object-glass and is presumed known. If the dimensions of the plate are known we can, by (1), calculate the area of the sky (in angular measure) which can be photographed. For example, the focal length of the Sheepshanks Equatorial of Cambridge Observatory is 589 cms., and the plates used are squares of side 16 cms. (The diameter of the object-glass is 30·5 cms.) Hence, by (1),

$$\tan \beta = \frac{16}{2 \times 589},$$

so that $\beta = 47'$ approximately. Thus a square region of the sky, $1\frac{1}{2}°$ by $1\frac{1}{2}°$, can be photographed on the plates.

Of great importance is the scale of the plate or the relation between a linear distance on the plate and the corresponding angular displacement in the sky. Thus for the plates just referred to, 1 cm. on the plate is equivalent to $5\frac{5}{6}'$, or 1 mm. is equivalent to 35″ approximately. We can express this somewhat differently; if two stars are separated by $5\frac{5}{6}'$ in the sky, the centres of their images on the photographic plate will be 1 cm. apart. Similar considerations apply equally to the reflecting telescope when used for photographic purposes.

159. *The tangent plane.*

In Fig. 106 the celestial sphere, with C as centre, is drawn. The tangent plane at A is drawn—this plane is at right angles to the radius CA and is therefore parallel to the photographic plate. It is to be remembered that A is the point on the celestial sphere towards which the optical axis of the telescope is directed. Produce CB to meet the tangent plane in D; then D will be called the projection of B on the tangent plane. The projection of any other point on the celestial sphere can be constructed in a similar manner by joining the centre C to the point under consideration and producing the radius to meet the tangent plane. Consider a star L whose projection on the tangent plane is N and whose image on the photographic plate is at M. If $\phi = O\hat{C}M = A\hat{C}L$, we have

$$\tan \phi = \frac{OM}{OC} = \frac{AN}{AC} \qquad \ldots\ldots(2).$$

It follows generally that the system of stellar images on the plate is *similar* to the system of the projections on the tangent plane, one system however being on a different linear scale from the other. Let AR', AS' be the positive directions of rectangular axes in the tangent plane; let OR, OS be parallel to AR', AS' respectively, in the plane of the plate, their positive directions being opposite to those of AR' and AS'. Let ξ', η' be the co-ordinates of the projection of a star on the tangent plane and ξ, η the co-ordinates of the image on the plate; then by the principle of similarity, we have

$$\frac{\xi'}{AC} = \frac{\xi}{OC} \qquad \ldots\ldots(3)$$

and $$\frac{\eta'}{AC} = \frac{\eta}{OC} \qquad \ldots\ldots(4).$$

160. *Standard co-ordinates.*

For the sake of geometrical simplicity we shall suppose the celestial sphere (centre C) and the tangent plane at A to be drawn as in Fig. 107. It is to be understood that A is the point on the celestial sphere towards which the telescope is pointed. If S is a star near A, its projection T on the tangent plane is obtained by joining C to S and producing CS to meet the

tangent plane in T. Draw the great circle arc AS; then, since
the plane of this great circle passes through C, it follows that all
radii joining C to points on AS lie in one plane and this plane
intersects the tangent plane in a straight line AT. More generally,
we can say that any great circle projects into a straight line in
the tangent plane. Let P be the north pole of the celestial
sphere. AP is the meridian of A and it projects into the straight
line AQ. We shall take AQ as the η'-axis of the tangent plane.
The ξ'-axis is taken to be AU, which is drawn perpendicular to

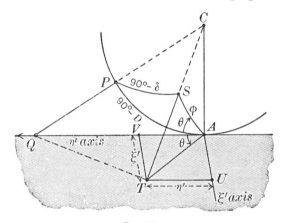

Fig. 107.

AQ, and its positive direction is taken to be eastwards of the
meridian AP so that increasing values of ξ' correspond to increasing values of the right ascension.

Since AT lies in the tangent plane, AT is perpendicular to
AC and is therefore the tangent at A to the great circle arc AS.
Similarly, AQ is the tangent to the great circle arc AP. Now
$Q\hat{A}T$ defines the spherical angle PAS; hence $Q\hat{A}T = P\hat{A}S$. Thus
we see that the angle between any two great circle arcs, intersecting at the tangential point A, is exactly reproduced on the
tangent plane as the angle between the two straight lines into
which the great circles project. This remark holds only for great
circles passing through the tangential point; for example, the
great circle SP projects into the straight line TQ and AP
projects into AQ; but $A\hat{Q}T$ (the angle between AQ and TQ) is

not equal to $S\hat{P}A$ (the angle between the great circles AP and SP).

Denote the arc AS by ϕ and $S\hat{A}P$ by θ; then $Q\hat{A}T = \theta$. Draw perpendiculars TU, TV to AU, AQ respectively. Then

$$VT = \xi' = AT \sin\theta \qquad \ldots\ldots(5)$$

and $$UT = \eta' = AT \cos\theta \qquad \ldots\ldots(6).$$

Now $$AT = AC \tan A\hat{C}T = AC \tan\phi.$$

Hence $$\frac{\xi'}{AC} = \tan\phi \sin\theta \qquad \ldots\ldots(7),$$

$$\frac{\eta'}{AC} = \tan\phi \cos\theta \qquad \ldots\ldots(8).$$

Hence, by (3) and (4),

$$\frac{\xi}{OC} = \tan\phi \sin\theta \qquad \ldots\ldots(9)$$

and $$\frac{\eta}{OC} = \tan\phi \cos\theta \qquad \ldots\ldots(10),$$

in which ξ and η are the co-ordinates of the image of S on the photographic plate with reference to rectangular axes through the centre O of the plate (Fig. 106), and drawn parallel, but oppositely directed, to the axes AU, AQ on the tangent plane. OC is the focal length of the telescope. Suppose that the focal length is known in millimetres and that the plate co-ordinates ξ and η are derived also in millimetres by processes which will be described later; then the values of ϕ and θ can be calculated from (9) and (10). As we shall see immediately, ϕ and θ are functions of the right ascensions and declinations of A and S; if the right ascension and declination of A are known, the right ascension and declination of S can then be deduced from the values of the co-ordinates ξ and η.

If we take the focal length OC to be the unit of length and ξ, η to be expressed in terms of this unit, we have from (9) and (10),

$$\xi = \tan\phi \sin\theta \qquad \ldots\ldots(11),$$

$$\eta = \tan\phi \cos\theta \qquad \ldots\ldots(12).$$

ξ and η are then called the *standard co-ordinates* of the star concerned. In this definition of standard co-ordinates—first

introduced by Professor H. H. Turner*—the following points
have to be noted: (i) the origin of the co-ordinate axes corresponds to a definite position, whose right ascension and declination are specified with respect to a standard mean equinox;
the epoch of this mean equinox is chosen to be 1900·0; (ii) the
ξ- and η-axes are correctly oriented for the epoch 1900·0; (iii) the
definition, being a purely geometrical one, excludes the effects of
instrumental imperfections and of refraction and aberration (all
of which will be considered later). The standard co-ordinates of
a particular star thus specify the position of the star uniquely,
and can therefore be used in place of right ascension and declination.

161. *Formulae for the standard co-ordinates.*

Let A, D be the right ascension and declination (referred to
1900·0) of the point A on the celestial sphere, and α, δ the
corresponding co-ordinates of the star S. We shall now show
how the relations between ξ, η and A, D, α and δ are obtained.
In the spherical triangle ASP (Fig. 107) we have: $AP = 90° - D$,
$SP = 90° - \delta$, $A\hat{P}S = \alpha - A$ (in the figure, S is eastwards of the
meridian AP), $AS = \phi$, $S\hat{A}P = \theta$. Then by formulae **A**, **B** and
C, we have

$$\cos \phi = \sin \delta \sin D + \cos \delta \cos D \cos (\alpha - A) \quad ......(13),$$

$$\sin \phi \sin \theta = \cos \delta \sin (\alpha - A) \quad ......(14),$$

$$\sin \phi \cos \theta = \sin \delta \cos D - \cos \delta \sin D \cos (\alpha - A) \quad ......(15).$$

Dividing (15) by (13) we obtain, using (12),

$$\eta = \frac{\cos D - \cot \delta \sin D \cos (\alpha - A)}{\sin D + \cot \delta \cos D \cos (\alpha - A)} \quad ......(16).$$

Define q as follows:

$$\cot q = \cot \delta \cos (\alpha - A) \quad ......(17).$$

Then

$$\eta = \frac{\cos D - \sin D \cot q}{\sin D + \cos D \cot q},$$

from which

$$\eta = \tan (q - D) \quad ......(18).$$

* *Monthly Notices*, vol. LIV, p. 13 (1893).

Again, dividing (14) by (13) and using (11), we have

$$\xi = \frac{\cot \delta \sin (\alpha - A)}{\sin D + \cos D \cot \delta \cos (\alpha - A)} \qquad \ldots\ldots(19)$$

$$= \frac{\cot q \tan (\alpha - A)}{\sin D + \cos D \cot q}$$

by (17), so that

$$\xi = \frac{\cos q \tan (\alpha - A)}{\cos (q - D)} \qquad \ldots\ldots(20).$$

The auxiliary quantity q, which has been introduced into (18) and (20) in order to simplify the logarithmic computations of ξ and η in the case when A, D, α and δ are all known, is readily seen to have a simple geometrical interpretation. Draw a great

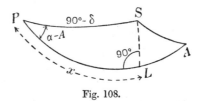

Fig. 108.

circle arc SL to cut AP at right angles in L (Fig. 108). Denote PL by x; then by formula D,

$$\cos x \cos (\alpha - A) = \sin x \tan \delta - \sin (\alpha - A) \cot 90°,$$

from which $\qquad \tan x = \cot \delta \cos (\alpha - A).$

Hence, by (17), $x = 90° - q$, so that q is the declination of L.

The formulae (18) and (20) enable the calculation of ξ and η to be made when A, D, α and δ are known.

We now derive the formulae which will give α and δ in terms of A, D, ξ and η. We have from (16),

$$\eta \sin D + \eta \cot \delta \cos D \cos (\alpha - A)$$
$$= \cos D - \cot \delta \sin D \cos (\alpha - A),$$

from which

$$\cot \delta \cos (\alpha - A) \{\eta \cos D + \sin D\} = \cos D - \eta \sin D,$$

and hence, $\qquad \cot \delta \cos (\alpha - A) = \dfrac{1 - \eta \tan D}{\eta + \tan D} \qquad \ldots\ldots(21).$

Again, from (19),

$$\cot \delta \sin (\alpha - A) = \xi \sin D + \xi \cos D \cot \delta \cos (\alpha - A)$$

$$= \xi \left\{ \sin D + \frac{\cos D \, (1 - \eta \tan D))}{\eta + \tan D} \right\},$$

by means of (21), whence we derive

$$\cot \delta \sin (\alpha - A) = \frac{\xi \sec D}{\eta + \tan D} \qquad \ldots\ldots(22).$$

Divide (22) by (21); then

$$\tan (\alpha - A) = \frac{\xi \sec D}{1 - \eta \tan D} \qquad \ldots\ldots(23),$$

from which $(\alpha - A)$ can be calculated and α obtained. When $(\alpha - A)$ has been found, δ can be obtained from (21) or (22).

In astronomical photography there are two fundamental processes employed directly or indirectly. The first is the calculation of the standard co-ordinates of one or more stars whose right ascensions and declinations are known; this process involves the use of equations (18) and (20). The second is the calculation of the right ascensions and declinations of stars from the values of their standard co-ordinates; this process is carried out by means of (23) and (21) or (22).

162. *The measurement and scale of photographic plates.*

For convenience we shall consider the plates taken with astrographic telescopes. These instruments, constructed according to a standard design, are in use in about a score of observatories scattered over the globe, the work to which they have been principally devoted being a complete photographic survey of the heavens. In this enterprise the co-operating observatories had definite zones (between certain parallels of declination) assigned to them; several observatories have finished their share of the work, but under existing world-conditions it is uncertain when the survey will be complete.

On each astrographic plate, a net-work system of parallel lines (Fig. 109) is photographed, either before or after the plate is exposed to the stars, so that on development the plate shows the stellar images and the réseau system of lines. The lines are equally spaced at intervals of five millimetres. We shall suppose that the central lines XOY and UOV correspond exactly to the

ξ- and η-axes already defined. Consider a star-image at S. The distance AS, parallel to OX, is measured by a machine and we shall suppose that $AS = 4\cdot14$ mms. The distance of S from the axis UOV (that is to say OC) is thus $10 + 4\cdot14$ or $14\cdot14$ mms. In a similar way, BS is measured and the distance OD obtained.

Now the standard co-ordinates are expressed in terms of the focal length regarded as the unit; hence the standard ξ co-ordinate

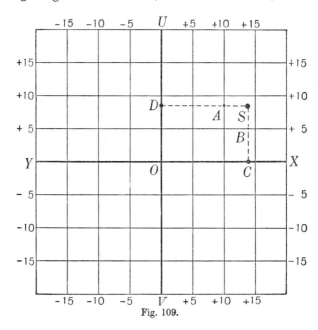

Fig. 109.

of S, for example, in Fig. 109 is OC (measured in mms.) divided by the focal length (measured in mms.). For simplicity, consider two stars (of declination δ_1 and δ_2) on the central meridian AP and let Y_1, Y_2 (along the η-axis) be their standard co-ordinates. Then, by (17) and (18),

$$Y_1 = \tan(\delta_1 - D) \quad \text{and} \quad Y_2 = \tan(\delta_2 - D).$$

Assuming that $(\delta_1 - D)$ and $(\delta_2 - D)$ are small angles and expressed in minutes of arc, we have

$$Y_1 = (\delta_1 - D)\sin 1' \quad \text{and} \quad Y_2 = (\delta_2 - D)\sin 1',$$

so that $\qquad Y_2 - Y_1 = (\delta_2 - \delta_1)\sin 1'.$

If d is the distance between the images measured in mms. and f is the focal length in the same unit, we have $Y_2 - Y_1 = d/f$, and consequently, $\delta_2 - \delta_1 = s.d$ ......(24),

where $s = (1/f) \operatorname{cosec} 1'$; the value of s can thus be regarded as known. s is the "scale" of the plate; it gives the number of minutes of arc corresponding to 1 mm. on the plate. Hence, by measuring the distance (in mms.) between the images of any two stars, we are enabled by (24) and the known value of s to deduce the angular separation of the two stars in the sky.* The astrographic telescope was designed to give the scale of 1 mm. to 1'.

In certain classes of work it is sometimes more convenient to measure small distances on the plate in terms of revolutions of the micrometer attached to the measuring machine. Proceeding as before, we measure the distance between two stellar images in terms of the micrometer scale (say it is 3·456 revolutions), and this number corresponds to the number of minutes (or seconds) of arc by which the stars are separated in the sky. For example, one revolution of the micrometer head of the Cambridge measuring machine corresponds to $17''\cdot58$ on a plate taken with the Sheepshanks telescope.

163. *The measured co-ordinates.*

In defining the standard co-ordinates of a star we have assumed (a) that the optical axis of the telescope passes through the origin of co-ordinates on the plate, (b) that the plate is perpendicular to the optical axis, (c) that the η-axis corresponds precisely to the projection of the central meridian, for the epoch 1900·0, on the tangent plane, (d) that the ξ-axis is perpendicular to the η-axis. In practice it is impossible to attain the geometrical perfection just indicated, and consequently the co-ordinates of a star-image, measured with reference to the axes on the plate, must be expected to differ (generally slightly) from the theoretical standard co-ordinates. But this is not all. Hitherto, in referring to star-images, we have ignored the effects of refraction and aberration. We have seen in previous chapters that owing to refraction and aberration the apparent position of a star on the celestial sphere is displaced by measurable or

* The focal length, f, and the scale of the measuring machine change slightly with temperature; allowance for these changes is made automatically by the "plate constants" a and e (section 167).

calculable amounts from its true position; consequently the
actual image of a star on the photographic plate will be somewhat displaced from the position it would occupy were these
effects inoperative. It will thus be realised that standard co-
ordinates are ideal co-ordinates, whereas the measured co-
ordinates of a star-image include the effects of geometrical (or
mechanical) imperfections and the effects of refraction and
aberration. At first sight the problem of deriving the standard
co-ordinates of a star from the measured co-ordinates of its
image on the photographic plate seems one of great difficulty;
actually, as we shall see, the solution in practice is extremely
simple. We shall now examine in detail the differences between
the standard and measured co-ordinates.

164. *Discussion of errors.*

We shall consider the errors individually. We shall denote by
ξ and η the true standard co-ordinates of a star and by x and y the
co-ordinates as influenced by the particular error concerned.

(*a*) *Error of orientation.*

In Fig. 110 let XOY and UOV be the axes of co-ordinates
correctly centred and oriented for the epoch 1900·0; let $X'OY'$

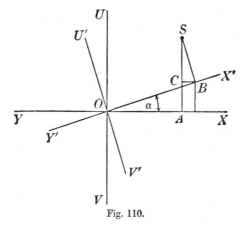

Fig. 110.

and $U'OV'$ be the axes on the plate (the central réseau lines)
correctly centred but erroneously oriented. Let $X'\hat{O}X = \alpha$.

Let S be a point whose standard co-ordinates are ξ and η—
referred to OX and OY—and whose co-ordinates referred to OX'

and OY' are x and y. Draw perpendiculars SA, SB to OX, OX' respectively. Then $OA = \xi$, $AS = \eta$, $OB = x$ and $BS = y$. We have

$$OA = OB \cos \alpha + BS \cos (90° + \alpha),$$

or

$$\xi = x \cos \alpha - y \sin \alpha,$$

whence

$$\xi - x = - 2x \sin^2 \frac{\alpha}{2} - y \sin \alpha.$$

Similarly,

$$\eta - y = x \sin \alpha - 2y \sin^2 \frac{\alpha}{2}.$$

We write these in the form:

$$\left. \begin{array}{l} \xi - x = a_1 x + b_1 y \\ \eta - y = d_1 x + e_1 y \end{array} \right\} \qquad \ldots\ldots (25),$$

where $a_1, \ldots e_1$ are simple functions of α. In practice α is always a small angle and in consequence the coefficients $a_1, \ldots e_1$ are also small. The formulae (25) are essentially linear in x and y.

(b) *Non-perpendicularity of axes.*

If the x-axis on the plate is not exactly perpendicular to the y-axis, it is clear that we shall again obtain linear formulae of the type (25) for the quantities $\xi - x$ and $\eta - y$.

A similar result is also obtained if the corresponding axes of the micrometer scale are not exactly perpendicular.

(c) *Centering error.*

Suppose firstly that during an exposure the direction of the optical axis corresponds to a given direction (A, D) referred to the mean equinox of 1900·0, and that ξ and η are the standard co-ordinates of a star (α, δ) with reference to the position (A, D) as centre. It is hardly to be expected that the straight line joining the centre of the object-glass to the origin of the impressed réseau co-ordinate axes will coincide exactly with the optical axis, and consequently the origin will correspond to slightly different values of A and D. Secondly, the optical axis may not be directed quite accurately to the position (A, D) for 1900·0. As a result, we must therefore assume that the origin of co-ordinates on the plate corresponds to a position $(A + \Delta A, D + \Delta D)$, where ΔA and ΔD may be supposed to be small quantities.

All other errors and influences being assumed absent, let x and y be the co-ordinates of an image with respect to the réseau axes. Then x and y may be taken to be the standard co-ordinates of the star concerned with reference to the position $(A + \Delta A, D + \Delta D)$ as centre. We shall denote $x - \xi$ and $y - \eta$ by $\Delta \xi$ and $\Delta \eta$ respectively.

Now, by (11) and (12),

$$\xi = \tan \phi \sin \theta, \quad \eta = \tan \phi \cos \theta,$$

where ϕ, θ are functions of A and D. Corresponding to incre-ments ΔA and ΔD, we shall have increments $\Delta \phi$, $\Delta \theta$. Hence we have

$$\Delta \xi = \Delta \phi \, (1 + \tan^2 \phi) \sin \theta + \Delta \theta \tan \phi \cos \theta,$$
$$\Delta \eta = \Delta \phi \, (1 + \tan^2 \phi) \cos \theta - \Delta \theta \tan \phi \sin \theta.$$

For a star at an angular distance of $1°$ from (A, D), $\tan \phi = 1/57$, and in the above formulae we can neglect such terms as have factors $\Delta \phi \tan^2 \phi$. We thus have

$$\left. \begin{array}{l} \Delta \xi = \Delta \phi \sin \theta + \eta \, \Delta \theta \\ \Delta \eta = \Delta \phi \cos \theta - \xi \, \Delta \theta \end{array} \right\} \qquad \ldots\ldots(26).$$

We have now to express $\Delta \phi$ and $\Delta \theta$ in terms of ΔA, ΔD.

From (13) we have

$$- \sin \phi \Delta \phi = \Delta D \, \{\sin \delta \cos D - \cos \delta \sin D \cos (\alpha - A)\}$$
$$+ \Delta A \cos \delta \cos D \sin (\alpha - A),$$

and, using (14) and (15), this becomes

$$\Delta \phi = - \Delta D \cos \theta - \Delta A \cos D \sin \theta \qquad \ldots\ldots(27).$$

From (14) we obtain

$$\Delta \theta \sin \phi \cos \theta + \Delta \phi \cos \phi \sin \theta = - \Delta A \cos \delta \cos (\alpha - A),$$

or, using (27),

$$\Delta \theta \sin \phi \cos \theta = \cos \phi \sin \theta \, (\Delta D \cos \theta + \Delta A \cos D \sin \theta)$$
$$- \Delta A \cos \delta \cos (\alpha - A).$$

Multiply (13) by $\cos D$ and (15) by $\sin D$ and subtract. Then

$$\cos \delta \cos (\alpha - A) = \cos \phi \cos D - \sin \phi \cos \theta \sin D.$$

Hence

$$\Delta \theta \sin \phi \cos \theta = \Delta D \cos \phi \sin \theta \cos \theta$$
$$+ \Delta A \, \{\cos \phi \sin^2 \theta \cos D - \cos \phi \cos D + \sin \phi \cos \theta \sin D\},$$

from which

$$\Delta \theta \sin \phi = \Delta D \cos \phi \sin \theta + \Delta A \, (\sin \phi \sin D - \cos \phi \cos \theta \cos D).$$

Multiplying this last equation by sec ϕ cos θ, we have

$$\eta \Delta\theta = \Delta D \sin \theta \cos \theta + \Delta A \,(\eta \sin D - \cos^2 \theta \cos D).$$

Inserting this expression for $\eta \Delta\theta$ and the expression for $\Delta\phi$ given by (27) in the first of (26), we obtain

$$\Delta\xi = - \Delta A \cos D + \eta \Delta A \sin D,$$

or $\qquad \xi - x = \Delta A \cos D - \eta \Delta A \sin D.$

As $(\xi - x)$ is of order ΔA, we can write this last equation with sufficient accuracy as

$$\left. \begin{aligned} \xi - x &= \Delta A \cos D - y \Delta A \sin D. \\ \text{Similarly,} \qquad \eta - y &= \Delta D + x \Delta A \sin D \end{aligned} \right\} \quad \ldots\ldots(28).$$

These formulae have the linear forms:

$$\xi - x = \qquad b_2 y + c_2,$$
$$\eta - y = d_2 x \qquad + f_2.$$

(d) *Error of tilt.*

This error is due to the non-perpendicularity of the optical axis to the plane of the plate. If i is the angle between the optical axis and the normal to the plate, the expressions for $\xi - x$, $\eta - y$ are of the form

$$\xi - x = \tan i \,(px^2 + qxy),$$
$$\eta - y = \tan i \,(pxy + qy^2).$$

As the angle i is in practice only a few minutes of arc and as the squares and products of x and y only are involved, the correction for tilt can generally be neglected.

The total effect of the various errors considered here is to give the displacements $(\xi - x)$ and $(\eta - y)$ in terms of essentially linear expressions; we can thus write the general formulae

$$\left. \begin{aligned} \xi - x &= ax + by + c \\ \eta - y &= dx + ey + f \end{aligned} \right\} \quad \ldots\ldots(29),$$

in which $a, b, \ldots f$ are small quantities depending on the small errors involved.

165. *Refraction.*

We now investigate the displacements of the stellar images due only to refraction. In Fig. 111 the tangent plane to the celestial sphere at A is drawn as in Fig. 107. Z is the zenith and

W is its projection on the tangent plane; let the co-ordinates of W be X, Y. Owing to refraction, a star S is seen at S', the displacement SS' being along the great circle arc joining S to Z. We shall use the formula [no. (7) of Chapter III]

$$SS' = k \tan ZS \qquad \ldots\ldots(30),$$

in which k is expressed in circular measure and ZS is written, without sensible loss of accuracy, in place of the observed zenith distance ZS'. The great circle $ZS'S$ projects into the straight line $WT'T$ on the tangent plane, T and T' being the

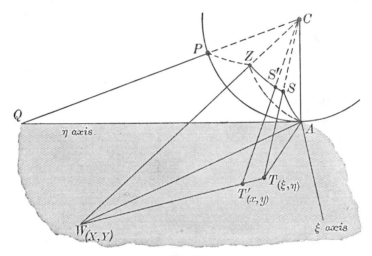

Fig. 111.

projections of S and S' respectively. Let ξ, η be the co-ordinates of T and x, y the co-ordinates of T'. If ξ, η and x, y are expressed in terms of AC as the unit of length, these quantities are the standard co-ordinates of the star and the measured co-ordinates of its image on the plate respectively.

Since the region to be photographed is generally no more than $2° \times 2°$, the different points on that part of the spherical surface concerned are actually very close to the corresponding projected points on the tangent plane; we accordingly assume that $SS' = TT'$. From (30) we have

$$TT' = k \tan ZS \qquad \ldots\ldots(31).$$

Since T, T', W are collinear, we have

$$\frac{x - \xi}{TT'} = \frac{X - \xi}{TW} \quad \text{and} \quad \frac{y - \eta}{TT'} = \frac{Y - \eta}{TW},$$

and writing $\Delta\xi$ for $(\xi - x)$ and $\Delta\eta$ for $(\eta - y)$ we obtain, from (31),

$$\Delta\xi = -\frac{k(X - \xi)}{TW} \tan ZS \qquad \ldots\ldots(32),$$

$$\Delta\eta = -\frac{k(Y - \eta)}{TW} \tan ZS \qquad \ldots\ldots(33).$$

We remind the reader that the different co-ordinates ξ, η, x, y, X, Y are all supposed to be expressed in terms of AC as the unit of length. In particular, ξ and η will thus be small quantities and in the sequel we shall neglect the much smaller quantities ξ^2, $\xi\eta$, η^2 and higher powers and products of ξ and η.

Now from the plane triangle TAW, we have

$$TW^2 = AT^2 + AW^2 - 2AT \cdot AW \cos TAW,$$

so that

$$(X - \xi)^2 + (Y - \eta)^2 = (\xi^2 + \eta^2) + (X^2 + Y^2)$$
$$- 2AT \cdot AW \cos TAW,$$

which gives us $\quad AT \cdot AW \cos TAW = X\xi + Y\eta \qquad \ldots\ldots(34).$

From the spherical triangle ZAS, we have by the cosine-formula **A,**

$$\cos ZS = \cos AS \cos AZ + \sin AS \sin AZ \cos ZAS$$

$$= \frac{AC}{CT} \cdot \frac{AC}{CW} + \frac{AT}{CT} \cdot \frac{AW}{CW} \cos ZAS.$$

Since $T\hat{A}W$ defines the spherical angle ZAS, we have from (34), putting $AC = 1$,

$$\cos ZS = \frac{1 + X\xi + Y\eta}{CT \cdot CW}.$$

But $\qquad\qquad CT^2 = 1 + \xi^2 + \eta^2$

$\qquad\qquad\qquad = 1$, when we neglect ξ^2 and η^2.

Hence $\qquad\qquad \cos ZS = \frac{1 + X\xi + Y\eta}{CW} \qquad \ldots\ldots(35),$

from which $\qquad \sin^2 ZS = \frac{CW^2 - (1 + X\xi + Y\eta)^2}{CW^2}$

$$= \frac{(CW^2 - 1) - 2(X\xi + Y\eta)}{CW^2}$$

$$= \frac{AW^2 - 2(X\xi + Y\eta)}{CW^2}.$$

Hence, using the binomial theorem, and neglecting ξ^2, etc., we obtain

$$\sin ZS = \frac{AW}{CW}\left(1 - \frac{X\xi + Y\eta}{AW^2}\right) \qquad \ldots\ldots(36).$$

From (35) and (36),

$$\tan ZS = \frac{AW\left(1 - \dfrac{X\xi + Y\eta}{AW^2}\right)}{1 + X\xi + Y\eta} \qquad \ldots\ldots(37).$$

Now
$$TW^2 = (X - \xi)^2 + (Y - \eta)^2$$
$$= X^2 + Y^2 - 2(X\xi + Y\eta), \text{ neglecting } \xi^2, \eta^2,$$
$$= AW^2 - 2(X\xi + Y\eta);$$
$$\therefore \ TW = AW\left(1 - \frac{X\xi + Y\eta}{AW^2}\right) \qquad \ldots\ldots(38).$$

Hence from (37) and (38),
$$\tan ZS = TW(1 + X\xi + Y\eta)^{-1}$$
$$= TW(1 - X\xi - Y\eta).$$

We thus obtain from (32),
$$\Delta\xi = -k(X - \xi)(1 - X\xi - Y\eta),$$
or
$$\Delta\xi = -kX + k\{(1 + X^2)\xi + XY\eta\} \qquad \ldots\ldots(39).$$

Since $(\xi - x)$ and k (expressed in circular measure) are both small, we can write x and y for ξ and η respectively on the right-hand side of (39) without introducing any appreciable error. Write $(\xi - x)$ for $\Delta\xi$; then (39) gives us
$$\Delta\xi \equiv \xi - x = -kX + k\{(1 + X^2)x + XYy\}\ldots\ldots(40).$$
Similarly, we obtain
$$\Delta\eta \equiv \eta - y = -kY + k\{XYx + (1 + Y^2)y\}\ldots\ldots(41).$$

The displacements due to refraction for the centre of the plate are $-kX$ and $-kY$, and these quantities, since they appear in the values of $(\xi - x)$ and $(\eta - y)$ for all the images on the plate, may be supposed to be incorporated in the undetermined constants c and f of equations (29). When $-kX$ and $-kY$ are omitted from (40) and (41) the remaining terms express the values of $(\xi - x)$ and $(\eta - y)$ for the *differential refraction*; these equations are then of the linear form

$$\xi - x = ax + by \qquad \ldots\ldots(42),$$
$$\eta - y = dx + ey \qquad \ldots\ldots(43),$$

in which, for example, $a = k(1 + X^2)$.

The coefficients a, b, d and e can be calculated if necessary. It is sufficient to consider the values of X and Y at the middle of the exposure. Let the sidereal time then be S. Then as PZ is the observer's meridian (Fig. 111), the hour angle of the vernal equinoctial point Υ is the right ascension of any point on the observer's meridian PZ; in particular, the right ascension of Z is S. Also $PZ = 90° - \phi$, where ϕ is the latitude, so that regarding Z as a particular point on the celestial sphere, its declination is ϕ. We now apply the formulae (17), (18) and (20). Let Q be defined by

$$\cot Q = \cot \phi \cos (S - A) \qquad \dots\dots(44);$$

then
$$Y = \tan (Q - D) \qquad \dots\dots(45),$$

$$X = \frac{\cos Q \tan (S - A)}{\cos (Q - D)} \qquad \dots\dots(46).$$

The value of k being known—it is $58{\cdot}2 \sin 1''$ in circular measure—we can then deduce the values of a, b, d and e in (42) and (43). In practice this calculation is hardly ever necessary, for it is sufficient to know that the differences between the standard co-ordinates and the corresponding co-ordinates of the image (as affected by refraction) are given with sufficient accuracy, as a general rule, by linear expressions in x and y with small coefficients a, b, etc.

However, when the altitude of the celestial region to be photographed is less than $30°$ or so, the simple formula (30) for the refraction is not sufficiently accurate; also the omission of quadratic terms can no longer be justified. Assuming that the refraction R is given by

$$R = A \tan ZS' + B \tan^3 ZS'$$

[see formulae (29) and (30) of Chapter III] and taking account of second order terms, we can write the formulae for $(\xi - x)$ and $(\eta - y)$ as follows:

$$\xi - x = \alpha x + \beta y + g x^2 + h x y + k y^2,$$
$$\eta - y = \gamma x + \delta y + l x^2 + m x y + n y^2.$$

These give the effect of differential refraction on the standard co-ordinates. The coefficients α, β, ... m, n are expressible in terms of A and B and the co-ordinates X and Y of the zenith. In practice, their values are derived without considering the theoretical formulae by which they can be expressed.

166. *Aberration.*

The investigation of the effect of aberration on the standard co-ordinates of a star is very similar to that in the previous section. From Chapter VIII, we know that if F is the position on the celestial sphere towards which the earth is moving at the time of the observation, the position of a star is displaced from its true position S to a position S' on the great circle arc SF, S' being nearer to F than S. The displacement SS' is given by

$$SS' = \kappa \sin FS \qquad \ldots\ldots(47),$$

where κ is the aberration constant whose value in circular measure is $20 \cdot 4 \sin 1''$. Confining ourselves to the effects of aberration only, we write, as before, ξ and η for the standard co-ordinates of a star and x and y for the co-ordinates of its image on the photographic plate. F is, of course, a definite point on the celestial sphere; we shall suppose that its projection on the tangent plane is W_1, with co-ordinates U and V. We shall also suppose that F, U and V correspond simply to the time of the middle of the exposure. Following the procedure of the previous section, we have the formulae corresponding to (32) and (33),

$$\Delta\xi = -\frac{\kappa(U-\xi)}{TW_1}\sin FS \qquad \ldots\ldots(48),$$

$$\Delta\eta = -\frac{\kappa(V-\eta)}{TW_1}\sin FS \qquad \ldots\ldots(49),$$

and from (36) and (38),

$$\sin FS = \frac{AW_1}{CW_1}\left(1 - \frac{U\xi + V\eta}{AW_1^2}\right) \qquad \ldots\ldots(50),$$

$$TW_1 = AW_1\left(1 - \frac{U\xi + V\eta}{AW_1^2}\right),$$

so that
$$\frac{\sin FS}{TW_1} = \frac{1}{CW_1}.$$

Now
$$CW_1^2 = 1 + U^2 + V^2.$$

We accordingly obtain

$$\Delta\xi = -\frac{\kappa U}{(1+U^2+V^2)^{\frac{1}{2}}} + \frac{\kappa\xi}{(1+U^2+V^2)^{\frac{1}{2}}}\ldots\ldots(51),$$

$$\Delta\eta = -\frac{\kappa V}{(1+U^2+V^2)^{\frac{1}{2}}} + \frac{\kappa\eta}{(1+U^2+V^2)^{\frac{1}{2}}}\ldots\ldots(52).$$

As in the previous section, we can write x for ξ and y for η without any sensible loss of accuracy on the right-hand sides of (51) and (52); also we can omit the constant terms (independent of ξ and η) on the right of these equations. We then obtain the expressions for *differential aberration* in the form

$$\xi - x = a_1 x \qquad \ldots\ldots(53),$$

$$\eta - y = d_1 y \qquad \ldots\ldots(54),$$

in which a_1 and d_1 are small (they have as a common factor $20 \cdot 4 \sin 1''$, which is of the order 10^{-4}). Again it is unnecessary, as a rule, to calculate the values of a_1 and d_1 from their theoretical expressions.

167. *The general relations between standard co-ordinates and measured co-ordinates. (Turner's method.)*

From the previous three sections we have seen that refraction, aberration and the instrumental errors taken separately produce in each co-ordinate a displacement of the image of a star on the plate from the position corresponding to the standard co-ordinates of the star, and that this displacement is given, generally with sufficient accuracy, as a linear expression in the co-ordinates. If we combine all the various effects we clearly obtain linear formulae for $(\xi - x)$ and $(\eta - y)$, which we can write in the general forms

$$\xi - x = ax + by + c \qquad \ldots\ldots(55),$$

$$\eta - y = dx + ey + f \qquad \ldots\ldots(56),$$

where ξ, η are the standard co-ordinates of the star and x, y are the measured co-ordinates of its image on the plate. In these equations a, b, etc. are small and dependent, in a composite way, on the instrumental errors, on refraction and on aberration. Thus in general x differs from ξ by a small quantity and we can write the equations, without loss of accuracy, in an alternative form, namely,

$$\xi - x = a\xi + b\eta + c \qquad \ldots\ldots(57),$$

$$\eta - y = d\xi + e\eta + f \qquad \ldots\ldots(58),$$

in which the quantities a, b, etc. are small. These quantities are called the *plate constants*. We shall now consider three practical applications: (i) the measurement of astrographic plates, (ii) the measurement of proper motions, (iii) the measurement of stellar parallaxes.

168. *The measurement of astrographic plates.*

The precise determination of the right ascensions and declinations of stars brighter than about the ninth magnitude is undertaken by meridian circle observers. Beyond this magnitude the stars are too faint to be observed, owing to the illumination necessary to show up the wires of the instrument. Actually, of course, the observation with this instrument (if it were practicable) of the multitudinous faint stars would be utterly beyond the astronomical resources of the world's observatories even if they were increased a thousandfold in number and employed solely for this purpose. Amongst the different investigations to which the photographic telescope is applied, the study of the positions of the stars is the one to which we first turn our attention.

We shall now suppose that a photograph has been taken of a region, the co-ordinates A, D of whose centre are known, and that a system of réseau lines has been impressed on the plate. With the measuring machine the co-ordinates x and y (with reference to the two central lines regarded as the axes of co-ordinates on the plate) of every stellar image can be derived. As the standard co-ordinates ξ and η are defined in terms of the focal length as unit of length, we shall suppose that x and y are defined in terms of the same unit; thus if the x-measure of an image on the plate is found to be p mms. and if f is the focal length in millimetres, we have $x = p/f$.

Having regard to the size of the astrographic field ($2° \times 2°$), we can take it as certain that any plate will contain the images of several stars whose right ascensions and declinations are known from meridian circle observations. Suppose for the moment that there are three such stars. Then by (18) and (20) the standard co-ordinates of these stars, which we shall designate *reference stars*, can be calculated. The measured co-ordinates of the images of the reference stars are also known. Then from (55) we have

$$\left.\begin{aligned}
\xi_1 - x_1 &= ax_1 + by_1 + c \\
\xi_2 - x_2 &= ax_2 + by_2 + c \\
\xi_3 - x_3 &= ax_3 + by_3 + c
\end{aligned}\right\} \qquad \ldots\ldots(59),$$

in which the suffixes 1, 2, 3 refer to the three stars. These three equations are sufficient to determine the three plate constants a, b and c. For any other star we have

$$\xi - x = ax + by + c,$$

and, as x and y are known from the measures of the image and the values of a, b and c have now been determined, we are therefore able to calculate the value of ξ—the standard ξ-co-ordinate of the star in question. In a similar way η is obtained, the values of the plate constants d, e and f having been calculated by means of the three reference stars. If it is desired, the calculation of the right ascension α and of the declination δ of this star can be carried out by means of (23) and (21). Actually the standard co-ordinates ξ and η of the star define its position uniquely with reference to the centre of the region (A, D) and, except for special objects, the calculation of the equatorial co-ordinates α and δ need not be undertaken.

In practice, more than three stars are chosen as reference stars. If there are N such stars, we shall have N equations of the form shown in (59). The constants a, b and c are then derived from a solution of the N equations according to the method of least squares (or an analogous process). This applies also to the determination of the plate constants d, e and f.

Our treatment of standard co-ordinates, immediately above, is not complete without further reference to the epoch to which they refer. The epoch chosen for the standard co-ordinates of star-places in the astrographic catalogue is 1900·0. Let us suppose that a plate was taken on 1904 March 4. For the epoch 1900·0, the standard co-ordinates of the reference stars can be found from the catalogues. The reduction of the measured co-ordinates of any other star leads to the standard co-ordinates, referred to the mean equator of 1900·0, of the star's position in the sky on the date 1904 March 4. If the proper motion and parallax of the star are negligible, the standard co-ordinates so obtained define the position of the star for the epoch 1900·0. The measured co-ordinates x and y of any one of the reference stars will contain the components μ_x and μ_y of proper motion, corresponding to the interval 1900·0 to 1904 March 4. (We denote by μ_x the component of proper motion along the ξ-axis and by μ_y the component along the η-axis; thus $\mu_x = \mu_a \cos \delta$ and $\mu_y = \mu_\delta$, all the quantities concerned being expressed in circular measure.) If μ_x and μ_y are known, the effects of μ_x and μ_y ought strictly to be removed from the values of x and y before the solution for the plate constants is made.

169. *Photographic observations of minor planets and comets.*

We shall suppose that a photograph of a minor planet (the general procedure is applicable also to comets) is taken at an instant t on a given date—in practice t is taken to be the time corresponding to the middle of the exposure. If we define the standard co-ordinates of selected reference stars with reference to the mean equinox at the beginning of the year in question, the measurement of the planet's image on the photographic plate and the subsequent reduction, as indicated in section 168, will give the position of the *image* referred to the mean equinox of the beginning of the year concerned. This position is called the planet's *mean place* or *astrographic place* corresponding to the time and date of the photograph. It is to be remembered that the reduction automatically removes, in particular, the effects of stellar aberration at the date in question. The mean place evidently includes the effect of the planet's parallax, which depends on the observer's position on the earth.

For ephemeris purposes, the *true mean place* is used. This is defined to be the position, referred to the mean equinox of the beginning of the year, corresponding to the *direction* of the planet with respect to the centre of the earth, which, it is to be remembered, is revolving around the sun. This last consideration involves the introduction of the corrections due to aberration.

The true mean place is derived from the mean place (i) by applying the corrections due to the planet's parallax, (ii) by adding the effects of aberration,

$$Cc + Dd \quad \text{in R.A.,}$$
$$Cc' + Dd' \quad \text{in declination}$$

[formulae (25) and (26), p. 184], and (iii) by ante-dating the observed time t of observation by τ, the interval required by light to travel from the planet to the earth (section 112).

170. *The measurement of proper motions.*

For the accurate determination of proper motions an adequate interval T between the two epochs of observation is necessary. Let us suppose that a region of the sky is photographed on 1904 March 4 and again on 1926 April 3. We can then take the interval T to be 22·1 years. After the measures have been reduced, according to the method of section 168, the first plate will give the standard co-ordinates ξ_1, η_1 (with reference to the mean equator for 1900·0 and to a definite central point with equatorial co-ordinates A and D for this epoch) of the position of a star X in the sky on 1904 March 4. If x_1, y_1 are the measured co-ordinates of its image on the first plate, we have

$$\xi_1 - x_1 = a_1 x_1 + b_1 y_1 + c_1 \qquad \ldots \ldots (60),$$

in which a_1, b_1 and c_1 are the constants of this plate, determined from a suitable number of reference stars. Now suppose that the second plate has been measured in a similar way. (We are assuming that a réseau system of lines has been impressed on each plate.) Generally, the constants of the second plate will be different from the constants of the earlier plate; let their values, determined from the reference stars, be a_2, b_2 and c_2. Let x_2, y_2 be the measured co-ordinates of the image of the star X on the second plate; let ξ_2, η_2 denote the standard co-ordinates (with reference to the mean equator for 1900·0 and to the same central point of the region) of the position of X in the sky on 1926 April 3. Then

$$\xi_2 - x_2 = a_2 x_2 + b_2 y_2 + c_2 \qquad \ldots \ldots (61).$$

If the star has no proper motion, ξ_2 and ξ_1 will clearly be the same. If the star has a proper motion with components μ_x and μ_y per annum, then, since the difference between ξ_2 and ξ_1 is the displacement (parallel to the ξ-axis) due to proper motion in T years, we have

$$\xi_2 - \xi_1 = T\mu_x \qquad \ldots \ldots (62).$$

Similarly,
$$\eta_2 - \eta_1 = T\mu_y \qquad \ldots \ldots (63).$$

The values of ξ_1, ξ_2, η_1 and η_2 are obtained from (60) and (61) and the corresponding equations in η. Thus the values of μ_x and μ_y are found from (62) and (63).

It is to be noticed that in these formulae the unit in which μ_x and μ_y are expressed is the same as that by which standard co-ordinates are defined. In practice, it is much more convenient to express the standard co-ordinates either in millimetres or in terms of the micrometer scale unit, or in minutes or seconds of arc. If we suppose that ξ, η, x and y are all expressed in seconds of arc, we obtain the values of μ_x and μ_y expressed in the familiar way, that is, in seconds of arc per annum. Then μ_y is the same as μ_δ (the proper motion in declination) and μ_x is $\mu_a \cos \delta$, where δ is the star's declination and μ_a is the proper motion (in seconds of arc) along the equator.

Instead of using the linear formulae (60) and (61) it may be necessary to include the quadratic terms mentioned at the end of section 165. From the equations of a sufficient number of reference stars the individual constants of the formulae for $(\xi - x)$ and $(\eta - y)$ are derived.

171. *The measurement of proper motions (Kapteyn's method).*

The method just described involves the use of réseau lines as intermediate axes of reference. For each plate, the measure of a co-ordinate x or y of an image requires a setting of the micrometer wire or scale on a neighbouring réseau line and then a setting on the image. For the second plate the procedure is the same. In the method now adopted in practice the work of measuring is reduced by half by eliminating the réseau system altogether. We shall first consider the plan suggested by Kapteyn, as the principle involved is most clearly seen in this application. A plate on which an exposure has been made is not developed, but is carefully stored. After a suitable interval of time, T, has elapsed the plate is again placed in the telescope and

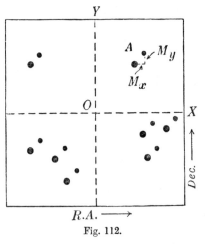

Fig. 112.

a new exposure made of the same region. To prevent the earlier and the later images of the same star from coinciding or over-

lapping, the plate-carrier is given a small displacement perpendicular to the optical axis. The plate is developed after the
later exposure. Each star is represented by two images, one
corresponding to the earlier epoch and the other to the later.
Fig. 112 illustrates this result; we shall suppose that the larger
dot at A represents the image of a particular star at the earlier
epoch, say, 1900 and the smaller dot the image at the second
epoch, say, 1920. If, at the earlier exposure, a bright star is
allowed to register a trail across the plate (this is effected by
stopping the driving mechanism of the telescope) the trail will
define with sufficient accuracy the ξ-axis; by means of the trail
the plate can then be oriented correctly in the measuring machine.

The procedure consists in measuring the components, M_x and
M_y, of the displacement between the 1900 and 1920 images.
Referred to hypothetical axes on the plate (represented by broken
lines in Fig. 112) we shall have, using (57), for the image of A
corresponding to the position of the star in 1900,

$$\xi_1 - x_1 = a_1\xi_1 + b_1\eta_1 + c_1 \qquad \ldots\ldots(64),$$

and, for its image in 1920,

$$\xi_2 - x_2 = a_2\xi_2 + b_2\eta_2 + c_2 \qquad \ldots\ldots(65).$$

(We cannot assume that $a_1 = a_2$, etc., because the effects of
refraction and aberration and the orientation of the plate, for
example, are hardly likely to be identical in 1900 and in 1920.)
The differences between the standard co-ordinates in 1900 and
1920 will be small—the differences are simply the effects of
proper motion in the interval—and as a_2, b_2 are small, we can
replace ξ_2 and η_2 by ξ_1 and η_1 on the right of (65) without
sensible error and thus obtain

$$\xi_2 - x_2 = a_2\xi_1 + b_2\eta_1 + c_2 \qquad \ldots\ldots(66).$$

From (64) and (66) we have by subtraction

$$\xi_2 - \xi_1 = x_2 - x_1 + (a_2 - a_1)\xi_1 + (b_2 - b_1)\eta_1 + (c_2 - c_1)\ldots(67).$$

In this equation, $\xi_2 - \xi_1 = T\mu_x$ and $(x_2 - x_1)$ is simply the
measured displacement M_x. Also we can replace ξ_1 by x_1 and
η_1 by y_1 on the right of (67) without altering the linear character
of the equation. We thus have the general expression for $T\mu_x$ in
the form

$$T\mu_x = M_x + ax_1 + by_1 + c \qquad \ldots\ldots(68).$$

There is a similar equation giving $T\mu_y$.

The plate constants a, b and c in (68) are determined by means of several stars—the *comparison stars*—scattered over the plate, whose proper motions are presumed to be small; faint stars are more likely to satisfy this condition than the brighter stars. If we assume that the proper motions of all the comparison stars are zero, then to each of these stars corresponds an equation of the type
$$ax + by + c + M_x = 0 \qquad \ldots\ldots(69),$$
where x, y are the measured co-ordinates of the star. The values of x, y need not be determined with great accuracy, as their factors a and b respectively are small quantities. If there are N comparison stars the solution of the N equations of the type (69) by the method of least squares yields the appropriate values of the plate constants a, b and c. A similar procedure is adopted for the N equations of the form
$$dx + ey + f + M_y = 0.$$
The substitution of the values of a, b and c in the general formula (68) gives for any other star the component μ_x of the proper motion. The value of μ_y is obtained similarly.

We have assumed that the proper motions of all the comparison stars are zero; by substituting the values of a, b and c appropriate to any comparison star in the equation (68), we can readily see how satisfactory (or otherwise) this assumption really is, for by this process we effectively determine the value of μ_x for each of the comparison stars. If it should prove that a comparison star has an appreciably large proper motion, it must be discarded and another star selected, if possible. This entails a new calculation of the plate constants. Actually, all the comparison stars cannot be expected to have zero proper motions; and the derived values of μ_x and μ_y for the stars in general are the values of their *relative proper motions*, that is, relative to the mean motion of the group of comparison stars.

172. *The reduction of relative to absolute proper motions.*

The proper motions of stars derived from meridian circle observations are *absolute proper motions*. Suppose that there is one star on the plate whose absolute proper motions μ_x' and μ_y' are known; the plate measures give the relative proper motions μ_x and μ_y. Then $\mu_x' - \mu_x$ and $\mu_y' - \mu_y$ are the corrections to be applied to all the relative proper motions to convert them into absolute proper motions. Meridian proper motions are generally not so accurate as one would wish, and therefore the above correction can only be obtained with sufficient accuracy if there are at least about ten stars with well-determined proper motions. This is a condition that is hardly likely to be fulfilled in general and consequently another method of deriving the correction has to be adopted.

We have seen on p. 263 that the components P_α and P_δ of parallactic motion, in seconds of arc per annum, are given by

$$P_\alpha = H \sin \lambda \sin \chi \sec \delta,$$
$$P_\delta = H \sin \lambda \cos \chi,$$

in which H is the secular parallax of the star, λ is its angular distance from the ant-apex of the solar motion and χ is the position angle of the ant-apex with reference to the star. Writing P_x for $P_\alpha \cos \delta$ and P_y for P_δ and denoting by H_m the secular parallax of a star of magnitude m, we have

$$P_x = H_m \sin \lambda \sin \chi \qquad \ldots\ldots(70),$$
$$P_y = H_m \sin \lambda \cos \chi \qquad \ldots\ldots(71).$$

For the stars in a photographic region we can take the average values of λ and χ to be the values at the centre of the plate. Consider a group of stars, N in number, all of magnitude m. Each star will have its own random motion in space but, as it is observed relative to the sun, there will be superimposed the parallactic motion. The observed proper motion will thus consist of the effects of the random motion and of the parallactic motion. Adding up all these effects for the N stars, the effects of the random motions on the proper motions will tend to cancel out, leaving N times the parallactic motion of a star of magnitude m (at the mean distance of the stars in the group) in the region of the sky concerned. Thus taking the average for the N stars, the components of their mean absolute proper motion ought to be very nearly the values of P_x and P_y in (70) and (71). Let $\bar{\mu}_x$, $\bar{\mu}_y$ denote the average values of the relative proper motions of the N stars as obtained from the measurement of the photographic plate. The corrections which we must apply to the components of the relative proper motions of all the stars photographed in order to convert the relative proper motions into absolute proper motions is evidently given by the quantities

$$P_x - \bar{\mu}_x \quad \text{and} \quad P_y - \bar{\mu}_y.$$

In practice several magnitude groups are formed and the weighted mean of the corrections derived from the several groups is taken as the final correction to be applied. The following table* illustrates the method. The centre of the photographic region is the star ϵ Cygni; the values of λ and χ are $147°$ and $102°$ respectively; then by (70),

$$P_x = 0 \cdot 53 H_m.$$

The values of H_m are taken from the table on p. 424. (The galactic latitude of ϵ Cygni is $-6°$.)

Group	Mag-nitude m	No. of stars N	H_m	P_x	NP_x	$N\bar{\mu}_x$	$N(P_x - \bar{\mu}_x)$
1	8·1	7	$0''{\cdot}0234$	$+ 0''{\cdot}01242$	$+ 0''{\cdot}087$	$+ 0''{\cdot}049$	$+ 0''{\cdot}038$
2	9·2	11	$0''{\cdot}0163$	$+ 0''{\cdot}00865$	$+ 0''{\cdot}095$	$+ 0''{\cdot}065$	$+ 0''{\cdot}030$
3	9·5	20	$0''{\cdot}0147$	$+ 0''{\cdot}00780$	$+ 0''{\cdot}156$	$+ 0''{\cdot}035$	$+ 0''{\cdot}121$
4	9·7	11	$0''{\cdot}0138$	$+ 0''{\cdot}00733$	$+ 0''{\cdot}080$	$+ 0''{\cdot}072$	$+ 0''{\cdot}008$
		49					$+ 0''{\cdot}197$

The correction to be applied to all the components μ_x of relative proper motion to convert them into absolute proper motion is thus $+ 0''{\cdot}197 \div 49$ or $+ 0''{\cdot}004$. The correction to be applied to the components μ_y is obtained in a similar way.

173. *The film-to-film method of measuring proper motions.*

Kapteyn's method of measuring proper motions, which we have described in section 171, suffers from several disadvantages, amongst which we mention the following: the photographic film is likely to deteriorate during the rather long interval between the early and late exposures, in which event the images of the later exposure will fall below the standard of quality expected in precise work of this kind; the images of the early exposure may be of poor quality owing to bad atmospherical conditions or indifferent "guiding" during the exposure, and this state of affairs is not brought to light until many years later. The method now commonly adopted is as follows. The photograph at the earlier epoch is taken in the usual way and, after a star has

---

* *Cambridge Astronomical Observations,* vol. XXVI, p. 18 (1928).

been allowed to register its trail, the plate is developed at once. If the plate is unsatisfactory another can be taken to replace it. After a suitable interval has elapsed, a photograph is again taken of the region, but in this case the plate is placed in its carrier with its film-side away from the object-glass. The stellar images are consequently made by the starlight after its passage through the glass of the plate. The plate—we shall call it the "reversed" plate—is developed in the usual way. The plate of the earlier epoch and the reversed plate are now bound together, film to film, by clips allowing for a small displacement of the pairs of images of the several stars. The appearance of the double images is then similar to that in Kapteyn's method as illustrated in Fig. 112. The superposition of the plates must be done with as little rotation of one plate relative to the other as possible. The measurement of the images and the derivation of the proper motions follow the same course as in Kapteyn's method. In practice, several plates are generally exposed on each region at each epoch and at least two separate exposures made on each plate. The material is then adequate to give precise values of the relative proper motions.

174. *The determination of the plate constants by the method of Christie and Dyson.*

We shall write equation (68) in the simple form

$$ax + by + c = M - p \qquad \ldots\ldots(72),$$

in which p is the quantity $T\mu_x$ and the suffix in M_x has been dropped. This equation holds strictly for every star in the region. In order to determine the plate constants by the simple method* proposed by Christie and Dyson, the comparison stars are selected in four equal groups, according to quadrants on the plate, so that the mean centres of position of the groups are as nearly symmetrical about the centre as possible. Assume that the co-ordinates of the group-centres are as in Fig. 113 and that there are N stars in each group.

The values of p in (72) are of course unknown. Let us assume, however, that the selected comparison stars have small relative proper motions, for example, less than $0''\!\cdot\!02$ per annum.

* *Monthly Notices, R.A.S.* vol. LV, p. 61.

A preliminary solution shows up stars which do not fulfil this condition; they can be removed from the appropriate groups and replaced by others.

Adding the N equations (72) for the comparison stars in group 1, we have

$$N(-aX+bY+c)=\underset{1}{\Sigma}p-\underset{1}{\Sigma}M \quad \ldots\ldots(73),$$

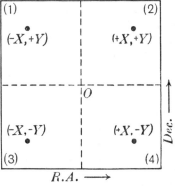

Fig. 113.

in which $\underset{1}{\Sigma}p$ denotes the sum of the proper motions (parallel to the x-axis) of the N stars in group 1 and $\underset{1}{\Sigma}M$ denotes the sum of the measured displacements between pairs of images of the N stars. We have similarly, for the other groups,

$$N(+aX+bY+c)=\underset{2}{\Sigma}p-\underset{2}{\Sigma}M \quad \ldots\ldots(74),$$

$$N(-aX-bY+c)=\underset{3}{\Sigma}p-\underset{3}{\Sigma}M \quad \ldots\ldots(75),$$

$$N(+aX-bY+c)=\underset{4}{\Sigma}p-\underset{4}{\Sigma}M \quad \ldots\ldots(76).$$

Add (73) and (75); then

$$-2NaX+2Nc=\underset{1}{\Sigma}p+\underset{3}{\Sigma}p-\underset{1}{\Sigma}M-\underset{3}{\Sigma}M \ldots\ldots(77).$$

Add (74) and (76); then

$$+2NaX+2Nc=\underset{2}{\Sigma}p+\underset{4}{\Sigma}p-\underset{2}{\Sigma}M-\underset{4}{\Sigma}M \ldots\ldots(78).$$

Subtracting (77) from (78) we obtain

$$4NaX=\{\underset{2}{\Sigma}p-\underset{1}{\Sigma}p+\underset{4}{\Sigma}p-\underset{3}{\Sigma}p\}+\{\underset{1}{\Sigma}M-\underset{2}{\Sigma}M+\underset{3}{\Sigma}M-\underset{4}{\Sigma}M\}$$
$$\ldots\ldots(79).$$

Similarly,

$$4NbY=\{\underset{1}{\Sigma}p+\underset{2}{\Sigma}p-\underset{3}{\Sigma}p-\underset{4}{\Sigma}p\}-\{\underset{1}{\Sigma}M+\underset{2}{\Sigma}M-\underset{3}{\Sigma}M-\underset{4}{\Sigma}M\}$$
$$\ldots\ldots(80),$$

$$4Nc=\{\underset{1}{\Sigma}p+\underset{2}{\Sigma}p+\underset{3}{\Sigma}p+\underset{4}{\Sigma}p\}-\{\underset{1}{\Sigma}M+\underset{2}{\Sigma}M+\underset{3}{\Sigma}M+\underset{4}{\Sigma}M\}$$
$$\ldots\ldots(81).$$

In these equations N, X, Y and $\underset{1}{\Sigma}M,\ldots\underset{4}{\Sigma}M$ are known and the plate constants a, b and c are now derived on the assumption that

$$\underset{2}{\Sigma} p - \underset{1}{\Sigma} p + \underset{4}{\Sigma} p - \underset{3}{\Sigma} p = 0,$$

$$\underset{1}{\Sigma} p + \underset{2}{\Sigma} p - \underset{3}{\Sigma} p - \underset{4}{\Sigma} p = 0,$$

$$\underset{1}{\Sigma} p + \underset{2}{\Sigma} p + \underset{3}{\Sigma} p + \underset{4}{\Sigma} p = 0,$$

from which $\underset{1}{\Sigma} p = \underset{4}{\Sigma} p; \quad \underset{2}{\Sigma} p = \underset{3}{\Sigma} p; \quad \underset{1}{\Sigma} p = -\underset{2}{\Sigma} p.$

In a group of N stars, some values of p may be expected to be positive and some negative; the value of $\underset{1}{\Sigma} p$, for example, consequently may be expected to be small and it will be practically negligible in comparison with the other quantities on the right of (79), ... (81). The third assumption mentioned above—that $\underset{1}{\Sigma} p + \underset{2}{\Sigma} p + \underset{3}{\Sigma} p + \underset{4}{\Sigma} p = 0$—is simply another way of expressing the fact that the proper motions derived in this way are *relative proper motions*. They can be converted into absolute proper motions by the method already described in section 172. A similar procedure is adopted for the components parallel to the y-axis. The values of the plate constants derived by this method are little inferior in accuracy to the values obtained by the rigorous, but more lengthy, method of least squares.

175. *The measurement of stellar parallaxes.*

In Chapter IX we have derived the displacements, in right ascension and declination, of a star due to parallax. We consider here only the displacement in right ascension, measures of which alone are made in practice. If α, δ are the heliocentric right ascension and declination of a star whose parallax Π is to be measured, and α_1 is the right ascension as viewed from the earth on a particular date T_1, then by (67), p. 222, we have

$$\alpha_1 - \alpha = \Pi F_1 \sec \delta \qquad \ldots\ldots(82),$$

in which F_1 is the parallax factor, the value of which can be calculated as the quantities on which it depends are all known. At a later date T_2, the heliocentric right ascension of the star will have increased owing to proper motion to $\alpha + T\mu_a$, T being the interval (expressed in terms of the year as unit) between T_1 and T_2 and μ_a being the proper motion in right ascension. If

α_2 is the geocentric right ascension at the date T_2 and F_2 the corresponding parallax factor, we shall have

$$\alpha_2 - (\alpha + T\mu_a) = \Pi F_2 \sec \delta \qquad \ldots\ldots(83).$$

From (82) and (83),

$$\alpha_2 - \alpha_1 = \Pi (F_2 - F_1) \sec \delta + T\mu_a \qquad \ldots\ldots(84).$$

Let ξ_1, η_1 be the standard co-ordinates (with reference to the centre of the region whose equatorial co-ordinates are A, D) corresponding to the geocentric co-ordinates α_1, δ_1 of the parallax star, and ξ_2, η_2 the standard co-ordinates corresponding to α_2, δ_2. Then by (23),

$$\tan (\alpha_1 - A) = \frac{\xi_1 \sec D}{1 - \eta_1 \tan D}.$$

As the parallax star is always chosen to be near the centre of the region, so that ξ_1, η_1 are very small and D differs hardly at all from δ_1 or δ, we can write, with all necessary accuracy,

$$\alpha_1 - A = \xi_1 \sec \delta.$$

Similarly,

$$\alpha_2 - A = \xi_2 \sec \delta.$$

Hence

$$\alpha_2 - \alpha_1 = (\xi_2 - \xi_1) \sec \delta \qquad \ldots\ldots(85).$$

Denote the proper motion along the parallel of declination δ by μ_x; then $\mu_x = \mu_a \cos \delta$ or $\mu_a = \mu_x \sec \delta$. Inserting this value of μ_a in (84) and using (85) we have

$$\xi_2 - \xi_1 = \Pi (F_2 - F_1) + T\mu_x \qquad \ldots\ldots(86).$$

Now let x_1, y_1 be the co-ordinates of the image with respect to rectangular axes on the plate taken on the date T_1; then by (57),

$$\xi_1 - x_1 = a_1\xi_1 + b_1\eta_1 + c_1.$$

Similarly,

$$\xi_2 - x_2 = a_2\xi_2 + b_2\eta_2 + c_2.$$

Since ξ_2 differs very little from ξ_1, the difference being due to parallax and proper motion, and as either differs little from x_1, y_1, we can write the difference between these two equations in the form $\quad \xi_2 - \xi_1 = x_2 - x_1 + ax_1 + by_1 + c.$

In practice the difference $(x_2 - x_1)$ between a pair of images of the parallax star is measured; denote it by M_x; then making use of (86) we obtain

$$\Pi (F_2 - F_1) + T\mu_x = M_x + ax + by + c \qquad \ldots\ldots(87),$$

on the right of which we have dropped the subscripts in x_1 and y_1.

The plate constants a, b and c are derived from the measures of several faint stars, whose parallaxes and proper motions may be regarded as negligible. Each comparison star thus contributes an equation of the type

$$ax + by + c + M_x = 0 \qquad \ldots\ldots(88).$$

The co-ordinates x, y of the comparison stars are supposed known. The solution of the several equations (88) gives the values of a, b and c.

Generally, the parallax star is chosen as the origin of the x and y co-ordinate axes, so that for this star $x = y = 0$ and (87) becomes

$$\Pi\,(F_2 - F_1) + T\mu_x = M_x + c \qquad \ldots\ldots(89).$$

This equation, derived from the measurement of a pair of plates taken on dates T_1 and T_2 such that the quantity $F_2 - F_1$ is as large as possible, contains two unknowns, the parallax Π and the component μ_x of proper motion (strictly, relative to the mean motion of the group of comparison stars).

Several other pairs of plates (six to ten in number) provide a corresponding number of equations of the type (89); these equations, when solved by the method of least squares, give the values of Π and μ_x. In practice M_x and c are found in terms of the unit of the micrometer attached to the measuring machine; thus by (89), Π and μ_x are found in terms of this unit. They are converted into seconds of arc by means of the known equivalent of the micrometer unit. [See also p. 411.]

Several methods are employed to obtain the measures M_x between the pairs of images of the parallax and comparison stars. Kapteyn's method, similar in principle to that described for the measurement of proper motions (section 171), has been tried. A second method, adopted at the Royal Observatory, Greenwich, depends on the use of reference lines and the principle is the same as if a system of réseau lines were marked on each plate. In some observatories a special kind of measuring machine—the stereo-comparator—enables the measurement of the quantities M_x to be made without the use of any intermediary lines of reference. This instrument is also used for the measurement of proper motions.

As an example, the measures of two pairs of plates, by Dr van

Maanen at Mt Wilson Observatory, of the star "Boss 3650" and other relevant details are summarised in the table below:

Plate no.	Date	Parallax factor F	$F_2 - F_1$	T	$M_x + c$
136	1914 June 15	$- 0·73$			
470	1915 Feb. 6	$+ 0·91$	$+ 1·64$	$0·64$	$- 31$
486	1915 Feb. 7	$+ 0·90$			
555	1915 June 4	$- 0·60$	$- 1·50$	$0·32$	$- 150$

The last column contains the values of $M_x + c$ in terms of the measuring-machine unit; one unit $= 0''·001649$. The first pair of plates gives the equation

$$+ 1·64\,\Pi + 0·64\,\mu_x = -\ 31,$$

and the second pair gives

$$- 1·50\,\Pi + 0·32\,\mu_x = -\ 150.$$

Eliminating μ_x, we obtain $\Pi = +\ 58$; we then obtain $\mu_x = -\ 197$. Multiplying these numbers by $0·001649$, we find

$$\Pi = +\ 0''·095,$$

$$\mu_x = -\ 0''·315.$$

From the measures of six pairs of plates, van Maanen finds

$$\Pi = +\ 0''·096 \pm 0''·003,$$

$$\mu_x = -\ 0''·302 \pm 0''·008.$$

Here $0''·096$ is the *most probable value* of Π derived from the measures; $\pm 0''·003$ is the *probable error* which indicates the degree of precision with which the value of Π has been obtained. Having regard to the inevitable errors of measurement and the like, we interpret the complete result for Π by saying that the probability that the *true* value of Π lies between the limits $0''·096 + 0''·003$ and $0''·096 - 0''·003$, that is between $0''·099$ and $0''·093$, is equal to the probability that it lies outside these limits.

EXERCISES

1. Two plates are centred at P and Q and $(x, 0)$ and $(\xi, 0)$ are the corresponding standard co-ordinates of a star whose image appears on each plate. Prove that

$$\xi = \frac{(x + c)}{1 - cx},$$

where c is the ξ-co-ordinate of P with respect to the plate centred at Q.

[The x and ξ axes are the projections on the respective plates of the great circle PQ.] [*Lond.* 1930.]

2. Show that, to a first approximation, the trail of a star of declination δ on a photographic plate is

$$y = \text{const.} + \tfrac{1}{2}x^2 \tan \delta,$$

and that the projection of a meridian distant $\Delta\alpha$ from the central meridian makes with the latter the angle $\tan^{-1}(\tan \Delta\alpha \sin D)$, where D is the declination of the plate-centre.

CHAPTER XIII

DETERMINATION OF POSITION AT SEA

176. *The sextant.*

In this chapter we shall consider the problem of determining a ship's position at sea from observations of heavenly bodies. The instrument used is the sextant, by means of which the altitude of a heavenly body above the sea-horizon can be measured. Fig. 114 is a diagram embodying the more important features of the

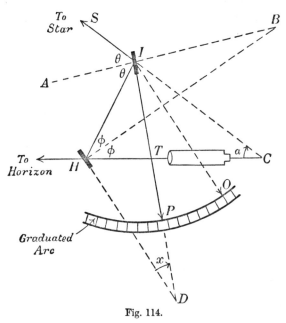

Fig. 114.

instrument. *I* (called the *index-glass*) represents a mirror perpendicular to the plane of the paper. Rigidly attached to the index-glass is an arm *IP*. The index-glass and the arm *IP* can rotate about an axis (perpendicular to the plane of the paper) at *I*. The centre of rotation at *I* is also the centre of a graduated arc *OP*. To any given position of the index-glass and of the arm *IP* corresponds a reading on the graduated arc. *H* is a small

rectangular piece of glass, fixed to the framework of the instrument and perpendicular to the plane of the paper; the upper half
of its surface is clear and the lower half is silvered, the latter
acting as a mirror. T is a small telescope attached to the framework. To find the altitude of a star (for example) above the sea-
horizon, the observer holds the instrument in a vertical plane
and points the telescope so as to see the horizon through the
upper half of H (called the *horizon-glass*). He then moves the
index-glass I by means of the arm IP until the image of the star
appears in the field of view. When the image appears on the line
of the horizon, he notes the reading on the graduated arc.

Let IS denote the direction of the star. A ray in the direction
SI is reflected by the mirror I along IH; it is then reflected by
the mirror portion of the horizon-glass H along HT, and the
star is thus observed in the telescope in the direction in which
the sea-horizon is seen. The altitude of the star is simply related
to the inclination of the index-glass I to the horizon-glass H
—in Fig. 114 the inclination is $I\hat{D}H$, which we denote by x. Let
AIB and HB be the normals to the mirrors I and H; then $I\hat{B}H$ is
evidently x. The laws of reflection give

$$S\hat{I}A = A\hat{I}H \equiv \theta,$$

and $$I\hat{H}B = B\hat{H}C \equiv \phi.$$

If a is the star's altitude above the sea-horizon, then $S\hat{C}H = a$.
From the triangle IHC, the exterior angle $SIH = I\hat{C}H + I\hat{H}C$,
so that $$2\theta = 2\phi + a \qquad \ldots\ldots(1).$$

Similarly, from the triangle IBH,

$$\theta = \phi + x \qquad \ldots\ldots(2).$$

Hence from (1) and (2), $$a = 2x \qquad \ldots\ldots(3),$$

or the star's altitude is twice the angle between the mirrors I and
H (or between their normals). The altitude a is zero when x is
zero, that is, when the mirrors I and H are parallel. In Fig. 114,
IO is parallel to the fixed direction HD; O is the zero-point of
the scale. The angle OIP can thus be found from the reading on
the graduated arc and, by (3), the star's altitude is twice this
angle. The arc of the sextant is generally about one-sixth of the
circumference of a circle (hence the name "sextant"), but instead

of having 60 divisions each representing one degree, the arc is divided into 120 divisions. In this way, the altitude is read directly from the scale without the necessity of applying the factor 2 of equation (3). With the aid of sub-divisions and a vernier, altitudes can be read with a first-class instrument to one-tenth of a minute of arc.

177. *The errors of a sextant.*

In this section we shall mention very briefly the errors* to which a sextant is liable.

(i) *Error of perpendicularity.* The index-glass I ought to be perpendicular to the plane of the graduated arc. (ii) *Side error.* The horizon-glass H ought to be perpendicular to the plane of the graduated arc. Errors (i) and (ii) can be removed by means of appropriate screws at I and H. (iii) *Collimation error.* The line of collimation of the telescope ought to be parallel to the plane of the arc. (iv) *Index error.* When I is parallel to H the pointer P ought to indicate $0°$ on the graduated arc. It is possible to remove the index error by means of a screw at H, but the usual custom is to determine the reading when I and H are parallel and to apply the appropriate correction to all altitude readings. The condition of parallelism of I and H is achieved most simply as follows (we assume that errors (i) and (ii) have first been eliminated). With the pointer set near $0°$, the telescope is pointed to a star. Two images of the star are then in the field of view; one is the direct image of the star as seen through the upper half of the horizon-glass H and the other is the image formed as the result of the reflections at I and H. With a slow-motion screw, the two images can be superimposed and the reading on the arc then corresponds to the position of I when it is parallel to H. As the instrument is not likely to be indefinitely in proper adjustment, the index error should be determined at frequent intervals. (v) *Centering error.* The pivot about which the arm IP revolves ought to be the centre of the graduated arc. In the best instruments this error is generally very small; it varies according to the particular position of the arm IP. In

---

* For the practical methods of eliminating or determining the errors, the reader is referred to more detailed treatises, e.g. *The Admiralty Manual of Navigation* and Doolittle's *Practical Astronomy as applied to Geodesy and Navigation.*

Great Britain, sextants are tested at the National Physical Laboratory before being put on the market, and the values of the centering errors determined there are noted on the sextant box.

178. *Corrections to the observed altitude.*

(i) *Dip of the horizon.* In making an observation, the observer's eye is at some distance above sea-level, and consequently the visible horizon will appear somewhat depressed below the horizontal plane, that is, the plane perpendicular to the direction of the observer's zenith. Thus the zenith distance of the visible horizon will be a little over 90°, and all altitudes measured from the visible or sea-horizon will require correction. In Fig. 115 let

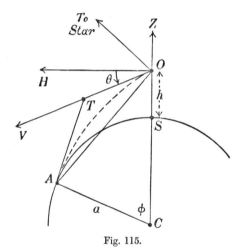

Fig. 115.

COZ be the direction of the zenith of an observer at O at a height h feet above sea-level. Consider the vertical plane ZOA containing the direction of the star observed. Let OH be perpendicular to OZ in this plane. Let A be the point of the sea-surface corresponding to the visible horizon. Owing to atmospherical refraction the path of a ray from A to O will be slightly curved and the direction in which A is actually observed will be along OV, the tangent at O to the dotted curve representing the path of the ray. The angle HOV (denoted by θ) is called the *dip of the horizon*. As A is the most distant point visible, the path of the ray will be tangential to the earth's surface at A. Consequently if AT is a tangent at A to the curved path of the ray, AT is perpendicular to CA. Since h is small compared with the earth's radius a, we assume that the curved path is approximately a circular arc; hence, as OV and AT are tangents to this arc, then $T\hat{O}A = O\hat{A}T$. Denote $O\hat{C}A$ by ϕ. It is usually assumed that $T\hat{O}A$ or $O\hat{A}T$ is a constant fraction of ϕ. We can then write

$$O\hat{A}T = \beta\phi \qquad \qquad \dots\dots(4),$$

where β is a constant numerical factor whose value is deduced to be $1/13$ approximately. We now have: $O\hat{A}C = 90° - \beta\phi$; $A\hat{O}C = 90° - (\theta + \beta\phi)$, and hence

$$90° - \beta\phi + 90° - (\theta + \beta\phi) + \phi = 180°,$$

from which $\qquad\qquad \phi\,(1 - 2\beta) = \theta \qquad\qquad \dots\dots(5).$

The earth's radius being denoted by a (in feet) we have, from the triangle AOC,

$$\frac{\sin(90° - \beta\phi)}{a + h} = \frac{\sin(90° - \theta - \beta\phi)}{a},$$

or $\qquad\qquad \dfrac{\cos \beta\phi}{\cos(\theta + \beta\phi)} = 1 + \dfrac{h}{a} \qquad\qquad \dots\dots(6),$

whence $\qquad\qquad \dfrac{2\sin(\theta/2)\sin\frac{1}{2}(\theta + 2\beta\phi)}{\cos(\theta + \beta\phi)} = \dfrac{h}{a}.$

As θ and ϕ are small angles, we can write this last equation as

$$\theta\,(\theta + 2\beta\phi) = 2h/a,$$

or, using (5), $\qquad\qquad \theta^2 = 2\,(1 - 2\beta)\,h/a \qquad\qquad \dots\dots(7).$

Inserting the value of β and expressing θ in minutes of arc, we have

$$\theta = \sqrt{\frac{22h}{13a}}\,\operatorname{cosec} 1'.$$

Now $a = 3960 \times 5280$ feet and $\operatorname{cosec} 1' = 3438$. Hence we obtain $\qquad\qquad \theta = 0\cdot98\,(h)^{\frac{1}{2}} \qquad\qquad \dots\dots(8),$

or, as a sufficiently good working rule: "The dip, in minutes of arc, is equal to the square root of the height, above sea-level, in feet". For example, if the observer's eye is 36 feet above sea-level the angle of dip, θ, is practically $6'\cdot0$.

As the observed altitude of the heavenly body is measured with respect to OV—the direction of the sea-horizon—the angle of dip is to be subtracted from the observed altitude; thus the altitude with respect to OH—the direction of the theoretical horizon—is obtained.

(ii) *Astronomical refraction.* We have seen in Chapter III that the effect of atmospherical refraction is to make the heavenly body appear nearer the zenith than it would be if the atmosphere were non-effective in deviating the rays of light in their passage through the air. The observed altitude is thus too great by the amount of refraction R given by formula (7) of Chapter III,

$$R = k \tan \zeta.$$

(This formula is sufficiently accurate for navigational purposes.)

(iii) *Semi-diameter* (S.D.). In observations of the sun, moon and the nearer major planets, the altitude of the centre of the disc cannot be accurately measured directly; the observation consists in measuring the altitude of the upper or lower limb of the body concerned, and applying to the observed altitude the value of the semi-diameter given in the almanacs. In this way, the altitude of the centre of the sun, moon or planet is derived.

(iv) *Parallax.* As we have seen in Chapter IX, the effect of parallax is to make the observed altitude less than it would be if the dimensions of the earth were negligible in comparison with the distance of the heavenly body observed. If P denotes the horizontal parallax of the body and a its altitude (previously corrected for dip, refraction and semi-diameter), the correction due to parallax is given with sufficient accuracy for navigational purposes by $P \cos a$, which is to be added to the observed altitude. This correction is important only in the case of the moon.

When the index error and the corrections (i) to (iv) have been applied, in the order indicated, to the observed altitude we obtain, by subtracting this corrected altitude from $90°$, the true zenith distance of the heavenly body.

As regards star observations, only the corrections (i) and (ii) are of practical moment. In nautical tables,* the sum of these

---

* For example, *Inman's Tables*; Bowditch's *Practical Navigator.*

corrections can be taken directly from a simple double-entry table.

As regards the sun, simple tables have been prepared to enable the total effect of corrections (i) to (iv) to be found by inspection.

Corrections (iii) and (iv) for the planets are small and in nautical observations are usually neglected.

Example. To find the true zenith distance of the moon's centre at U.T. 10^h, 1931 March 24; given $h = 25$ feet; observed altitude of the moon's lower limb $= 32°\ 20'\cdot0$ (corrected for index error). From the *Nautical Almanac*, S.D. $= 15'\cdot2$; $P = 55'\cdot8$.

Obs. altitude (a)		$32°\ 20'\cdot0$
Dip	$-$	$5'\cdot0$
Refraction	$-$	$1'\cdot5$
Semi-diameter	$+$	$15'\cdot2$
Parallax ($P \cos a$) $+$		$47'\cdot0$

Hence the corrected altitude $= 33°\ 15'\cdot7$

and thus the true zenith distance is $56°\ 44'\cdot3$.

179.   *The position circle.*

An essential part of the navigator's observation concerns the exact U.T. at which the observation of altitude was made. For this purpose a reliable chronometer keeping U.T. as nearly as possible is necessary. The daily radio time-signals enable the navigator to determine the error of the chronometer at suitable intervals; consequently, when he measures the altitude of a heavenly body he knows the exact U.T. at which the observation was made. The various corrections in section 178 being applied, a complete observation yields two definite pieces of information:

(i)   the true zenith distance of the heavenly body,

(ii)   the U.T. of the observation.

Now, from the U.T. it is easy to find the position on the earth's surface at which the body concerned is exactly overhead at the moment of observation. Let U be this point in Fig. 116, which represents the earth with its centre at C. At the moment of observation the body is thus in the direction CUS. CP is the direction of the north pole and therefore $S\hat{C}P$ is the north polar distance of the heavenly body, that is, the great circle arc PU

is $(90° - δ)$, in which $δ$ is the declination of the body at the
U.T. of the observation. But PU is the colatitude of U and
thus the latitude of U is simply the declination of the heavenly
body. Again, if PGQ is the Greenwich meridian, $G\hat{P}U$ is the
longitude (measured west
of Greenwich) of U. But
$G\hat{P}U$ is also equal to
the angle between the
celestial meridian of the
heavenly body and the
celestial meridian corre-
sponding to the Green-
wich meridian PGQ. But
this latter angle is simply
the hour angle of the
heavenly body, at the
particular U.T., with re-
spect to the Greenwich
meridian. Thus the longi-
tude of U (measured westwards) is the hour angle

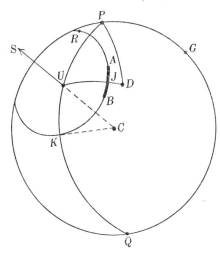

Fig. 116.

of the heavenly body at Greenwich at the U.T. concerned. This
hour angle can be calculated from the data in the almanacs.
Accordingly, the position of U on the earth's surface can be
definitely specified. U is known as the *geographical position* (or
the sub-solar or sub-stellar point, in the case of the sun or a star
respectively) of the body observed at the particular U.T.

Consider now the observation of altitude from which the true
zenith distance z is derived. If the observer were actually at
such a point as K on the earth's surface, the direction of his
zenith would be along the radius CK; also the direction of the
heavenly body at the time of observation is along the radius
CU; hence $K\hat{C}U$ or the great circle arc UK is the true zenith
distance z. Now since the U.T. of the observation is known,
the geographical position U is a definite point on the earth's
surface; the true zenith distance z is also known, and therefore
the observer must be situated somewhere on a small circle KJR
of which U is the pole, every point of this small circle being at

an angular distance z from U. This small circle is called the *position circle*. It is to be noted that a single observation of a heavenly body leads only to a certain small circle, on which the observer is situated. If we suppose that a similar observation of another heavenly body is made at the same U.T., a second position circle will be derived on which the observer must be situated. Therefore his actual position must be at one or other of the two points of intersection of the two position circles. As the approximate position of the ship is always known, there is no difficulty in deciding which of the two points is the correct position.

The approximate position of the ship is obtained by the navigator by the process of *dead reckoning*. He knows with fair accuracy the ship's various speeds and courses since leaving harbour, or since the last determination of position. He can also estimate, perhaps with moderate success, the effects of winds and currents on his progress. Using these data, he plots on a chart the ship's position hour by hour, and when he makes an astronomical observation he obtains from his chart the position according to dead reckoning—the D.R. position, so called. We shall suppose that in Fig. 116 D represents the D.R. position of the ship at the time of the altitude observation. It is to be noted that the D.R. position is, at best, only an approximate estimate; after 24 hours or more without astronomical observations it is not unlikely that the D.R. position may be in error by 10 miles or more.

180. *The position line (St Hilaire's method).*

As we have seen, the altitude observation of a heavenly body yields the information that at the time of observation the ship is situated on a certain small circle KJR (Fig. 116); at this moment the estimated position of the ship is at D. It is clear that the only part of the position circle with which the navigator need concern himself is that part in the immediate neighbourhood of D. His object then is to represent on his chart this part AJB of the position circle (shown with a heavy line in Fig. 116). Now the latitude and longitude of U and of D are known; hence the length of the great circle arc UD can be calculated. But this arc is simply the angle between the radii CU and CD, and as CD produced gives the direction of the zenith at the particular point D, the arc UD is thus the zenith distance of the heavenly body at the U.T. of the observation for a hypothetical observer situated at D. We shall call this zenith distance UD the *calculated zenith distance*. Now the length of the arc UJ is known from the observation—it is the true zenith distance z. Hence by subtraction we obtain the length of the arc DJ. This arc DJ is known as the *intercept*. Expressed in minutes of arc, it gives the distance in nautical miles of the D.R. position D from the nearest point J on the position circle. The arc DJ is perpendicular to the position circle at J, for U is the pole of KJR.

Also, the spherical angle UDP is easily seen to be the azimuth of the heavenly body, at the U.T. concerned, for an observer at D; it can be calculated or found by inspection in such tables as Burdwood's *Azimuth Tables*. In Fig. 116 the azimuth of J is the same as the azimuth of U, which we now suppose to be known. Thus, under the circumstances depicted in Fig. 116, where the calculated zenith distance UD is greater than the true zenith distance UJ, the navigator can draw from the D.R. position on his chart a straight line in the direction given by the azimuth of the heavenly body; he then marks off along this line a distance equal to the intercept, and through the point so obtained he draws a straight line, called the *position line*,* perpendicular to the line of azimuth. The position line represents on his chart (which we will consider in greater detail in section 182) the portion AJB of the position circle KJR (Fig. 116). It is clear that if D is within the position circle, that is, if the calculated zenith distance is less than the true zenith distance, the intercept will be marked off in the direction opposite to that given by the azimuth.

181. *Example of the calculation of the intercept.*†

The observed altitude of the sun's lower limb was $17° 27'\cdot0$ at U.T. $16^{\text{h}} 31^{\text{m}} 2^{\text{s}}$ on 1931 March 10, the height of the observer's eye above sea-level being 25 feet and the index error of the sextant being $-2'\cdot0$. The ship's estimated position D by dead reckoning was: Lat. $41° 15'$ N, Long. $7° 28'$ W. To calculate the intercept.

* Sometimes called the *Sumner line*.

† We retain U.T. in this chapter although G.M.T. is used in navigation circles, e.g. in the *Nautical Almanac* (for use in navigation).

The longitude 7° 28′ W is equivalent, in time-measure, to 29ᵐ 52ˢ W.

For the U.T. and date, we have from the almanac:

Sun's declination   4° 18′·0 S
Equation of time   − 10ᵐ 33ˢ
Sun's semi-diameter   16′·1

In Fig. 117, Z is the zenith of the ship's estimated position. Thus $PZ = 41° 45′$ (the colatitude). Also X is the position of

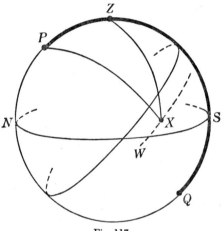

Fig. 117.

the sun's centre on the celestial sphere and PX is 94° 18′·0. The angle ZPX is the hour angle of the sun (H.A.T.S.). We shall use the haversine formula (section 13) to compute ZX (the calculated zenith distance):

U.T.	16ʰ 31ᵐ	2ˢ	(March 10)		
	− 12				
G.M.A.T.	4 31	2			
Long. (W)	− 29	52			
H.A.M.S. at D	4 1	10			
Equation of time	− 10	33			
H.A.T.S.	3ʰ 50ᵐ 37ˢ		log hav	9·366 40	
PX	94° 18′·0		log sin	9·998 78	
PZ	41 45 ·0		log sin	9·823 40	
	52° 33′·0		log hav θ	9·188 58	

$$\text{hav } \theta \quad 0{\cdot}154\ 37$$
$$\text{hav } 52^\circ\ 33'{\cdot}0 \quad 0{\cdot}195\ 97$$
$$\overline{\text{hav } ZX \quad 0{\cdot}350\ 34}$$

$\therefore\ ZX = \underline{72^\circ\ 35'{\cdot}0}$ (the calculated zenith distance).

The reader may verify this calculation by means of the cosine-formula. We now find the true zenith distance.

Observed alt. of sun's lower limb ...	$17^\circ\ 27'{\cdot}0$
Index error	$-\ \ 2\ {\cdot}0$
Dip (for height of eye, 25 feet) ...	$-\ \ 5\ {\cdot}0$
Refraction	$-\ \ 3\ {\cdot}1$
Semi-diameter	$+\ 16\ {\cdot}1$
$\therefore$ Corrected observed altitude ... $=$	$17^\circ\ 33'{\cdot}0$
Thus the true zenith distance is ...	$72^\circ\ 27'{\cdot}0$
But the calculated zenith distance is	$72^\circ\ 35'{\cdot}0$

Hence the intercept is 8′·0.

The sun's azimuth is 118° W or S 62° W. (We leave this calculation as an exercise to the reader.)

Since the calculated zenith distance is greater than the true zenith distance (72° 27′·0), the position line on the chart (Fig. 118) is obtained by drawing from D (the estimated position)

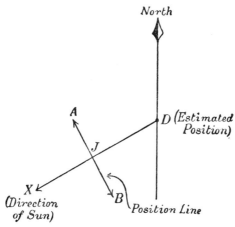

Fig. 118.

a line DX in the direction S 62° W, and taking a point J such that DJ represents eight nautical miles; the straight line AB

drawn through J at right angles to DX is the position line on which the ship's position must be at the time of observation. This position line is also shown later in Fig. 122.

182. *Mercator's chart.*

Unless a ship is manœuvring, its course is generally constant for several hours at least; in these circumstances, its track on the earth's surface is called a *rhumb line* or a *loxodrome*, which can be more precisely defined as a curve on the earth's surface such that the tangent at any point of the curve cuts the meridian through that point at a constant angle. In Fig. 119, $AXZB$ is a rhumb line cutting all the meridians between A and B at a

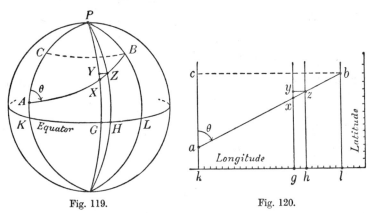

Fig. 119.   Fig. 120.

constant angle θ. It is evident that a projection of the earth's surface on which rhumb lines are represented by straight lines would be of inestimable convenience to the navigator, for then the track of his ship would be *represented* on the chart by a straight line. Such a chart is Mercator's projection. We now state the two main principles of its construction. (i) All rhumb lines on the surface of the earth are represented by straight lines on the chart. (ii) The angle between any two intersecting rhumb lines is correctly represented on the chart; for example, if two rhumb lines intersect at an angle of 30°, then on the chart the angle between the two corresponding straight lines is also 30°. Now the equator is a rhumb line, and all meridians of longitude are rhumb lines intersecting the equator at right angles. Hence on the chart the equator will be represented by a straight line (*kl* in Fig. 120) and the meridians KAP and LBP, for example, will be represented by parallel straight lines *ka*, *lb* perpendicular to *kl*. On the chart the meridians are equally spaced, so that the length of *kl* is proportional to the difference of longitude between K and L (or between A and B). Again, all parallels of latitude are rhumb lines intersecting the meridians at right angles; hence on the chart parallels of latitude such as CB will be represented by straight lines parallel to *kl* (for example, *cb* in Fig. 120) and perpendicular to *ka* and *lb*.

Consider two neighbouring points X and Z on the rhumb line AB (Fig. 119). Draw YZ, a parallel of latitude. We can regard XYZ as a small plane triangle in which $Y\hat{X}Z = \theta$ (the constant course along the rhumb line). If ΔL denotes the angle XPZ (the difference of longitude between X and Z) and ϕ the latitude of Z, then $YZ = \Delta L \cos \phi$. If the latitude of X is $\phi - \Delta\phi$, then $XY = \Delta\phi$. Now $YZ = XY \tan \theta$, so that

$$\Delta\phi \sec \phi = \Delta L \cot \theta \qquad \ldots\ldots(9),$$

in which we suppose that $\Delta\phi$ and ΔL are expressed in circular measure. Now on the chart the meridians of X and Z are represented by the straight lines *gx* and *hz*, and the rhumb line AB is represented by the straight line *ab* making the angle θ with the meridians. The distance *gh* represents the difference of longitude ΔL between the meridians GXP and HZP. Let us suppose, for example, that the scale of the chart is chosen so that one minute of arc of longitude is represented by one millimetre on the chart; then if ΔL is equivalent to n minutes of arc, the length of *gh* will be n mms. But since ΔL is expressed in circular measure, we have $\Delta L = n \sin 1'$. Hence (9) becomes

$$\Delta\phi \sec \phi = n \cot \theta \sin 1' \qquad \ldots\ldots(10).$$

In the small triangle *xyz* (Fig. 120), which represents on the chart the small triangle XYZ on the sphere, we have $y\hat{x}z = \theta$ and $yz \equiv gh = n$ mms. Denote the length of *xy* in millimetres by Δy. Then

$$\Delta y = n \cot \theta \qquad \ldots\ldots(11).$$

From (10) and (11), putting $\sin 1' = 1/3438$, we obtain

$$\Delta y = 3438 \sec \phi \Delta\phi \qquad \ldots\ldots(12).$$

If the difference of latitude between X and Z is $1'$, so that $\Delta\phi$ (in circular measure) is $1/3438$, then by (12) $\Delta y = \sec \phi$ mms.

It is thus seen that as ϕ increases the actual distance (measured in millimetres) between the parallels of latitude ϕ and $(\phi + 1')$ on the chart increases from the equator northwards or southwards, becoming infinite when the north and south poles are reached; for this reason, the north and south poles cannot be represented on the Mercator chart. On the chart, the latitude scale is marked on one or more lines parallel to lb.

From (12), by integration, we have for the length of lb (expressed in millimetres), which we denote by y_2,

$$y_2 = 3438 \log_e \tan\left(\frac{\pi}{4} + \frac{\phi''}{2}\right) \qquad \ldots\ldots(13),$$

in which ϕ'' is the latitude of B. Similarly, if $y_1 = ka$ and ϕ' is the latitude of A,

$$y_1 = 3438 \log_e \tan\left(\frac{\pi}{4} + \frac{\phi'}{2}\right) \qquad \ldots\ldots(14).$$

These formulae give the distances on the chart between the equator and the parallels of latitude, on which A and B are situated, in millimetres, that is to say, in terms of the distance, as unit, on the longitude scale corresponding to one minute of arc of longitude; this latter scale is engraved on kl or on one or more lines parallel to kl.

We now consider the problem of deriving, from measures made on the chart, the distance in nautical miles between any two positions on the earth measured along the rhumb line joining them. Consider the neighbouring points X and Z; on the chart they are represented by x and z. The triangles XYZ and xyz are similar; hence

$$XZ : XY = xz : xy = \sec\theta : 1 \qquad \ldots\ldots(15).$$

If $XY = 1'$ (one nautical mile), then by (15) $XZ = \sec\theta$ nautical miles. Now in the triangle xyz the distance xy *represents* one nautical mile and since, by (15), $xz = xy \sec\theta$, the measurement of xz will give the number of nautical miles between X and Z, provided xz is measured in terms of xy as the unit. The latitude scale thus provides the unit by which a distance measured on the chart gives correctly in nautical miles the rhumb-line distance between the corresponding points on the earth. As this unit varies in length according to $\sec\phi$, the process of finding the number of nautical miles represented by a line such as ab on the chart is theoretically complicated, for it involves the division of ab into a large number of sections and deriving the number of nautical miles in each section by reference to the unit of the latitude scale immediately opposite. In practice, however, it is sufficiently accurate to divide ab into sections of 30 or 40 nautical miles in length and to measure each in terms of the latitude unit opposite the mid-point of the section concerned.

183. *Determination of the ship's position from two observations of altitude.*

We consider now the general problem of determining the ship's position from two observations, making due allowance for the "run" of the ship between the observations, which we shall suppose to be made at times t_1 and t_2. Let D (Fig. 121) on the

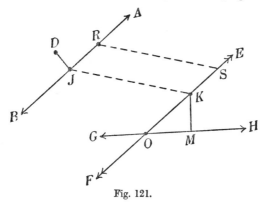

Fig. 121.

chart be the estimated (D.R.) position at t_1. Computing the zenith distance of the heavenly body observed at t_1 for the latitude and longitude of D, and using the corrected observed altitude, we obtain the intercept. Then having found the azimuth of the body we can draw on the chart the position line at t_1, according to the procedure of sections 180 and 181. Let AB be the deduced position line at t_1; then the position of the ship a t_1 is on the line AB. Let JK be parallel to the ship's course and equal in length to the distance travelled in the interval $(t_2 - t_1)$. Through K draw EF parallel to AB. Take any point R on AB and draw RS parallel to JK to meet EF in S. If the ship were actually at J at time t_1, then its position at t_2 would be at K.

Similarly, if its position at t_1 were at R, its position at t_2 would be at S. Hence it is evident that, as the ship's position at t_1 is somewhere on AB, its position at t_2 must lie on EF. This line EF is called the *transferred position line*.

We now utilise the second observation made at t_2. We can use K as the estimated position at this time and, calculating the intercept as before, we obtain the position line GH on which the ship's position must lie at t_2. But the first observation and the application of the ship's run place the ship on EF at t_2. Hence its position at t_2 is at O, the intersection of EF and GH.

The ship's true position at t_1 (if it is required) is obtained by drawing through O a line parallel to JK; the point of intersection of this line with AB is the ship's position at t_1.

It is evident that we can utilise observations of the same body made at t_1 and t_2 (for example, the sun), provided the change of azimuth in the interval is such that the two derived position lines do not intersect at too small an angle.

We illustrate the principles of this section in the following example involving observations of the sun and a star.

184. *Example of finding the ship's position from two observations.*

The estimated position of a ship steaming N 84° E at 8 knots was Lat. 48° 15′ N, Long. 7° 28′ W at U.T. 16ʰ 31ᵐ 2ˢ on 1931 March 10. The following observations were made:

At U.T. 16ʰ 31ᵐ 2ˢ, observed altitude of sun's lower limb was 17° 27′·0.

At U.T. 18ʰ 46ᵐ 10ˢ, observed altitude of Betelgeuse was 48° 55′·0.

To find the ship's position at U.T. 18ʰ 46ᵐ, given that the index error of the sextant is − 2′·0 and that the height of eye is 25 feet.

The first observation is that considered in detail in section 181. We now plot the results on the chart (Fig. 122). D is the estimated position given; DJ is the intercept 8′·0 drawn in the direction S 62° W and AB is the position line at U.T. 16ʰ 31ᵐ.

The interval between the two observations is $2\frac{1}{4}$ hours; the run is therefore 18′·0. We draw JK in the direction N 84° E (the ship's course) and make JK equal to eighteen divisions of the latitude scale opposite. EF is drawn through K parallel to AB; EF is the transferred position line on which the ship must be situated at U.T. 18^h 46^m. We take K as the estimated position at this latter time, and use this position to derive the position line given by the observation of Betelgeuse. From the chart we find that K is the point, Lat. 48° 13' N, Long. 7° 12'·5 W.

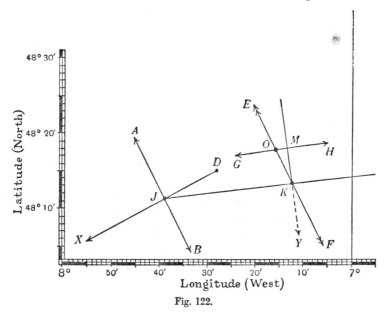

Fig. 122.

We first calculate the hour angle of the star as follows (its right ascension and declination are 5^h 51^m 27^s and + 7° 23'·8 respectively):

	U.T.	18^h	46^m	10^s	(March 10)
		− 12			
G.M.A.T.		6	46	10	
Long. (W)		− 28	50		
H.A.M.S.		6	17	20	
R.A.M.S.*		23	9	54	
Sid. time		5	27	14	
R.A. star		5	51	27	
H.A.		23	35	47	

* The almanacs give the sidereal time at midnight for every day in the year; the tabulated quantities + 12^h evidently give the values of the R.A.M.S. at each midnight.

If X denotes the position of Betelgeuse on the celestial sphere
and Z the zenith of the point K, we have

$$PX = 82° 36'·2; \quad PZ = 41° 47'·0; \quad Z\hat{P}X \equiv 24^{h} - \text{H.A.} = 0^{h} 24^{m} 13^{s}.$$

We calculate ZX by the cosine-formula:

log cos PX	9·109 71	log sin PX	9·996 37
log cos PZ	9·872 55	log sin PZ	9·823 68
		log cos $Z\hat{P}X$	9·997 57
	8·982 26		9·817 62

Hence $\cos ZX = 0·096\ 00 + 0·657\ 08 = 0·753\ 08.$

∴ $ZX = 41° 8'·5.$

The azimuth is found by calculation to be 171° E, so that the
bearing of the star is S 9° E—in the direction KY in Fig. 122.

We now correct the observed altitude:

Observed alt. of star...	...	...	...	48° 55'·0
Index error	...	...	...	...	− 2 ·0
Dip	...	...	...	...	− 5 ·0
Refraction	...	...	...	...	− 0 ·9
∴ corrected observed alt. ...				= 48° 47'·1
Thus the true zenith distance is		...		41° 12'·9
But the calculated zenith distance is		41° 8'·5

Hence the intercept is 4'·4.

As the calculated zenith distance is smaller than the true
zenith distance, the point K is evidently within the position
circle given by the observation of Betelgeuse. Hence to obtain
the position line on the chart we draw KM equal to the intercept
(4'·4), and in the direction *opposite* to that given by the star's
azimuth. The line GH drawn through M perpendicular to KM is
the position line resulting from the observation of the star.

The ship's position at U.T. $18^{h} 46^{m}$ is given by O, the intersection of GH with EF. From the chart it is found that O is the
position: Lat. 48° 17'·8 N, Long. 7° 16' W.

185. *Special methods.*

When the pole star is observed the position line can be obtained very easily by means of simple tables based on the formula in Exercise 20, p. 55. These tables are given in the
Nautical Almanac (abridged), the *American Ephemeris* and

Inman's Tables. Since the bearing of the pole star is in general very nearly north, the resulting position line is practically a parallel of latitude; thus an observation of the pole star yields the ship's latitude.

When the hour angle of the heavenly body observed is within 30 or 40 minutes from the meridian, simple tables based on the formula of Exercise 21, p. 55, facilitate the calculation of the zenith distance. Such observations are known as *ex-meridian* observations.

In the general problem, the arithmetical computations can be greatly lightened by means of special tables, amongst which may be mentioned *Altitude Tables** by R. de Aquino and *Position Line Tables (Sine Method)** by W. M. Smart and F. N. Shearme, the general principles of which are used in other tables. The position circle on the earth is deducible from two items of observation: (i) the observed altitude of the heavenly body concerned, (ii) the U.T. at which the observation is made. As the position circle is independent of the estimated position of the ship, so also is the position line on the chart, and although we employ the estimated position (D) to derive the position line, any other point C within 30 or 40 miles of D would serve equally well. In the sine method, a particular point C is selected in a way that can be best illustrated by means of an example. Suppose that the ship's estimated position D is: Lat. 48° 39' N, Long. 7° 18' W, and that the hour angle of the heavenly body computed for D is $2^h 18^m 32^s$. We choose the latitude of C to be the integral number of degrees nearest to that of D and its longitude such that the hour angle for C is the nearest multiple of 4^m. In this example, the latitude of C is 49° N and its longitude such that the hour angle for C is $2^h 20^m$; hence C must be $1^m 28^s$ or 22'·0 *east* of D, so that its longitude is 6° 56' W. The point C is plotted on the chart. The choice of C in the general case—its latitude a whole number of degrees and its longitude such that the hour angle of the heavenly body is a multiple of 4^m—allows the tabulation in a compact form of certain quantities which are used in the subsequent calculations.

In Fig. 123 let Z be the zenith of C, X the heavenly body (declination δ) and H the hour angle for the longitude of C. We

* Published by J. D. Potter, 145 Minories, London, E. 1.

shall first suppose that this hour angle is within 6^h of the meridian. Draw the great circle arc ZY to cut the great circle PX at right angles. Let PY and ZY be denoted by U and p respectively, and let ϕ denote the latitude of C. From the right-angled triangle PZY, in which $PZ = 90° - \phi$, we have

$$\tan U = \cot \phi \cos H \qquad \ldots\ldots(16).$$
$$\cos p = \sin \phi \sec U \qquad \ldots\ldots(17).$$

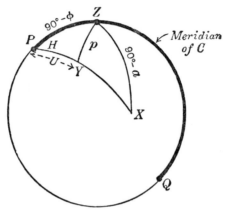

Fig. 123.

Let the zenith distance ZX be denoted by $(90° - a)$; we call a the calculated altitude. From the triangle ZYX, in which $YX = PX - U = 90° - \delta - U$, we have

$$\sin a = \sin (\delta + U) \cos p \qquad \ldots\ldots(18).$$

The values of U, calculated by means of (16), for each degree of latitude and at intervals of 4^m in the hour angle, from $0^h\,0^m$ to $6^h\,0^m$ or from $18^h\,0^m$ to $24^h\,0^m$, are given in the tables. Adjacent to each of these entries, the corresponding values of log cos p, given by (17), are tabulated. The computation of the calculated altitude a follows simply from (18). The difference between a so derived and the observed altitude (corrected) is the intercept, which, of course, has to be drawn from C on the chart. In this way the position line can be drawn with a minimum of arithmetical computation.

When the hour angle (H) of the heavenly body is more than 6^h from the meridian, a slight modification is necessary. Let H_1

denote $(12^h - H)$ or $(H - 12^h)$. The quantities log cos p and U are taken from the tables for the appropriate value of H_1 and the formula corresponding to (18) is

$$\sin a = \sin (\delta - U) \cos p.$$

186. *Equation of a great circle on Mercator's chart.*

The great circle distance between two points on the earth's surface is shorter than the rhumb-line distance, and the difference becomes of economic importance in long oceanic voyages. In such instances it is the practice to follow the appropriate great circle as nearly as possible. Suppose that the great circle is represented on the chart. It is divided into a suitable number of sections, and between the beginning and end of a section the ship follows the appropriate rhumb-line course (represented by the straight line joining the ends of the section), which is easily obtained from the chart. When the ship reaches the end of a section, its course is altered to that corresponding to the next section; and so on.

To find the equation of the curve representing the great circle joining two points U and V (Fig. 124) we proceed as follows. Let the great circle UV cut the equator in O and let the longitude (measured east) of O be L_0, PG being the Greenwich meridian. Let the inclination of UV to the equator (in the direction of increasing east longitudes) be denoted by i; i can have all values between $0°$ and $180°$. Let L and ϕ denote the east longitude and north latitude respectively of any point X on the great circle UV. Let x and y be the co-ordinates

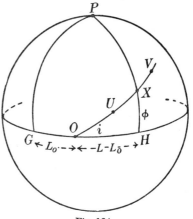

Fig. 124.

of X on the chart; we shall suppose that x and y are expressed in terms of the radian as unit. Then we have

$$x = L \qquad \qquad \dots\dots(19),$$

where L is expressed in circular measure, and, by (13),

$$y = \log_e \tan \left(\frac{\pi}{4} + \frac{\phi}{2} \right) \qquad \ldots\ldots(20).$$

From the spherical triangle XOH, in which H is the intersection with the equator of the meridian through X, we have: $OH = L - L_0$, $HX = \phi$, $X\hat{O}H = i$ and $X\hat{H}O = 90°$. Formula **D** gives

$$\tan \phi = \sin (L - L_0) \tan i \qquad \ldots\ldots(21).$$

But (20) can be written

$$e^y = \tan \left(\frac{\pi}{4} + \frac{\phi}{2} \right) = \frac{1 + \sin \phi}{\cos \phi} = \frac{\cos \phi}{1 - \sin \phi}.$$

Hence

$$\sec \phi + \tan \phi = e^y,$$

$$\sec \phi - \tan \phi = e^{-y},$$

from which

$$\tan \phi = \tfrac{1}{2} (e^y - e^{-y}) = \sinh y \qquad \ldots\ldots(22).$$

Hence, using (19) and (22) with (21), we obtain the equation of the great circle in the form

$$\sinh y = \sin (x - L_0) \tan i \qquad \ldots\ldots(23),$$

in which we regard L_0 and i as two constants associated with the great circle. The values of L_0 and i can now be expressed in terms of the longitudes L_1 and L_2 of U and V respectively, and the corresponding latitudes ϕ_1 and ϕ_2. From (21), we have

$$\left. \begin{aligned} \tan \phi_1 &= \sin (L_1 - L_0) \tan i \\ \tan \phi_2 &= \sin (L_2 - L_0) \tan i \end{aligned} \right\} \qquad \ldots\ldots(24).$$

By division,

$$\frac{\tan \phi_1}{\tan \phi_2} = \frac{\sin (L_1 - L_0)}{\sin (L_2 - L_0)},$$

from which we obtain

$$\frac{\tan \phi_2 - \tan \phi_1}{\tan \phi_2 + \tan \phi_1} = \frac{\sin (L_2 - L_0) - \sin (L_1 - L_0)}{\sin (L_2 - L_0) + \sin (L_1 - L_0)},$$

giving, on further simplification,

$$\tan \left(\frac{L_2 + L_1}{2} - L_0 \right) = \tan \left(\frac{L_2 - L_1}{2} \right) \cdot \frac{\sin (\phi_2 + \phi_1)}{\sin (\phi_2 - \phi_1)}$$

$$\ldots\ldots(25).$$

The value of L_0 is calculated from (25). $\tan i$ is then obtained from one of the formulae (24).

To find the position of a point on the curve with any assumed longitude L, we calculate the corresponding latitude ϕ by means of (21). The x and y co-ordinates are then given by (19) and (20).

The plotting of a great circle track on the Mercator chart can be much simplified by means of the Gnomonic Projection.

EXERCISES

1. Using the effect of refraction as given in section 178, p. 318, prove that the distance, in nautical miles, of the horizon for an observer h feet above sea-level is

$$\left(\frac{26}{11}\frac{h}{a}\right)^{\frac{1}{2}} \text{cosec } 1',$$

where a is the earth's radius in feet.

2. An observer, on the mast of a ship, 80 feet above sea-level, can just see a light which is 100 feet above sea-level. Show that his distance from the light is $21\frac{3}{4}$ nautical miles.

3. Show that in a place whose latitude is ϕ sunrise at the equinoxes will be visible at the top of a mountain h feet high about $4 \sqrt{h} \sec \phi$ seconds sooner than at its foot. [*Coll. Exam.*]

4. The path of the setting sun makes an angle θ with the horizon. Prove that if the sun's declination is δ a mountain in latitude ϕ, of height $\dfrac{1}{n}$ of the earth's radius, will have its summit illuminated

$$12 \sqrt{2} \text{ cosec } \theta \sec \delta/\pi \sqrt{n} \text{ hours}$$

after the sun has set on the plain at its base.

Determine to the nearest minute the value of this expression at the summer solstice for a mountain three miles high in latitude $45°$. [*Coll. Exam.*]

5. The angle of depression of the sun's upper limb at setting is observed from an eminence to be d, and from a neighbouring eminence h feet higher to be $d + \Delta d$, where Δd is expressed in seconds of arc. Prove that the earth's radius in feet is approximately

$$\frac{h \cos d \cot d}{\Delta d \sin 1''} \cdot$$

6. If s is the length of the rhumb line joining two points in latitudes ϕ_1 and ϕ_2 and θ is the angle which the rhumb line makes with the meridians, prove that

$$s = a (\phi_2 - \phi_1) \sec \theta,$$

where a is the earth's radius. [*Ball.*]

7. Show that on Mercator's projection the equation of a small circle on the earth (assumed spherical) is of the form

$$\cosh y/c \sec x/c = \text{const.},$$

provided that the small circle does not include a pole of the earth and x, y are measured on the chart from a certain origin.

Obtain the radius of curvature of this curve, at a point where the tangent lies N and S, in the form

$$\rho = c \sin \theta (\cos^2 \delta - \sin^2 \theta)^{-\frac{1}{2}},$$

where δ is the latitude of the centre of the above small circle and θ is its angular radius.

Determine the error made in neglecting the curvature of the "position line" when determining longitude from a timed observation of altitude $30°$ in latitude $51° 30'$, the error in the assumed latitude being $100'$; the observation is made near the prime vertical and the assumed latitude corresponds to an ideal observation exactly on the prime vertical. [*Lond.* 1926.]

8. An observation of the altitude of a star at a place of known latitude ϕ is made when the star is on the prime vertical. Show that an error in ϕ will have no sensible effect on the calculated longitude.

9. The latitudes and west longitudes of two places F and G are (ϕ_1, λ_1) and (ϕ_2, λ_2) respectively. When simultaneous observations of the same heavenly body (declination δ) are made at F and G, the corrected altitude is a in each case. Prove that

$$\sin^2 (\lambda_2 - \lambda_1) \cos^2 \delta = P_1{}^2 - 2P_1 P_2 \cos (\lambda_2 - \lambda_1) + P_2{}^2$$
$$= S_1{}^2 - 2S_1 S_2 \cos (\lambda_2 - \lambda_1) + S_2{}^2,$$

where $\qquad P_i = \sin a \sec \phi_i - \sin \delta \tan \phi_i$ $\Big\}$ $\quad (i = 1, 2),$

and $\qquad S_i = \cos a \sin A_i$

A_1 and A_2 being the azimuths at F and G respectively. [*Lond.* 1930.]

10. At U.T. $15^h 17^m 48^s$ on a certain date the sun's observed altitude (corrected) was $89° 34'$. Assuming that the ship's latitude was known to be $23° 3'$, that the sun's declination was $23° 13'$ and the equation of time $+ 0^m 12^s$, prove that the ship's longitude was either $49° 4'$ W or $49° 56'$ W.

11. When the latitude and longitude are found by simultaneous observations of the altitudes a_1, a_2 of two known stars, prove that the two possible places of observation will have the same longitude if

$$\sin a_1/\sin a_2 = \sin \delta_1/\sin \delta_2,$$

where δ_1, δ_2 are the declinations of the stars. [*Coll. Exam.*]

12. Assuming that an observation of the sun's altitude is made correctly but that the U.T. used in the subsequent calculations is in error by Δt seconds. show that the resultant position line is displaced through $\frac{1}{4}\Delta t \sin A \cos \phi$ nautical miles, where A is the sun's azimuth and ϕ is the latitude.

[*Lond.* 1930.]

13. Two altitudes a_1, a_2 of the sun are taken, at an interval of time $2h$, and the position lines cut orthogonally. Show that

$$\sin a_1 \sin a_2 = 1 - 2 \sin^2 h \cos^2 \delta.$$ [*Coll. Exam.*]

14. In determining the hour angle of a star a sailor makes errors Δa in his altitude and $\Delta\phi$ in his assumed latitude. Show that the error in hour angle will be given by

$$\Delta H = \Delta\phi \cot A \sec \phi - \Delta a \sec \phi \operatorname{cosec} A,$$

where ϕ is the latitude and A is the azimuth. [*M.T.* 1923.]

15. At 6 p.m. a star X is observed (all corrections having been applied) as follows: altitude, 33° 20′; true bearing, 220°.

At 7 p.m. a star Y is observed similarly, as follows: altitude, 63° 50′; true bearing, 290°.

The calculated altitudes are *both* computed for the position Lat. 60° N, Long. 30° W: they are 33° 15′ and 63° 56′ respectively. If the ship is steaming 070° at 10 knots find, by plotting on squared paper, the position of the ship at 6 p.m. and at 7 p.m. [*Lond.* 1928.]

16. An intercept p is drawn in the direction θ east of north from the position: east longitude, L; north latitude, ϕ. If the true coordinates of the ship are $L + \Delta L$, $\phi + \Delta\phi$ prove that

$$\Delta L \cos \phi \sin \theta + \Delta\phi \cos \theta - p = 0$$

where ΔL, $\Delta\phi$ and p are expressed in minutes of arc.

17. A ship's navigator observes that the altitude of a known star as it crosses the southern meridian is $a°$, and the time of the transit is t hours after Greenwich mean noon. The star's declination is $\delta°$ and its right ascension is a hours. From the almanac he finds that the mean sun's right ascension was σ hours at Greenwich mean noon on the day of the observation. Show that the ship is in latitude $90° + \delta - a$ and west longitude $15\left(\sigma - a + t\,\dfrac{366\frac{1}{4}}{365\frac{1}{4}}\right)$, both measured in degrees. [*M.T.* 1921.]

CHAPTER XIV

BINARY STAR ORBITS

187. *Visual binary stars.*

The name "double star" is ordinarily applied to a pair of stars seen very close together in the telescope. This apparent closeness is due to one of two causes: (i), the two stars may be at greatly differing distances from the earth, but nearly in the same direction as viewed from the earth; (ii), they may actually be close together in space, forming a system in which their mutual gravitational attraction might be expected to be exhibited by orbital motion. In 1803, Sir W. Herschel first demonstrated, from observations carried on over a quarter of a century, that certain double stars showed relative orbital motion. A large number of stars of this character have since been discovered, and to these the name *binary stars* is given. The class of double stars in (i) above are called *optical double stars*; we do not consider these further.

Generally, the two members of a binary star are of unequal brightness. The brighter star is called the *primary* and the fainter is called the *companion.*

In Fig. 125, let A denote the primary and B the companion. Let AN define the direction of the north celestial pole—AN is then part of the meridian through A. The angle NAB, denoted by θ, is the *position angle* of B with respect to A. The position angle is measured from 0° to 360° eastwards, in the direction indicated by the arrow. The angular distance between A and B is called simply the distance AB or, sometimes, the separation, and is generally denoted by ρ. Thus ρ and θ define the position of B with respect to A. If A and B form a binary system then,

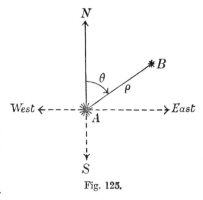

Fig. 125.

owing to the mutual gravitational attraction of the two stars—
which we assume to be given by the same law as in the planetary
system—the companion will describe an elliptic orbit relative to
the primary. This is the *true orbit* and its plane is the true orbital
plane. In general, this plane will be distinct from the plane
perpendicular to the line of sight and consequently the observed
orbit, called the *apparent orbit*, will be the projection of the true
orbit on the plane perpendicular to the line of sight. This latter
plane is the plane of the apparent orbit. The observations furnish
the details concerning the apparent orbit, and it will be our
purpose to show how the elements of the true orbit can be
deduced. The importance of this subject in astronomy lies in the
fact that valuable information concerning the masses of the
stars can be obtained.

188. *The micrometer.*

The instrument used in practice for measuring* distance and
position angle is the micrometer. This is attached to the eye-end
of the telescope. In the field of view of the micrometer eye-piece
are generally three spider-wires, one of which, XY (Fig. 126),

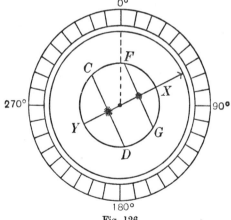

Fig. 126.

is fixed centrally and two, CD and FG, are perpendicular to XY.
The plate to which these wires are fitted can be rotated about
the telescopic axis. By adjustment, the images of the two stars

---

* For the more open binaries, the measures can also be obtained photographically.

A and *B* lie on *X Y*. A graduated circle enables the position angle θ to be derived, provided the reading corresponding to the direction *ON* (position angle 0°) can first be obtained. This is effected in practice as follows. The telescope is directed to any star and its image placed on *X Y*. If the telescope is stopped, the diurnal motion will carry the star along a parallel of declination; if *X Y* is perpendicular to the meridian *ON*, the star will thus appear to travel along *X Y*. The procedure then is to rotate the plate carrying the wires until this condition is fulfilled. In this position, the reading on the graduated scale corresponds to position angle 90° or 270°; hence the reading corresponding to position angle 0° is deduced. Alternatively, by rotating the graduated circle about the optical axis of the telescope, the scale can be made to correspond exactly with position angle; in this case, no correction is necessary to the readings.

The distance ρ between *A* and *B* is derived by means of the two wires *CD* and *FG*, each of which can be moved perpendicularly to *X Y* by fine screws to which micrometer heads are attached. When the instrument is adjusted, as shown in Fig. 126, the observation for distance consists in placing *CD* over *A* and *FG* over *B*. The micrometer readings furnish the value of ρ, expressed in terms of the number of revolutions of the micrometer heads. From observations made on two stars whose angular distance is accurately known, the value of one revolution of a micrometer can be expressed in seconds of arc.

The complete observation of a binary star leads to the value of the distance ρ, expressed in seconds of arc, and the value of the position angle θ.

189. *The elements of the true orbit of a visual binary.*

Let a sphere be drawn with the primary star *S* as centre (Fig. 127). The straight line joining the earth to *S* cuts the sphere at *E*. The plane of the great circle, of which *E* is the pole, represents the plane of the apparent orbit. Let *K* be the other pole. We suppose that the great circle *HLG* defines the plane of the true orbit of the companion around *S*, the angle *GLD* being the inclination *i*. The straight line *MSL*, which is the intersection of the two planes considered, is the line of nodes. Let the radius *SN* define the direction of position angle 0°. The true orbit of the companion relative to the primary S is shown shaded in Fig. 127. When the companion is at F in its true orbit, its position in the apparent orbit will be obtained by drawing a perpendicular from F to the plane of the great circle NLD. This perpendicular will lie in the plane of the great circle KGD, where G is the point on the sphere obtained by producing SF. Thus the observed position of the companion will lie on the radius SD and therefore its position angle θ is the arc ND. (We assume position angles to be measured in the sense NLD.)

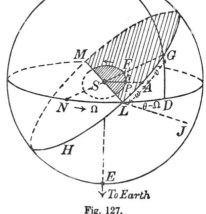

Fig. 127.

If r is the radius vector SF, the distance ρ is given by

$$\rho = r \cos GD \ \ ...(1).$$

The position angle of that node which is less than $180°$ from N is denoted by Ω; in Fig. 127, $NL = \Omega$ and consequently $LD = \theta - \Omega$.

Let P be the point on the true orbit at which the companion is nearest the primary; this point is called *periastron*; the other extremity of the major axis is called *apastron*. The semi-major axis (expressed in seconds of arc) of the true orbit will be denoted by a and the eccentricity by e. Let v be the true anomaly when the companion is at F. Then assuming that the companion moves in the direction of the arrow near P, the angle FSP is v and therefore $AG = v$. Let ω denote the arc LA. Then $LG = v + \omega$. Hence we obtain, from the triangle LGD right-angled at D,

$$\cos LG = \cos LD \cos GD,$$

or $$\cos GD = \cos (v + \omega) \sec (\theta - \Omega),$$

or, using (1), $$\rho = r \cos (v + \omega) \sec (\theta - \Omega) \quad ......(2).$$

Also, from the triangle LGD, we obtain

$$\tan (\theta - \Omega) = \tan (v + \omega) \cos i \quad ......(3).$$

12-2

We denote the orbital period by T (in years). The mean angular motion n of the companion around S in the true orbit is given by

$$n = 2\pi/T \qquad \ldots\ldots(4).$$

Let the time of periastron passage be denoted by τ. Then the mean anomaly M at time t is given by

$$M = n\,(t - \tau) = E - e\sin E \qquad \ldots\ldots(5),$$

where E is the eccentric anomaly. The true anomaly v is related to E by

$$\tan\frac{v}{2} = \left(\frac{1+e}{1-e}\right)^{\frac{1}{2}}\tan E/2 \qquad \ldots\ldots(6).$$

We have also $\qquad n^2a^3 = \dfrac{4\pi^2a^3}{T^2} = G\,(m_1 + m_2) \qquad \ldots\ldots(7),$

where m_1, m_2 are the masses of the two stars and G is the constant of gravitation. We regard T as an element, since the sum of the masses is defined in terms of a and T.

The elements of the true orbit are: a, e, i, Ω, ω, τ and T. If all the elements are known, the value of E at time t can be derived from (4) and (5), and then the true anomaly v is found by means of (6). Then from (3) and (2), the values of θ and ρ can be finally obtained. In this way, the observed quantities ρ and θ at any time can be compared with the calculated values based on the elements of the true orbit. The problem with which we shall be more intimately concerned is the derivation of the elements of the true orbit from the observations which enable the apparent orbit to be drawn.

It is to be remarked that although the value of the inclination may be known, the orbital plane is not uniquely determined. So far as observations with the micrometer are concerned, it is uncertain whether the true orbital plane is that given by the great circle HLG, or that given by the great circle of which an arc LJ only is shown in Fig. 127, the angle DLJ being also i. Now as regards the orbital plane HLG, in which the angular motion of the companion is in the direction from L towards A, it is evident that as the companion passes through L the component of its orbital motion in the line of sight (that is, perpendicular to the plane of the apparent orbit) is directed away from the earth. In this instance, the inclination is denoted by $+\,i$. If the companion's angular motion is in the direction LJ, the inclination is denoted by $-\,i$.

When the position angles increase with the time, as in the two orbits just considered whose planes are defined by LG and LJ, the motion is *direct*. When the position angles decrease, as they would do if the angular motion were in the direction GL or JL, the motion is *retrograde*.

190. *The apparent orbit of a visual binary.*

As we are considering the orbit of the companion relative to the primary, the latter will be situated at a focus of the true elliptic orbit. Now the apparent orbit is the projection of the true orbit on the plane perpendicular to the line of sight and is also an ellipse. But it does not necessarily follow that the focus of the true ellipse (that is, S) projects into a focus of the apparent ellipse. Let the ellipse in Fig. 128 represent the apparent orbit and S the primary. From what has just been said, S is generally not at a focus of this ellipse. If SN denotes the direction defining position angle $0°$ and SM that of $\theta = 90°$, the

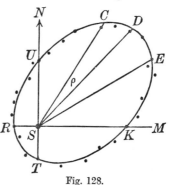

Fig. 128.

general equation of the ellipse referred to SN and SM as x and y axes respectively is

$$Ax^2 + 2Hxy + By^2 + 2Gx + 2Fy + 1 = 0 \quad ......(8),$$

in which there are five independent constants A, B, ... F, defining the particular ellipse concerned. If the companion is at C, an observation gives ρ and θ, from which the rectangular co-ordinates x and y of C are given by

$$x = \rho \cos \theta, \quad y = \rho \sin \theta.$$

Five such observations spread over the orbit are in theory sufficient to determine the five constants A, B, ... F of (8), but, owing to the unavoidable errors in measuring ρ and θ, the ellipse cannot be determined accurately in this way. A large number of observations, spread over many years, give a series of points such as C, D, E, ... from which a preliminary ellipse can be drawn. The test as to the accuracy of this ellipse is afforded by the following consideration. In the true orbit the areas swept out by the moving radius vector are, by Kepler's second law, proportional to the relevant intervals of time, and the ratio of any two such areas is not altered by projection. Hence, in the apparent orbit, if t_1, t_2, t_3 are the times at which the companion is successively at C, D and E, the ratio of area CSD to area DSE is equal to the ratio of $(t_2 - t_1)$ to $(t_3 - t_2)$. The latter ratio is, of course, accurately known. This test can be rapidly carried out with a planimeter, and the preliminary ellipse is modified until the requirements indicated are fulfilled. We assume now that the apparent ellipse has been satisfactorily drawn.

The constants A, B, ... F of the general equation (8) can be conveniently derived as follows. Let the co-ordinate axes cut the apparent ellipse (drawn to a convenient scale) in K, R, U and T. Let the co-ordinates of U and T be $(x_1, 0)$ and $(- x_2, 0)$. Since these co-ordinates satisfy (8), we have

$$A x_1{}^2 + 2G x_1 + 1 = 0,$$
$$A x_2{}^2 - 2G x_2 + 1 = 0,$$

from which A and G are determined. A similar procedure for the points K and R enables B and F to be found. If (ξ, η) are the measured co-ordinates of a point D on the apparent ellipse, H is found from the relation

$$- 2H\xi\eta = A\xi^2 + B\eta^2 + 2G\xi + 2F\eta + 1.$$

Amongst the many methods of deriving the elements of the true ellipse, there are the two well-known methods of Kowalsky and Zwiers, which we shall describe in detail. In each method the apparent orbit forms the basis of the subsequent procedure.

191. *Kowalsky's method of determining the elements of a visual binary orbit.*

The constants A, B, ... F of the equation (8) for the apparent orbit are first derived according to the method given, or otherwise, the x-axis (Fig. 129) being SN (in the direction of position angle 0°) and the y-axis being SR (position angle 90°). K is the pole of the great circle NLR. We take SK as the z-axis.

Let SA be the direction of periastron and C the pole of the plane of the true orbit. Then since S is the focus of the true orbit of the companion relative to the primary, the equation of the orbital ellipse referred to rectangular axes SA and SB in the plane of the true orbit is

$$\frac{(\xi + ae)^2}{a^2} + \frac{\eta^2}{b^2} = 1 \qquad \ldots \ldots (9),$$

in which $b^2 = a^2 (1 - e^2)$. The apparent orbit, whose equation is (8), is the projection of the ellipse, given by (9), on the plane NLR.

Let (l_1, m_1, n_1) be the direction-cosines of SA with respect to

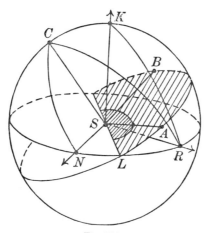

Fig. 129.

the axes SN, SR and SK. Then if A is joined by great circle arcs to N, R and K, we have

$$l_1 = \cos AN, \quad m_1 = \cos AR, \quad n_1 = \cos AK.$$

Similarly, let (l_2, m_2, n_2) and (l_3, m_3, n_3) be the direction-cosines of SB and SC with respect to SN, SR and SK. Then

$$l_2 = \cos BN, \quad m_2 = \cos BR, \quad n_2 = \cos BK,$$

and $\qquad l_3 = \cos CN, \quad m_3 = \cos CR, \quad n_3 = \cos CK.$

From the spherical triangles ANL, ARL, AKL we obtain

$$\left.\begin{aligned} l_1 &= \cos \Omega \cos \omega - \sin \Omega \sin \omega \cos i \\ m_1 &= \sin \Omega \cos \omega + \cos \Omega \sin \omega \cos i \\ n_1 &= \sin \omega \sin i \end{aligned}\right\} \quad \ldots \ldots (10).$$

From the triangles BLN, BLR and BLK—or simply writing $(90° + \omega)$ for ω in (10)—we obtain the direction-cosines of SB:

$$\left.\begin{aligned}
l_2 &= -\cos\Omega\sin\omega - \sin\Omega\cos\omega\cos i \\
m_2 &= -\sin\Omega\sin\omega + \cos\Omega\cos\omega\cos i \\
n_2 &= \cos\omega\sin i
\end{aligned}\right\} \quad \dots\dots(11).$$

From the triangles CLN, CLR, CLK we obtain

$$\left.\begin{aligned}
l_3 &= \sin\Omega\sin i \\
m_3 &= -\cos\Omega\sin i \\
n_3 &= \cos i
\end{aligned}\right\} \quad \dots\dots(12).$$

Amongst the well-known relations connecting the various direction-cosines, we shall use the following:

$$\begin{aligned}
l_1 m_2 - l_2 m_1 &= n_3 &\dots\dots(13),\\
l_1{}^2 + l_2{}^2 + l_3{}^2 &= 1 &\dots\dots(14),\\
m_1{}^2 + m_2{}^2 + m_3{}^2 &= 1 &\dots\dots(15),\\
l_1 m_1 + l_2 m_2 + l_3 m_3 &= 0 &\dots\dots(16).
\end{aligned}$$

Let (ξ, η) denote the co-ordinates of any point on the true ellipse with reference to the axes SA, SB, and let (x, y) be the co-ordinates of the projection of this point on the plane NLR with reference to the axes SN, SR. Then from the circumstances of projection, we have

$$x = l_1\xi + l_2\eta,$$
$$y = m_1\xi + m_2\eta,$$

from which
$$(l_1 m_2 - l_2 m_1)\,\xi = m_2 x - l_2 y,$$

or, using (13),
$$\xi = \frac{m_2 x - l_2 y}{n_3} \quad \dots\dots(17).$$

Similarly,
$$\eta = -\frac{m_1 x - l_1 y}{n_3} \quad \dots\dots(18).$$

Now these values of ξ and η satisfy (9); hence

$$\frac{(m_2 x - l_2 y + aen_3)^2}{a^2 n_3{}^2} + \frac{(m_1 x - l_1 y)^2}{b^2 n_3{}^2} = 1 \quad \dots\dots(19).$$

But x and y satisfy (8); hence (19) and (8) must represent the same ellipse on the plane of the apparent orbit. It follows that the coefficients of x^2, xy, y^2, ... in (8) must be severally proportional to the coefficients of x^2, xy, y^2, ... in (19). Let f denote the common ratio. Then we have

$$A = \frac{f}{n_3^2}\left(\frac{m_2^2}{a^2} + \frac{m_1^2}{b^2}\right)$$

$$B = \frac{f}{n_3^2}\left(\frac{l_2^2}{a^2} + \frac{l_1^2}{b^2}\right)$$

$$H = -\frac{f}{n_3^2}\left(\frac{l_2 m_2}{a^2} + \frac{l_1 m_1}{b^2}\right)$$

$$G = \frac{f e m_2}{a n_3}$$

$$F = -\frac{f e l_2}{a n_3}$$

$$\quad\quad\quad\quad \cdots\cdots(20).$$

and

$$1 = -f(1 - e^2)$$

From these equations we obtain, after a simple reduction,

$$F^2 - G^2 + A - B = \frac{(l_1^2 - m_1^2 + l_2^2 - m_2^2)}{n_3^2 a^2 (1 - e^2)^2}$$

$$= \frac{m_3^2 - l_3^2}{n_3^2 a^2 (1 - e^2)^2},$$

by the aid of (14) and (15). Inserting the values of l_3, m_3 and n_3 given by (12), we have

$$F^2 - G^2 + A - B = \frac{\cos 2\Omega \tan^2 i}{p^2} \quad \cdots\cdots(21),$$

where

$$p = a(1 - e^2) \quad \cdots\cdots(22).$$

Again, we have $FG - H = -\dfrac{(l_1 m_1 + l_2 m_2)}{n_3^2 a^2 (1 - e^2)^2}$

$$= \frac{l_3 m_3}{n_3^2 p^2},$$

by the aid of (16) and (22). Hence, from (12),

$$FG - H = -\frac{\sin 2\Omega \tan^2 i}{2p^2} \quad \cdots\cdots(23).$$

From (21) and (23), there results

$$(F^2 - G^2 + A - B)\sin 2\Omega + 2(FG - H)\cos 2\Omega = 0 \ \cdots(24),$$

from which Ω can be determined. It is to be remembered that the value of Ω, according to our convention, lies between $0°$ and $180°$.

Using the value of Ω just found, the value of $\dfrac{\tan^2 i}{p^2}$ is found by either (21) or (23).

Again, $\quad F^2 + G^2 - (A + B) = \dfrac{(l_1{}^2 + m_1{}^2 + l_2{}^2 + m_2{}^2)}{n_3{}^2 a^2 (1 - e^2)^2}$

$$= \dfrac{(2 - n_1{}^2 - n_2{}^2)}{p^2 \cos^2 i},$$

using (14) and (15). But by (10) and (11),

$$2 - n_1{}^2 - n_2{}^2 = 2 - \sin^2 i$$

$$= 2 \cos^2 i + \sin^2 i.$$

Hence $\qquad F^2 + G^2 - (A + B) = \dfrac{2}{p^2} + \dfrac{\tan^2 i}{p^2} \qquad \ldots\ldots(25).$

But $\dfrac{\tan^2 i}{p^2}$ has already been determined; hence the value of p^2 can be determined from (25). When p has been found, the value of $\tan^2 i$—and hence the inclination i—can be calculated (as mentioned on p. 344 the sign of i is indeterminate).

Expressing f as $-a/p$ by (22), we have from (20),

$$G = - \frac{e}{p} \frac{m_2}{\cos i},$$

$$F = \frac{e}{p} \frac{l_2}{\cos i}.$$

Inserting the values of m_2 and l_2 from (11), we have

$$\sin \Omega \sin \omega - \cos \Omega \cos \omega \cos i = \frac{Gp \cos i}{e} \qquad \ldots\ldots(26),$$

$$\cos \Omega \sin \omega + \sin \Omega \cos \omega \cos i = - \frac{Fp \cos i}{e} \qquad \ldots(27).$$

Multiply (26) by $\sin \Omega$ and (27) by $\cos \Omega$ and add. Then

$$\sin \omega = \frac{p}{e} \cos i \, (G \sin \Omega - F \cos \Omega) \qquad \ldots\ldots(28).$$

Multiply (26) by $\cos \Omega$ and (27) by $\sin \Omega$ and subtract. Then

$$\cos \omega \cos i = - \frac{p}{e} \cos i \, (G \cos \Omega + F \sin \Omega) \ldots\ldots(29).$$

Hence, from (28) and (29),

$$\tan \omega = \frac{(F \cos \Omega - G \sin \Omega) \cos i}{F \sin \Omega + G \cos \Omega} \qquad \ldots\ldots(30).$$

This formula allows ω to be calculated (i and Ω being known). The eccentricity e can then be found from (28) or (29), for all the quantities ω, p, i, Ω have now been evaluated. The semi-major axis a can now be derived from (22). We have thus shown how the elements a, e, i, Ω and ω can be obtained from the apparent ellipse.

There remains the problem of finding the period T and τ (the time of periastron passage). For any point on the apparent ellipse corresponding to time t, we have the values of ρ and θ. Now by (3),
$$\tan (v + \omega) = \tan (\theta - \Omega) \sec i,$$
from which the true anomaly v can be found. The eccentric anomaly E can then be calculated from (6). The mean anomaly M is now found from
$$M = E - e \sin E.$$
But $M = n (t - \tau)$, or
$$M = \frac{360°}{T} (t - \tau) \qquad \ldots\ldots(31).$$

Hence, for any time t, we have an equation (31) involving two unknowns T and τ. Theoretically two equations of the type (31) are sufficient to determine T and τ. In practice, however, their values are generally determined by a least-square solution of several equations of type (31).

192. *Zwier's method of determining the elements of a visual binary orbit.*

The apparent ellipse is again supposed drawn—it is the ellipse whose area is shown shaded in Fig. 130. The determination of the elements can be carried out by (a) a graphical method, or (b) an analytical method based on the former.

(a) *Graphical method.* Let C be the centre of the apparent ellipse and S the primary star. The straight line joining C and S cuts the apparent ellipse in A_1, which is the projection of periastron in the true orbit, for C is the projection of the centre of the true orbit. Since ratios are unaltered by projection, we have
$$CS : CA_1 = ae : a,$$
or
$$\frac{CS}{CA_1} = e,$$
where e is the eccentricity of the true ellipse. Thus by determining C, the centre of the apparent ellipse, the eccentricity is

at once obtained. This procedure can be utilised as a check on the value of e given by the formulae of Kowalsky's method.

Let $B_1 CE_1$ be the diameter of the apparent ellipse conjugate to the diameter $A_1 CD_1$. It can be obtained readily by drawing any chord UW parallel to $A_1 D_1$ bisecting it at V and joining C to V. Let X be any point on the apparent ellipse. Draw XR parallel to CB_1 meeting CD_1 in R. Produce RX to T so that

$$RT/RX = 1/(1 - e^2)^{\frac{1}{2}} \equiv k \quad \ldots\ldots(32).$$

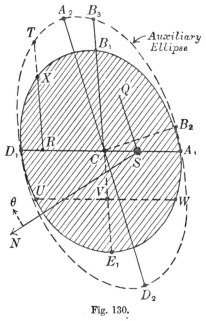

Fig. 130.

If this construction is supposed made for every point on the apparent ellipse, the curve $D_1 A_2 B_3 A_1 D_2$ will be the result. This curve is an ellipse, as will be shown, known as the *auxiliary ellipse*. In particular, we have

$$CB_3 = k . CB_1 \quad \ldots\ldots(33),$$

where k is defined by (32).

In the true ellipse, the minor axis is the conjugate of the major axis and this property holds for the corresponding projected lines. Hence CB_1 is the projection of the semi-minor axis of the true ellipse. Now consider the eccentric circle of the true ellipse. If a chord of the true ellipse, parallel to the minor axis, is produced both ways to meet the eccentric circle, the ratio of the chord of the eccentric circle to the chord of the true ellipse is $1/(1 - e^2)^{\frac{1}{2}}$ or k. It follows from the construction of Fig. 130 that the curve $D_1 A_2 B_3 A_1 D_2$ is the projection of the eccentric circle on the plane of the apparent orbit and is consequently an ellipse; it touches the apparent ellipse at A_1 and D_1.

Let CA_2, CB_2 be the semi-major and semi-minor axes respectively of the auxiliary ellipse, and denote them by α and β

respectively. We must then have

$$\frac{\beta}{\alpha} = \cos i \qquad \qquad \ldots\ldots(34).$$

Hence, by drawing the axes CA_2 and CB_2, the value of the inclination i can be determined.

Also, the semi-major axis a is equal to the radius of the eccentric circle, being the only radius not shortened by projection. But the radius of the eccentric circle is a, the semi-major axis of the true orbit; hence

$$a = \alpha \qquad \qquad \ldots\ldots(35).$$

Thus a is obtained from the auxiliary ellipse.

Now the diameter parallel to the line of nodes is not altered by projection; hence the major axis A_2CD_2 is parallel to the line of nodes. If SN is the direction given by position angle $\theta = 0°$ and SQ is parallel to CA_2, the position angle NSQ, measured in the direction of the arrow near N, is the longitude Ω of the node (Ω is between $0°$ and $180°$). Thus Ω is obtained from the auxiliary ellipse.

In Fig. 131 let SA denote the direction of periastron, and let KAG be part of a great circle through A and the pole K of the great circle NLG which defines the plane of the apparent orbit. Let λ denote LG. Then since $LA = \omega$, $A\hat{L}G = i$, $A\hat{G}L = 90°$ we have, by the four-parts formula **D**,

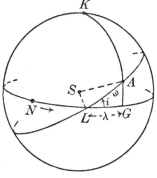

Fig. 131.

$$\tan \lambda = \tan \omega \cos i \qquad \ldots\ldots(36).$$

But SL defines the line of nodes and SG defines the projection of periastron. Hence, from Fig. 130, the angle A_2CA_1 or QSA_1 is λ. When this angle is measured, the element ω is obtained from (36).

We have thus found, from the auxiliary ellipse, the elements a, e, i, Ω and ω. The remaining elements T and τ are found as indicated in Kowalsky's method.

(b) *Analytical method.* In practice the auxiliary ellipse is not actually drawn, but its properties are used. In Fig. 130 denote the various semi-diameters as follows:

$$CA_1 = a_1; \quad CB_1 = b_1; \quad CB_3 = b_2; \quad CA_2 = \alpha; \quad CB_2 = \beta.$$

Let θ_1 be the position angle of A_1 and θ_2 the position angle of B_1. Then $B_1\hat{C}A_1$ is $(\theta_1 - \theta_2)$.

Now CB_1 and CA_1 are conjugate semi-diameters of the apparent ellipse and, by the construction of the auxiliary ellipse, it is evident that CA_1 and CB_3 are conjugate semi-diameters of the auxiliary ellipse. We have the well-known relations connecting pairs of conjugate semi-diameters:

$$\alpha^2 + \beta^2 = a_1{}^2 + b_2{}^2 \qquad \ldots\ldots(37),$$
$$\alpha\beta = a_1 b_2 \sin(\theta_1 - \theta_2) \qquad \ldots\ldots(38).$$

But, by (33), $\qquad b_2 = kb_1,$

and we can write

$$(\alpha + \beta)^2 = a_1{}^2 + 2ka_1 b_1 \sin(\theta_1 - \theta_2) + k^2 b_1{}^2 \equiv g^2 \ldots(39),$$

and

$$(\alpha - \beta)^2 = a_1{}^2 - 2ka_1 b_1 \sin(\theta_1 - \theta_2) + k^2 b_1{}^2 \equiv h^2 \ldots(40),$$

whence $\qquad \alpha = \tfrac{1}{2}(g + h) \quad \text{and} \quad \beta = \tfrac{1}{2}(g - h) \qquad \ldots\ldots(41).$

The apparent ellipse is supposed drawn; the diameter conjugate to CA_1 is also actually drawn. We can thus obtain the values of e, a_1, b_1 and $(\theta_1 - \theta_2)$ by measurement; since e is now known, k can be calculated from (32). Thus the quantities g and h can be calculated and by (41) the values of α and β are deduced.

Consider now the angle A_2CA_1 which we denoted previously by λ; it is the angle between the semi-major axis CA_2 of the auxiliary ellipse and the radius CA_1. Now the co-ordinates of A_1 referred to CA_2 and CB_2 as axes are $(a_1 \cos \lambda, a_1 \sin \lambda)$, and hence we have

$$\frac{a_1{}^2 \cos^2 \lambda}{\alpha^2} + \frac{a_1{}^2 \sin^2 \lambda}{\beta^2} = 1,$$

from which $\qquad \tan \lambda = \pm \dfrac{\beta}{\alpha}\left(\dfrac{\alpha^2 - a_1{}^2}{a_1{}^2 - \beta^2}\right)^{\frac{1}{2}},$

or, using (34), $\qquad \tan \lambda = \pm \cos i \left(\dfrac{\alpha^2 - a_1{}^2}{a_1{}^2 - \beta^2}\right)^{\frac{1}{2}} \qquad \ldots\ldots(42).$

As the sign of $\tan \lambda$ is the same as the sign of $\tan(\theta_1 - \theta_2)$, there are then two possible values of λ, differing by 180°, satisfying (42). Now $N\hat{S}Q = \Omega$, and since θ_1 is the position angle of A_1 and $A_2\hat{C}A_1 = \lambda$, we have

$$\Omega = \theta_1 - \lambda \qquad \ldots\ldots(43).$$

Now Ω is defined to lie between $0°$ and $180°$ and since θ_1 is known from measurement, the appropriate value of λ satisfying (42) is indicated by (43); this last equation enables Ω to be found. The value of ω is then found from (36), which is

$$\tan \lambda = \tan \omega \cos i.$$

The remaining elements T and τ are derived as before.

193. *The masses of the stars.*

The two elements of primary concern in this connection are the period T (in years) and the semi-major axis a. From the period, we deduce the mean angular motion n given by

$$n = 2\pi/T \qquad \ldots\ldots(44).$$

It is to be remembered that the micrometrical measures of the distances ρ between the companion and the primary are expressed in seconds of arc, and that consequently we obtain a expressed also in seconds of arc. Let Π be the parallax of the binary. Then if d is its distance from us measured in astronomical units, the value of Π in seconds of arc is given by

$$\Pi = \frac{1}{d} \operatorname{cosec} 1'' \qquad \ldots\ldots(45).$$

Let a_1 denote the semi-major axis of the true orbit expressed in astronomical units. Then a (in seconds of arc) is given by

$$a = \frac{a_1}{d} \operatorname{cosec} 1'' \qquad \ldots\ldots(46).$$

Hence, from (45) and (46),

$$a_1 = \frac{a}{\Pi} \qquad \ldots\ldots(47).$$

Now the relation between the mean angular velocity n and the linear semi-major axis a_1 is

$$n^2 a_1{}^3 = G\,(m_1 + m_2),$$

where G is the constant of gravitation and m_1, m_2 are the masses of the two stars. Hence, by (44),

$$\frac{a_1{}^3}{T^2} = \frac{G}{4\pi^2}\,(m_1 + m_2) \qquad \ldots\ldots(48).$$

Consider now the earth's orbit around the sun. We have a similar equation to (48), namely,

$$\frac{a_0{}^3}{T_0{}^2} = \frac{Gm}{4\pi^2} \qquad \dots\dots(49),$$

where m is the sun's mass (we neglect the earth's mass) and a_0, T_0 refer to the earth's orbit. But $a_0 = 1$ astronomical unit and $T_0 = 1$ year. If the sun's mass m is taken as the unit of mass, then G expressed in terms of these units is given by

$$G = 4\pi^2 \qquad \dots\dots(50).$$

Hence, for the binary, equation (48) becomes

$$\frac{a_1{}^3}{T^2} = m_1 + m_2,$$

where the stellar masses are now expressed in terms of the sun's mass as unit. Then, using (47), we derive

$$m_1 + m_2 = \frac{a^3}{\Pi^3 T^2} \qquad \dots\dots(51).$$

This equation enables the sum of the masses of a binary star to be determined when the parallax and its orbit are known.

It is found that, for the majority of visual binaries, the mass of the system is about twice the mass of the sun. When other evidence is reviewed, it is exceptional to find the mass of a star outside the range $\frac{1}{5}m$ to $50m$, where m denotes the sun's mass.

It is to be noted that only the sum of the masses can be calculated from (51). It is only in comparatively rare instances, when additional observational material of a different character is available, that the individual masses of primary and companion can be deduced.

194. *Dynamical parallaxes.*

The formula (51) can be used to determine the approximate value of Π when this quantity cannot be measured by the usual method. We can write

$$\Pi = \frac{a}{T^{\frac{2}{3}} (m_1 + m_2)^{\frac{1}{3}}} \qquad \dots\dots(52),$$

in which a and T are supposed to be derived from the observations of the visual orbit. If we put $m_1 + m_2 = 2$, in accordance with the average result for *known* visual binaries, we obtain what is called the dynamical parallax. In any given instance, of course, the value of $(m_1 + m_2)$ is actually unknown, but as the cube root of this quantity is involved in the formula (52), the resultant uncertainty in the calculated value of Π, due to the assumption concerning the mass of the system, is comparatively small. Dynamical parallaxes supply valuable information for certain investigations in which the distances of the stars are required.

195. Spectroscopic binaries.

A binary of this class consists of two members too close together to be resolved by the telescope into a double star; the duplicity of the system is inferred from spectroscopic observations of radial velocity. In a binary system which has its centre of gravity at G (Fig. 132), each star describes an elliptic orbit around G. In general, the orbital plane will be inclined to the line of sight. The orbital elements are defined as for visual binaries (Ω, the longitude of the node, is an exception, as it cannot be defined at all, owing to the absence of a definite reference point on the plane perpendicular to the line of sight).

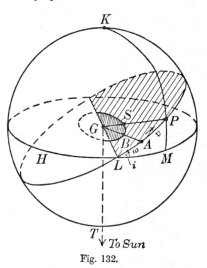

Fig. 132.

On page 213, reference has been made to the spectroscopic method of measuring the radial velocities of the stars and, in particular, the formula for the variable part of the radial velocity due to the earth's orbital motion round the sun was obtained. We shall suppose here that this variable part is removed from the observed radial velocity of a star; the value thus obtained represents the star's radial velocity relative to the sun.

In Fig. 132 let the straight line joining G and the sun intersect the sphere centred at G in T. The great circle of which T is the pole is HLM and we can refer to this plane, without ambiguity, as the plane of the apparent orbit. Let K be the other pole of HLM.

Let r (in kilometres) denote the radius vector GS, where S is the position of one of the stars in its orbit with respect to G. Let GB be the radius vector when the star is nearest G. Then $S\hat{G}B$ or $P\hat{G}A$ is the true anomaly v. (We suppose that the direction of motion is from A towards P.) L is the ascending node and we put $LA = \omega$ as before.

Let z denote the distance of S from the plane of HLM, reckoned positive when S is on the same side of this plane as K. Then

$$z = r \sin PM,$$

where KPM is the great circle arc through P. But from the triangle PLM, in which $LP = v + \omega$, $P\hat{L}M = i$ and $P\hat{M}L = 90°$, we have

$$\sin PM = \sin (v + \omega) \sin i.$$

Hence

$$z = r \sin (v + \omega) \sin i \qquad \ldots\ldots(53),$$

in which z is expressed in kilometres. The rate at which z varies with the time, that is, $\dfrac{dz}{dt}$, gives the radial velocity of the star S relative to the centre of gravity G of the system.

In general, the system itself will have a radial velocity with respect to the sun; denote it by V. Then V is the radial velocity of G, regarded positive when the motion is one of recession from the sun. Hence the radial velocity R of the star S with reference to the sun is given by

$$R = V + \frac{dz}{dt} \qquad \ldots\ldots(54).$$

The value of R is deduced from the spectroscopic observations. When both stars are sufficiently bright to register their spectra on the photographic plate, the radial velocities of both stars can be inferred. For the present, however, we shall consider the type of spectroscopic binary, only one of whose components enables radial velocity measures to be made.

Now we have the following formulae of elliptic motion applicable to the motion of the star S around G:

$$r = \frac{a(1 - e^2)}{1 + e \cos v} \qquad \ldots\ldots(55),$$

$$r^2 \frac{dv}{dt} = h = \{\mu a (1-e^2)\}^{\frac{1}{2}} \qquad \text{......(56),}$$

$$\mu = n^2 a^3 \qquad \text{......(57),}$$

where a is the semi-major axis in kilometres and n is the mean angular motion. Now from (53),

$$\frac{dz}{dt} = \frac{dr}{dt} \sin(v+\omega) \sin i + r \cos(v+\omega) \sin i \frac{dv}{dt},$$

and we find easily from the preceding formulae that

$$\frac{dr}{dt} = \frac{nae \sin v}{(1-e^2)^{\frac{1}{2}}}; \quad r\frac{dv}{dt} = \frac{na(1+e\cos v)}{(1-e^2)^{\frac{1}{2}}}.$$

Hence we obtain

$$\frac{dz}{dt} = \frac{na \sin i}{(1-e^2)^{\frac{1}{2}}} [\cos(v+\omega) + e\cos\omega] \qquad \text{......(58).}$$

Thus from (54) we can express the radial velocity R in terms of V, v and the elements of the orbit.

196. *The velocity curve.*

The orbital periods of spectroscopic binaries are generally several days only, and as observations may be carried on over several months, or even years, the orbital periods can be found with great accuracy. From all the available observations a curve is drawn, giving the relation between the radial velocity R and the time (through an interval equal to the period). Fig. 133

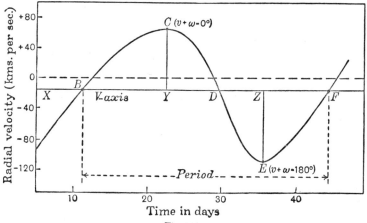

Fig. 133.

shows a typical velocity curve for which the period is $33\frac{1}{3}$ days; it refers to the star H.R. 8800, the spectroscopic observations of which were made at the Dominion Observatory, Victoria, B.C. (*Publications*, vol. I, p. 243). The curve between B and F represents the complete cycle of changes in the radial velocity R. At C the radial velocity has its maximum positive value (about 60 kms. per second) and at E its maximum negative value (about -115 kms. per second).

Actually, it is found (we shall explain this later) that the binary is approaching the sun with a speed of 15 kms. per second. Thus $V = -15$ kms. per second. The line XYZ parallel to the time axis, in accordance with this value of V, is called the V-axis; ordinates measured from this line to the curve give the appropriate values of $\dfrac{dz}{dt}$. For the present, we shall suppose that the V-axis has been drawn.

There are many methods of deriving the orbital elements from the velocity curve; we shall consider only one of these, which is extensively used in practice.

197. *Method of Lehmann-Filhés.*

We shall write (58) as follows:

$$\frac{dz}{dt} = K\left[\cos(v+\omega) + e\cos\omega\right] \qquad \ldots\ldots(59),$$

where

$$K = \frac{na\sin i}{(1-e^2)^{\frac{1}{2}}} \qquad \ldots\ldots(60).$$

Now dz/dt is a maximum when $(v+\omega)$ is $0°$ and, as the radial velocity is a maximum at C, the ordinate YC (which we denote by α), measured from the V-axis, is given by

$$\alpha = K(1 + e\cos\omega) \qquad \ldots\ldots(61).$$

Also, the radial velocity has its greatest negative value at E, and this occurs when $v + \omega = 180°$. If β denotes the length of the ordinate EZ, we have, numerically,

$$\beta = K(1 - e\cos\omega) \qquad \ldots\ldots(62).$$

From (61) and (62),

$$K = \tfrac{1}{2}(\alpha + \beta) \qquad \ldots\ldots(63),$$

$$e\cos\omega = \frac{\alpha - \beta}{\alpha + \beta} \qquad \ldots\ldots(64).$$

The measurement of α and β enables the values of K and $(e \cos \omega)$ to be derived.

The V-axis is determined from the consideration that the area BCD above this axis is equal to the area DEF below. This is easily seen as follows. Any ordinate is dz/dt and the abscissa is the time t. The area BCD is thus given by the integral

$$q \int \frac{dz}{dt} \, dt,$$

where the limits are the values of t corresponding to B and D, and q is a constant depending on the units in terms of which the area is measured. Hence this area is simply

$$q \, (z_D - z_B),$$

where z_D denotes the value of z corresponding to the point D on the velocity curve. Similarly the area DEF is

$$q \, (z_F - z_D).$$

But $z_F = z_B$, since BF corresponds to a complete cycle, and therefore the numerical expressions for the two areas are equal. The drawing of the V-axis is then largely a matter of trial and error; a first attempt is made and the areas above and below this line are measured by means of a planimeter. After one or more trials the position of the V-axis can be finally obtained to satisfy the conditions of area. In Fig. 133 the position of the V-axis is found according to the principles described, and the value of V obtained is -15 kms. per second. This is the radial velocity of the binary (H.R. 8800) relative to the sun.

Again, the area CYD, which we denote by Δ_1, is given by

$$\Delta_1 = z_D - z_C \qquad \ldots\ldots(65).$$

Now at C, the value of $(v + \omega)$ is $0°$ and, from (53), $z_C = 0$. Hence

$$\Delta_1 = z_D \qquad \ldots\ldots(66).$$

In the same way, the area DZE is equal to $(z_E - z_D)$, and as z_E is zero, the area DZE is equal to the area CYD, that is, Δ_1. We can prove in a similar manner that the areas BCY and ZEF are equal.

If v_1 and r_1 are the values of the true anomaly and radius vector in the true orbit at the point corresponding to D in Fig. 133, then by (53) and (66),

$$\Delta_1 = r_1 \sin (v_1 + \omega) \sin i \qquad \ldots\ldots(67).$$

But, at D, the value of dz/dt is zero; hence by (59),

$$\cos (v_1 + \omega) = - e \cos \omega$$

$$= - \frac{\alpha - \beta}{\alpha + \beta} \quad \text{by (64)}.$$

Hence $\qquad \sin (v_1 + \omega) = \pm \frac{2\sqrt{\alpha\beta}}{\alpha + \beta}.$

Now in passing along the velocity curve from C towards E, the radial velocity changes from being positive to negative just before and after the point D. Hence, referring to Fig. 132, we see that the point D on the velocity curve (Fig. 133) must correspond to the point on the true orbit for which z has a maximum positive value. Hence by (53), $\sin (v_1 + \omega)$ is positive and therefore

$$\sin (v_1 + \omega) = + \frac{2\sqrt{\alpha\beta}}{\alpha + \beta} \qquad \dots\dots(68).$$

Consider now the area ZEF, which we denote by Δ_2. Then

$$\Delta_2 = z_F - z_E,$$

or, since $z_E = 0$, we obtain $\Delta_2 = z_F$.

If v_2 and r_2 refer to the point on the true orbit corresponding to F on the velocity curve, we have

$$\Delta_2 = r_2 \sin (v_2 + \omega) \sin i \qquad \dots\dots(69).$$

But, at F, dz/dt is zero; thus

$$\cos (v_2 + \omega) = - e \cos \omega = - \frac{\alpha - \beta}{\alpha + \beta},$$

and now $\qquad \sin (v_2 + \omega) = - \frac{2\sqrt{\alpha\beta}}{\alpha + \beta} \qquad \dots\dots(70),$

the negative sign being taken in (70), since F corresponds to the point on the true orbit for which z has the maximum negative value. Hence, using (67), (68), (69) and (70), we find

$$\frac{\Delta_1}{\Delta_2} = - \frac{r_1}{r_2}.$$

The procedure has been such that Δ_2 is expressed mathematically as a negative quantity. If Δ_1 and Δ_2 are both regarded as positive quantities, we can write

$$\frac{\Delta_1}{\Delta_2} = \frac{r_1}{r_2} \qquad \dots\dots(71).$$

But $r_1 = \dfrac{a\,(1 - e^2)}{1 + e \cos v_1}$, with a similar expression for r_2; hence

$$\frac{\Delta_1}{\Delta_2} = \frac{1 + e \cos v_2}{1 + e \cos v_1}.$$

We can write this last equation as follows:

$$\frac{\Delta_1}{\Delta_2} = \frac{1 + e \cos (\overline{v_2 + \omega} - \omega)}{1 + e \cos (\overline{v_1 + \omega} - \omega)}$$

$$= \frac{1 + e \cos \omega \cos (v_2 + \omega) + e \sin \omega \sin (v_2 + \omega)}{1 + e \cos \omega \cos (v_1 + \omega) + e \sin \omega \sin (v_1 + \omega)},$$

which, by the use of (64), (68) and (70), reduces to

$$\frac{\Delta_1}{\Delta_2} = \frac{2\alpha\beta - \sqrt{\alpha\beta}\,(\alpha + \beta)\,e \sin \omega}{2\alpha\beta + \sqrt{\alpha\beta}\,(\alpha + \beta)\,e \sin \omega},$$

from which it is easily found that

$$e \sin \omega = \frac{2\sqrt{\alpha\beta}}{\alpha + \beta} \cdot \frac{\Delta_2 - \Delta_1}{\Delta_2 + \Delta_1} \qquad \ldots\ldots(72).$$

By measuring the areas Δ_1 and Δ_2 (which are both positive quantities) the value of $e \sin \omega$ can be deduced from (72). But we had also

$$e \cos \omega = \frac{\alpha - \beta}{\alpha + \beta};$$

hence this equation and (72) determine e and ω.

To determine the time τ of perihelion passage we note that, then, $v = 0$, and if $\dot{z}_1$ is the corresponding value of dz/dt the value of $\dot{z}_1$ is given from (59) by

$$\dot{z}_1 = K\,(1 + e) \cos \omega \qquad \ldots\ldots(73).$$

The value of $\dot{z}_1$ can now be found since K, e and ω are all supposed known. There are two ordinates of the velocity curve given by (73). However, the ambiguity is resolved by noting that, at C, $v + \omega = 0$, and at E, $v + \omega = 180°$. Thus if ω is found to be, say, $60°$, the value of v at C is $300°$ and at E is $120°$; in this instance periastron will correspond to the ordinate between C and D on the velocity curve and so the time τ (abscissa) of periastron passage, measured from the epoch corresponding, say, to B, is found.

Again, we have from (60) and (63),

$$K = \frac{na \sin i}{(1 - e^2)^{\frac{1}{2}}} = \tfrac{1}{2}\,(\alpha + \beta) \qquad \ldots\ldots(74).$$

Also if the period T is supposed known, $n = 2\pi/T$, and from (74),

$$a \sin i = \frac{T(\alpha + \beta)(1 - e^2)^{\frac{1}{2}}}{4\pi}.$$

The period T is usually expressed in days and α and β are measures of velocities expressed in kilometres per mean solar second. With these units, the last equation becomes (since 1 day = 86,400 seconds)

$$a \sin i = \frac{21600T}{\pi}(\alpha + \beta)(1 - e^2)^{\frac{1}{2}} \qquad \ldots\ldots(75).$$

This formula gives the value of the quantity $(a \sin i)$ expressed in kilometres, T being expressed in days. Unless the inclination i can be derived by other means, the semi-major axis a, with respect to the centre of gravity of the system, cannot be determined.

198. *Two spectra visible.*

When both members of the binary are sufficiently bright, the velocity curve associated with each star can be analysed by the previous, or any other, method. Let now a_1 denote the semi-major axis of one orbit, with respect to the centre of gravity of the system, and a_2 the semi-major axis of the other orbit. Then the semi-major axis of the orbit of one star relative to the other star is $(a_1 + a_2)$, which we denote by a_0. If a_0 is expressed in astronomical units and the period T in days, the sum of the masses of the two stars (expressed in terms of the sun's mass as unit) is given by

$$m_1 + m_2 = a_0{}^3 \left(\frac{365\frac{1}{4}}{T}\right)^2 \qquad \ldots\ldots(76).$$

But $(a_1 \sin i)$ can be found by means of (75) in terms of kilometres, and from the analysis of the second velocity curve the value of $(a_2 \sin i)$ can be similarly found. Since $a_0 = a_1 + a_2$ and one astronomical unit = 149,500,000 kms., formula (76) becomes, on multiplying both sides by $\sin^3 i$,

$$(m_1 + m_2)\sin^3 i = \left(\frac{a_1 \sin i + a_2 \sin i}{149,500,000}\right)^3 \left(\frac{365\frac{1}{4}}{T}\right)^2,$$

in which a_1, a_2 are expressed in kilometres and T in days, or, using (75),

$$(m_1 + m_2)\sin^3 i = 1\cdot298 \times 10^{-8}\, T\,(1 - e^2)^{\frac{3}{2}}\,\{(\alpha_1 + \beta_1) + (\alpha_2 + \beta_2)\}^3,$$

where α_1, β_1 and α_2, β_2 refer to the two velocity curves. On writing $K_1 = \frac{1}{2}(\alpha_1 + \beta_1)$ and $K_2 = \frac{1}{2}(\alpha_2 + \beta_2)$, this last equation is:

$$(m_1 + m_2) \sin^3 i = 10\cdot38 \times 10^{-8}\, T\, (1 - e^2)^{\frac{3}{2}} (K_1 + K_2)^3 \quad ...(77).$$

The value of $(m_1 + m_2) \sin^3 i$ can thus be obtained.

Now, since G is the centre of gravity of the system,

$$m_1 a_1 = m_2 a_2 \qquad \qquad ......(78),$$

or $\qquad\qquad m_1\,(a_1 \sin i) = m_2\,(a_2 \sin i).$

Hence, by (75), $\qquad \dfrac{m_1}{m_2} = \dfrac{\alpha_2 + \beta_2}{\alpha_1 + \beta_1} = \dfrac{K_2}{K_1} \qquad ......(79).$

Thus the ratio of the masses can be derived. The formula (77) for the sum of the masses involves the unknown value of the inclination, and therefore the individual masses cannot be found unless i can be derived by other means.

199. *The mass-function.*

We return now to the case when only one spectrum is visible. From (78), we have

$$\frac{a_1}{a_1 + a_2} = \frac{a_1}{a_0} = \frac{m_2}{m_1 + m_2};$$

$$\therefore\; (a_1 \sin i)^3 = \frac{m_2^3 a_0^3 \sin^3 i}{(m_1 + m_2)^3} \qquad ......(80).$$

But from (76), expressing a_0 in kilometres,

$$m_1 + m_2 = \left(\frac{a_0}{149,500,000}\right)^3 \left(\frac{365\frac{1}{4}}{T}\right)^2,$$

or $\qquad\qquad m_1 + m_2 = 3\cdot993 \times 10^{-20} \dfrac{a_0^3}{T^2} \qquad ......(81).$

Hence from (80) and (81), eliminating a_0, we derive

$$\frac{m_2^3 \sin^3 i}{(m_1 + m_2)^2} = \frac{3\cdot993 \times 10^{-20} (a_1 \sin i)^3}{T^2} \qquad ......(82).$$

The quantity on the right can be evaluated. The function on the left is the *mass-function*. It is generally given by computers of spectroscopic orbits.

200. *The masses of spectroscopic binaries.*

When one spectrum only is photographed, the information concerning the masses of the stars is contained in the value of the mass-function. When both spectra are visible, we have seen that the velocity curves enable the ratio of m_1 to m_2 to be determined, and also the quantity $(m_1 + m_2) \sin^3 i$. Thus for any individual binary, the actual values of the masses cannot be found owing to the presence of the unknown inclination i in the formulae. Nevertheless, valuable information regarding the masses is derived from statistical considerations. Consider a large number of spectroscopic binaries of the same spectral type distributed over the sky. We assume that the masses of such binaries are much alike. If we take the average mass of a system to be M and the number to be N, we can write, forming the sum $\Sigma \, (m_1 + m_2) \sin^3 i$, from the known values of this quantity,

$$\Sigma \, (m_1 + m_2) \sin^3 i = NMS,$$

where S denotes the mean value of $\sin^3 i$, with reference to the random inclinations over the sky. The value of S is generally taken to be 2/3. In this way we can derive an estimate of the average mass M of a spectroscopic binary of given spectral type.

The most massive spectroscopic binary known is the star B.D. $+ \, 6° \, 1309$; for this star, $m_1 \sin^3 i = 75$ and $m_2 \sin^3 i = 63$ (in terms of the sun's mass). The individual masses are thus at least 75 and 63 times the sun's mass.

201. *Eclipsing binary stars.*

If the line of sight is in or near the plane of the orbit of a binary star, it is evident that during each orbital period the component A will pass wholly or partially in front of the component B, thus eclipsing the latter. The effect will be noted in the diminution of the light of the binary. In the same way the component B will cause a total or partial eclipse of A. The analysis of the light curve of such a star cannot be given here, and the reader is referred to *Astrophysical Journal*, vol. xxxv, p. 333; vol. xxxvi, pp. 243, 390, where a detailed account of the method of analysis is given. It may be added, however, that the light curve yields the inclination, i, and if both components can be observed spectroscopically, the mass of each is obtained.

EXERCISES

1. The greatest apparent diameter of a visual binary orbit is $1''\!\cdot\!8$; the parallax is $0''\!\cdot\!072$ and the period 50 years. If the orbit is circular, show that the mass of the binary is 25/32 times the sun's mass. [*Lond.* 1925.]

2. The semi-major axis of the orbit of Krüger 60 is $2''\!\cdot\!46$; the period is $44\cdot3$ years; the parallax of the system is $0''\!\cdot\!257$. Calculate the mass of the binary in terms of the sun's mass. [*Lond.* 1928.]

3. The greatest angular separation of the components of τ Cygni is $0''\!\cdot\!91$, the period is $47\cdot0$ years and the parallax is $0''\!\cdot\!05$. Assuming that the orbit is circular, calculate the mass of the binary.

4. Calculate the dynamical parallax of the binary β 7642, for which $a = 2''\!\cdot\!87$ and $T = 317\cdot5$ years.

The trigonometrical parallax is $0''\!\cdot\!088$; comment on the difference between the two values of the parallax. [*Lond.* 1926.]

5. Explain how the shape and position of the true orbit of a visual binary can be deduced from observations of position angle and distance extended over a whole period.

If a, b be the semi-axes of the apparent orbit, h, k the co-ordinates of the primary referred to these semi-axes, show that the inclination i of the orbit to the plane at right angles to the line of sight is given by

$$\sin^2 i = \frac{2\lambda}{a^2 + b^2 - h^2 - k^2 + \lambda},$$

where
$$\lambda^2 = (h^2 + k^2)^2 + (a^2 - b^2)^2 + 2(a^2 - b^2)(k^2 - h^2).$$
[*Lond.* 1922.]

6. Prove that if the complete apparent orbit of a visual binary has been obtained, and on any chord POQ through the primary O we take a point R such that OR is a harmonic mean between PO and OQ, the locus of R is an ellipse, of which the length of the major axis gives the latus rectum of the true orbit, the direction of this major axis gives the direction of the line of nodes, and the ratio of the minor to the major axis is the cosine of the inclination. [*Lond.* 1923.]

7. The true period of an eclipsing binary is 3 days, and its velocity in the line of sight (away from the sun) is 30 kms. per second. Show that its apparent period is greater than the true one by 26 seconds. [*Lond.* 1923.]

8. A variable star has ecliptic co-ordinates λ, β. If T is the time of a maximum for an observer on the earth and T_0 the corresponding time with respect to the sun, show that, c being the velocity of light in kilometres per second and a the radius of the earth's orbit in kilometres,

$$T_0 = T - \frac{a}{c}\cos \beta \cos (\lambda - \odot),$$

where $\odot$ is the sun's longitude. [*Lond.* 1927.]

CHAPTER XV

OCCULTATIONS AND ECLIPSES

202. *Occultations of stars by the moon.*

As the moon's sidereal period of orbital revolution around the earth is about $27\frac{1}{3}$ days, it moves eastwards with reference to the stars at an average rate of rather more than half a degree per hour. In its passage over the stellar background it is continually interposing its disc between us and the stars, and the sudden disappearance of a star in this way is called the *occultation* of the star by the moon. After an interval, which depends on a variety of factors, the star reappears. The disappearance and reappearance of the star are generally referred to as *immersion* and *emersion* respectively. The disappearance of the star and its reappearance are instantaneous phenomena and, if the time of one or the other is noted accurately, there is obtained at that instant a definite relation between the moon's position in the sky and the position of the observer, it being assumed that the star's position is known accurately. Formerly, occultations were utilised for the determination of longitude, but the introduction of radio time-signals has rendered the occultation method obsolete.

If the moon's position is known accurately, the particulars of the occultation of a star at any place can be predicted and, under these circumstances, it is to be expected that prediction and observation would agree. Now the moon's position at any instant is predicted in the almanacs according to the relevant theory in celestial mechanics and it is found, partly as a result of the observation of occultations, that the mean longitude of the moon, as deduced from such observations, is in defect of the theoretical value at the present time (1944) by $1''$. This discrepancy is usually attributed to changes in the earth's rotational period, and as this period is the basis of all time-measurement, the importance of careful observations of the moon's position and, in particular, of the observations of occultations hardly needs to be emphasised.

203. *The geometrical conditions for an occultation.*

Consider Fig. 134, in which the earth (regarded as a spheroid) and the moon (regarded as a sphere) are shown with their centres at E and M respectively. Let MS be the straight line joining the moon's centre and the star concerned at a particular instant. Since the star can be regarded as at an infinite distance, the rays from the star lying within a circular cylinder, whose axis is MS and whose cross-sectional radius is equal to the moon's radius, will be stopped by the moon. Suppose that this cylinder intersects the earth's surface in the curve FGH (only part of the complete curve is shown in Fig. 134); then, at the particular

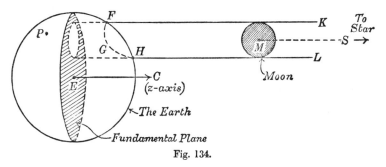

Fig. 134.

instant concerned, the star will be about to disappear or reappear behind the lunar disc, as viewed from points on the curve FGH, and at all places on the earth's surface within this curve the star will be invisible. For an occultation of a particular star to be visible at some point on the earth's surface, the cylinder must evidently intersect the earth's surface.

The plane passing through the earth's centre and perpendicular to the line MS is called the *fundamental plane*; in the subsequent discussion the normal EC to this plane is taken to be the z-axis. It is to be noted that, owing to the Earth's rotation, the fundamental plane and the z-axis are continually altering relative to fixed axes in the Earth.

As viewed from any point on the earth, the moon will be displaced, owing to parallax, away from the zenith by an angle which may be just over 1° (corresponding to the maximum or horizontal parallax); also its semi-diameter is 16'. Hence, for a star with the same right ascension as the moon at any instant,

an occultation may be possible at some place if the star's declination does not differ by more than about $1\frac{1}{3}°$ from the moon's
tabulated declination. By comparing the moon's tabulated
positions in the almanac with the positions of stars, the selection
of stars, according to the criterion just mentioned, is made for
which occultations are possible.

204. *Bessel's method of investigating an occultation.*

This is the method used in the almanacs for predicting the
circumstances of an occultation. We shall use the following
symbols:

α, δ—the apparent R.A. and Dec. of the star.
α_1, δ_1—the apparent R.A. and Dec. of the moon's centre.
P_1—the equatorial horizontal parallax of the moon.

r_1—the moon's geocentric
distance.
k—the moon's radius.   } All expressed in terms of the
ρ—the observer's geocentric   earth's equatorial radius as
distance.   the unit of length.

ϕ'—the observer's geocentric latitude.

In Fig. 135 let E (the earth's centre) be the centre of the
celestial sphere shown. The radius EC is drawn parallel to the
straight line joining the moon's centre to the star; EC is the
z-axis and the plane DBA, of which EC is the normal and which
passes through the centre E, is the fundamental plane. Let EP
be the earth's axis of rotation, P being the north pole. Since C is
the pole of the fundamental plane, the plane of any great circle
drawn through C is perpendicular to the fundamental plane; in
particular, the great circle joining C and P defines a plane perpendicular to the fundamental plane. Let it intersect the latter
in EB and the equator in F; $BPCF$ is defined to be the x-plane.
Since the star is regarded as at an infinite distance, EC is the
direction of the star, as viewed from the earth, and PCF is the
meridian through the star. The earth's rotation carries this
meridian westwards, that is, in the direction FD. The positive
direction of the x-axis is defined to be EA—the eastward drawn
normal of the plane $BPCF$. It is evident that A is on the celestial
equator and the R.A. of this point is, accordingly, $90° + \alpha$ (since

$FA = 90°$ and $\gamma F = \alpha$, where γ is the vernal equinox). Also FC is δ (the star's declination). EB is the y-axis.

Let EX be the geocentric direction of the moon's centre. Let (x, y, z) denote the rectangular co-ordinates of the moon's centre,

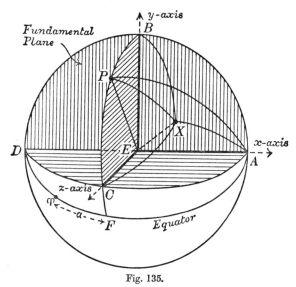

Fig. 135.

in terms of the earth's equatorial radius as unit, with respect to the co-ordinate system described. Joining X to A, B and C by great circle arcs, we have

$$\frac{x}{r_1} = \cos AX, \quad \frac{y}{r_1} = \cos BX, \quad \frac{z}{r_1} = \cos CX.$$

In the spherical triangle APX, we have:

$$PX = 90° - \delta_1, \quad PA = 90°, \quad X\hat{P}A = (90° + \alpha) - \alpha_1.$$

We thus obtain, using the cosine-formula,

$$x = r_1 \cos \delta_1 \sin (\alpha_1 - \alpha) \qquad \dots\dots(1).$$

In the spherical triangle BPX, we have:

$$BP \equiv FC = \delta, \quad PX = 90° - \delta_1,$$
$$B\hat{P}X \equiv B\hat{P}A + A\hat{P}X = 180° + \alpha - \alpha_1.$$

Hence, by the cosine-formula,

$$y = r_1 [\sin \delta_1 \cos \delta - \cos \delta_1 \sin \delta \cos (\alpha_1 - \alpha)]\dots\dots(2).$$

We do not require the corresponding expression for z.

Since the fundamental plane is perpendicular to the straight line joining the moon's centre to the star, x and y are the co-ordinates, at any instant, on the fundamental plane of the centre of the shadow cast by the moon with reference to the light from the star or, in other words, x and y are the co-ordinates of the point of intersection of the axis of the cylinder with the fundamental plane.

Also r_1 and P_1 are related by the formula

$$\frac{1}{r_1} = \sin P_1 \qquad \ldots\ldots(3).$$

For an occultation, $(\alpha_1 - \alpha)$ and $(\delta_1 - \delta)$ may be regarded as small angles and, if $(\alpha_1 - \alpha)$ is expressed in seconds of time and $(\delta_1 - \delta)$ and P_1 in seconds of arc, the formulae (1) and (2) can be written, using (3), with sufficient accuracy for purposes of prediction, as

$$x = \frac{15}{P_1}(\alpha_1 - \alpha)\cos \delta_1 \qquad \ldots\ldots(4),$$

$$y = \frac{(\delta_1 - \delta)}{P_1} \qquad \ldots\ldots(5).$$

We now consider the co-ordinates (ξ, η) of the observer with reference to the fundamental plane. To avoid complications in Fig. 135, we shall now suppose that EX defines the geocentric direction of the observer, so that PX is the observer's meridian. As before,

$$\frac{\xi}{\rho} = \cos AX, \quad \frac{\eta}{\rho} = \cos BX.$$

Now in the triangle APX, $PX = 90° - \phi'$, $PA = 90°$ and $X\hat{P}A = C\hat{P}A - C\hat{P}X$. If h denotes the hour angle of the star at the instant concerned, $C\hat{P}X = h$, and therefore $X\hat{P}A = 90° - h$. Hence

$$\xi = \rho \cos \phi' \sin h \qquad \ldots\ldots(6).$$

Similarly, in the triangle BPX, we have $BP = \delta$, $PX = 90° - \phi'$, and $B\hat{P}X = 90° + (90° - h) = 180° - h$. Hence

$$\eta = \rho\,[\cos \delta \sin \phi' - \sin \delta \cos \phi' \cos h] \qquad \ldots\ldots(7).$$

Referring to Fig. 134 we see that, for immersion or emersion, the point of projection of the observer on the fundamental plane must lie on the circle of radius k in which the cylinder cuts the fundamental plane. This condition is thus expressed by

$$(x - \xi)^2 + (y - \eta)^2 = k^2 \qquad \ldots\ldots(8).$$

For the calculation of occultations, the value of k (the moon's radius expressed in terms of the earth's equatorial radius as unit) is taken to be 0·2725.

205. *The Besselian elements of an occultation.*

The quantities, relating to the occultation of any given star, tabulated in the almanacs are:

T_0—the U.T. of conjunction in R.A. of the moon and the star.

H—the hour angle of the star at Greenwich at the instant T_0.

Y—the value of the y-co-ordinate [see formula (5)] at the instant T_0. At this instant, the value of the x-co-ordinate is evidently zero.

x', y'—the rate of change of x and y *per hour* of mean solar time.

For example, from (5), $y' = \dfrac{60\Delta\delta_1}{P_1}$,

where $\Delta\delta_1$ is the rate of change, *per minute*, in the moon's declination; $\Delta\delta_1$ is given in the almanacs. In addition, the limiting parallels of latitude are given, outside which an occultation is impossible. Owing, however, to the changing declination of the moon, the point of intersection with the earth's surface of the axis MS of the shadow cylinder (Fig. 134) will not move along a parallel of latitude; it follows that an occultation may not be visible even if the observer is within the limiting parallels.

206. *The prediction of an occultation at any place.*

We shall assume that an approximate estimate of the U.T. of an occultation, visible at a particular point on the earth's surface, has first been obtained. We shall refer later to the methods of deriving this estimated U.T., which we shall suppose to be $(T_0 + t)$.

If λ is the longitude of the observer west of Greenwich, the hour angle h of the star is given by

$$h = H - \lambda + t' \qquad \ldots\ldots(9),$$

where t' is the number of sidereal hours equivalent to t hours of mean solar time. Hence the values of ξ and η at U.T. $(T_0 + t)$

13   SA

can be calculated from (6) and (7) and the known terrestrial co-ordinates of the observer, ρ and ϕ'. The star's declination is given in the almanacs.

Also, since $x = 0$ at time T_0 (the moon and the star are then in conjunction as regards right ascension) and $y = Y$ at time T_0, then at time $(T_0 + t)$ we have

$$x = x't, \quad y = Y + y't \qquad \ldots\ldots(10),$$

t being expressed in hours.

If $(T_0 + t)$ is actually the true time of the occultation, the equation $\qquad (x - \xi)^2 + (y - \eta)^2 = k^2$

must be satisfied by the values of ξ, η, x, y just calculated. In general, however, the estimated time is likely to differ by a minute or two from the true time, and we proceed as follows. We shall suppose that $(T_0 + t)$ is the estimated time and $(T_0 + t + \Delta t)$ is the accurate time concerned, where Δt is expressed in hours. We calculate x, y and ξ, η for the U.T. $(T_0 + t)$, as already indicated; we denote these values by x_1, y_1, ξ_1 and η_1 respectively. The values of x, y, ξ and η corresponding to the true time $(T_0 + t + \Delta t)$ of the occultation are then derived from

$$\begin{aligned} x &= x_1 + x'.\Delta t, \quad y = y_1 + y'.\Delta t \\ \xi &= \xi_1 + \xi'.\Delta t, \quad \eta = \eta_1 + \eta'.\Delta t \end{aligned} \left.\right\} \qquad \ldots\ldots(11),$$

where ξ', η' denote the rate of change in ξ and η per mean solar hour. Since (8) must be satisfied for the occultation, we have

$$[x_1 - \xi_1 + \Delta t \, (x' - \xi')]^2 + [y_1 - \eta_1 + \Delta t \, (y' - \eta')]^2 = k^2,$$

or, writing $\qquad x_1 - \xi_1 = f, \quad y_1 - \eta_1 = g \qquad \ldots\ldots(12),$

and neglecting squares of Δt, we deduce

$$\Delta t = \frac{k^2 - f^2 - g^2}{2 \, (fx' + gy' - f\xi' - g\eta')} \qquad \ldots\ldots(13).$$

In this formula, $k = 0.2725$; f and g are derived from the quantities x_1, ξ_1, y_1 and η_1 computed for G.C.T. $(T_0 + t)$; also x' and y' are tabulated in the almanacs. To calculate Δt, it is necessary to know the values of ξ' and η'. Now by (6) and (7),

$$\xi' \equiv \frac{d\xi}{dt} = \rho \cos \phi' \cos h \frac{dh}{dt} \qquad \ldots\ldots(14),$$

$$n' \equiv \frac{d\eta}{dt} = \rho \cos \phi' \sin \delta \sin h \frac{dh}{dt} \qquad \ldots\ldots(15),$$

in which dh/dt means the change in the star's hour angle (expressed in circular measure) in one hour of mean solar time. Now h increases from 0 to 2π in 23^h 56^m of mean solar time,

$$\therefore \quad \frac{dh}{dt} = \frac{2\pi}{23\cdot 93} = 0\cdot 2625.$$

Write $$Q = \rho \cos \phi' \cos h \qquad \ldots\ldots(16);$$

then, from (6), (14), (15) and (16),

$$f\xi' + g\eta' = 0\cdot 2625 \,(fQ + g\xi \sin \delta) \qquad \ldots\ldots(17)$$
$$= 0\cdot 2625 a_0, \quad \text{say.}$$

Hence $$\Delta t = \frac{k^2 - f^2 - g^2}{2\,[fx' + gy' - 0\cdot 2625 a_0]} \qquad \ldots\ldots(18),$$

from which Δt can be derived.

If necessary, greater accuracy can be attained by regarding $(T_0 + t + \Delta t)$ as the approximate time of the phenomenon, and repeating the calculations in the way just described.

There are several graphical methods* which enable an approximate estimate of the time of the phenomenon to be made. We can only refer very briefly to a method suggested by L. J. Comrie. Referred to the projection, on the fundamental plane, of the observer's position as origin, the co-ordinates of the centre of the shadow are (f, g). Draw a circle of radius k and compute the values of f and g for three times t_1, t_2 and t_3 in the neighbourhood of T_0. In actual practice t_1, t_2 and t_3 are always taken to be three consecutive values of the series -2^h, -1^h, 0^h, $+1^\mathrm{h}$, $+2^\mathrm{h}$, in which 0^h corresponds to T_0. Plotting the positions defined by these values of the co-ordinates (f, g), we obtain the track of the centre of the shadow with reference to the observer's position. This track or curve is approximately a straight line. If an occultation is possible, the curve will intersect the circle of radius k in two points in general, and the corresponding times can be deduced from the data with which the curve was drawn. These times are the approximate times of immersion and emersion. Formula (18) can then be used to derive the time of occurrence of the occultation.

The position angle χ (measured eastwards from the north

* For example, see W. F. Rigge, *The Graphic Construction of Eclipses and Occultations* (Loyola Univ. Press, Chicago, 1924).

point of the moon's disc) corresponding to immersion or emersion is given from the figure by

$$\sin \chi = -\frac{f}{k}, \quad \cos \chi = -\frac{g}{k} \qquad \ldots\ldots(19),$$

where the values of f and g now refer to one or other of the points of intersection of the track and the circle.

In the *Nautical Almanac* predictions of occultations visible at Greenwich and Edinburgh are given. The approximate time of the phenomenon at a place $\Delta\lambda$ degrees west and $\Delta\phi$ degrees north of Edinburgh, say, can be derived from the formula

Approx. time = (Predicted U.T. for Edinburgh) $+ a\Delta\lambda + b\Delta\phi$,

where a and b are quantities tabulated for each occultation. This approximate time serves as the basis for the accurate prediction according to the method already described.

207. *The reduction of occultations.*

As we have remarked in section 202, the observations of occultations are now applied solely to the important problem of determining the errors in the moon's longitude.

Let T denote the U.T. at the observed instant of an occultation. From (4) and (5) we have

$$x = \frac{15}{P_1} (\alpha_1 - \alpha) \cos \delta_1 \qquad \ldots\ldots(20),$$

$$y = \frac{\delta_1 - \delta}{P_1} \qquad \ldots\ldots(21),$$

in which α_1, δ_1 and P_1 are the moon's right ascension, declination and horizontal parallax corresponding to time T; $(\alpha_1 - \alpha)$ is expressed in seconds of time and $(\delta_1 - \delta)$ in seconds of arc. Instead of (21), a more accurate formula is usually employed; it is derived as follows. From (2) and (3) we have

$$y = \frac{\sin (\delta_1 - \delta)}{\sin P_1} + \frac{2}{\sin P_1} \cos \delta_1 \sin \delta \sin^2 \left(\frac{\alpha_1 - \alpha}{2} \right),$$

or—$(\delta_1 - \delta)$, $(\alpha_1 - \alpha)$ and P_1 being expressed as indicated above—

$$y = \frac{\delta_1 - \delta}{P_1} + \frac{(\alpha_1 - \alpha) \cos \delta_1 \sin \delta}{2P_1} \cdot 15^2 (\alpha_1 - \alpha) \sin 1'',$$

or, using (20), $\quad y = \dfrac{\delta_1 - \delta}{P_1} + \dfrac{x (\alpha_1 - \alpha) \sin \delta}{27500} \qquad \ldots\ldots(22).$

The formulae (20) and (22) enable us to calculate the quantities x and y corresponding to time T.

If h is the hour angle of the star at T, we calculate the values of ξ and η by means of (6) and (7).

We then derive f and g for the time T, where

$$f = x - \xi, \quad g = y - \eta \qquad \ldots\ldots(23).$$

In Fig. 136 let X be the star's position at time T. We assume that, owing to an error in mean longitude, the moon's tabulated right ascension and declination are in error. Let M be the centre of the moon, according to the tabulated values of α_1 and δ_1 at time T. The angular distance XM, calculated from the erroneous values of α_1 and δ_1, will thus be somewhat different from the observed angular distance, which is the moon's semi-diameter S. Denote the arc XM by S'.

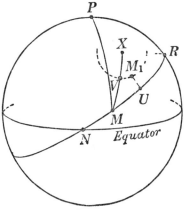

Fig. 136.

Denoting the angle PMX —the position angle of the star with respect to M—by χ, we have from (19) and (23)

$$\tan \chi = f/g = (x - \xi)/(y - \eta) \qquad \ldots\ldots(24)$$

from which χ is calculated.

Also, $\quad - r_1 S' = (x - \xi)\, \mathrm{cosec}\, \chi = (y - \eta)\, \sec \chi,$

where S' is expressed in circular measure. We then calculate S'.

In Fig. 136 let NMR be the great circle determined by the moon's orbital plane. Let ρ denote the position angle PMR. Hence the angle XMR is $\rho - \chi$.

ρ is calculated as follows. Let $\Delta\alpha_1$ and $\Delta\delta_1$ denote the variation of α_1 (in seconds of time) and of δ_1 (in seconds of arc) in 1 minute. These quantities are obtained from the almanac. Then the components of the moon's motion along its orbit are, in circular measure, $15\Delta\alpha_1 \sin 1'' \cos \delta$ (along the parallel of declination through M) and $\Delta\delta_1 \sin 1''$ (along the meridian MP). Hence

$$\tan \rho = \frac{15\Delta\alpha_1 \cos \delta}{\Delta\delta_1} \qquad \ldots\ldots(25).$$

To summarise up to this point: In Fig. 136 we know for the observed instant T of the actual occultation (i) the position of the star X, (ii) the angular distance XM ($= S'$) and (iii) the angle XMR ($= \rho - \chi$). Now at the time, T, of occultation the angular distance of the star X from the moon's centre is S (the moon's semi-diameter). Since we know the position of X definitely, the moon's centre must lie somewhere on the small circle (cutting XM at V) whose centre is X and whose angular radius is S. Let M_1 be the actual position of the moon's centre. Draw M_1U perpendicular to NMR. Then MU represents the correction to the moon's tabulated longitude (measured along NMR) and UM_1 the corresponding correction in latitude. Denote MU by $\Delta\lambda$ and UM_1 by $\Delta\beta$. These quantities are small compared with S (or XV) and so the arc VM (or $S' - S$) is small. The moon's true centre M_1 is, accordingly, close to V so that M_1V is perpendicular to XM. Projecting MU and UM_1 on MV we obtain

$$S' - S = \Delta\lambda \cos (\rho - \chi) + \Delta\beta \sin (\rho - \chi) \quad \ldots\ldots(26).$$

Several observations of the same occultation made at different places provide the corresponding number of equations of the form of (26). The solution, by least squares, leads to the values of the corrections $\Delta\lambda$ and $\Delta\beta$.

208. *Eclipses of the moon.*

A lunar eclipse occurs when the moon passes into the shadow (with reference to the light from the sun) cast by the earth. The phenomenon can only take place when the earth is directly between the sun and the moon, that is to say, when the moon is in opposition; this corresponds to "full moon".

If the moon's orbital plane coincided with the plane of the ecliptic, lunar eclipses would occur at every full moon. As, however, the orbital plane is inclined at an angle of about 5° to the plane of the ecliptic, the conditions for the occurrence of a lunar eclipse require that the moon must be on or near the ecliptic when full, that is, the moon must be at or near one of the nodes.

When the whole of the moon's disc is obscured, the eclipse is said to be a *total eclipse*; when only a part at most is obscured, the eclipse is a *partial eclipse*.

In Fig. 137 let S and E be the centres of the sun and earth respectively. A cone can be drawn whose generators are the external tangents to the solar and terrestrial globes; the vertex of this cone is at V, and that part of the cone between XV and YV (dark shading) is called the *umbra*. In the figure the moon's orbit is shown and, if the moon passes completely within the umbra, a total eclipse occurs.

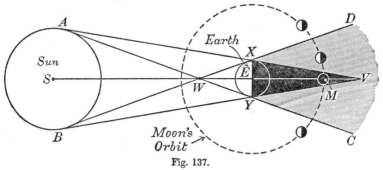

Fig. 137.

Another cone can be drawn whose generators are the internal tangents to the solar and terrestrial globes; its vertex is W. The parts of this cone represented by DXV and CYV (lightly shaded) receive a varying amount of illumination from the sun; along XD and YC the illumination is complete and along XV and YV it is zero. The lightly shaded portions of the internal cone form the *penumbra*, and when the moon passes through the penumbra towards the umbra its brightness gradually decreases until it is completely obscured when it is wholly immersed in the umbra.

209. *The angular radius of the shadow cone at the moon's geocentric distance.*

In the remainder of this chapter, we shall use the following notation:

P—equatorial horizontal parallax of sun.

P_1—equatorial horizontal parallax of moon.

S—sun's semi-diameter.

S_1—moon's semi-diameter.

r—geocentric distance of sun.

r_1—geocentric distance of moon.

s—angular radius (with respect to the earth's centre) of the umbral cone at the moon's distance.

We assume that r and r_1 are expressed in terms of the earth's equatorial radius as the unit of distance.

Consider Fig. 138. The radius of the umbral cone at the moon's distance when a total eclipse is occurring is MN. The angular

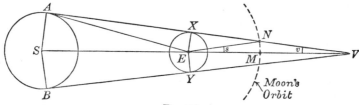

Fig. 138.

radius, s, with respect to E is $N\hat{E}M$. Denote $X\hat{V}E$ by v. Then we have
$$X\hat{N}E = s + v.$$
But $X\hat{N}E$ is the angle subtended at the moon by the earth's radius and, regarding the earth as a sphere, we see that $X\hat{N}E = P_1$. Hence
$$P_1 = s + v \qquad \dots\dots(27).$$
Again, $S\hat{E}A = X\hat{A}E + v$. But SEA is the angle subtended at the earth by the sun's radius and is therefore the sun's semi-diameter S. Also, $X\hat{A}E$ is evidently the sun's parallax P. Hence
$$S = P + v \qquad \dots\dots(28).$$
From (27) and (28) we have
$$s = P + P_1 - S \qquad \dots\dots(29).$$
In a similar manner it is easily shown that the angular radius s' of the penumbral cone is given by
$$s' = P + P_1 + S \qquad \dots\dots(30).$$
The formulae for s and s' in (29) and (30) give the angular radii of the geometrical shadows concerned. It is found, however, that the earth's atmosphere increases, owing to absorption, the theoretical radii by about 2 per cent. For the prediction of lunar eclipses we use the expressions
$$s = \tfrac{51}{50}(P + P_1 - S) \qquad \dots\dots(31),$$
$$s' = \tfrac{51}{50}(P + P_1 + S) \qquad \dots\dots(32).$$
As an example, we find from (31), using the following values $P = 9''$, $P_1 = 57'$, $S = 16'$, that, very nearly,
$$s = 42'.$$

The angular diameter of the umbral cone is thus 84'. If the eclipse is *central*, that is, if the axis EV of the shadow cone passes through the moon's centre, totality will last while the moon's centre moves through the angle $84' - 2S_1$ or $52'$ if we take $S_1 = 16'$. Now the interval between two consecutive full moons is about 30 days; hence the moon's angular motion relative to the shadow is about 12° per day or 30' per hour. Hence the central eclipse which we have been considering will last for about $\frac{52}{30}$ hour or about $1\frac{3}{4}$ hours.

The value of s varies from a maximum of 44'·6 to a minimum of 37'·8; the former value occurs when the moon is nearest the earth (that is, in perigee) at the same time as the earth is furthest from the sun (that is, in aphelion); the minimum value occurs when the moon is in apogee and the earth in perihelion.

The length EV of the earth's shadow is easily obtained. From (28), $v = S - P$, so that
$$EV = EX \operatorname{cosec} (S - P).$$
Putting $S = 16'$, $P = 9''$ and $EX = 3960$ miles, we find that
$$EV = 859,000 \text{ miles.}$$

210. *The ecliptic limits.*

We consider now the effect of the inclination to the ecliptic of the moon's orbital plane in restricting the number of eclipses.

In Fig. 139 let NM be the great circle, on the celestial sphere centred at E, which defines the plane of the moon's orbit. Let M and C be the centres of the moon and the shadow respectively, when an eclipse is about to take place. Let C_1 be the position of the shadow's centre when the moon is at the node N. Denote NC_1 by ξ and let t be the time required by the moon to go from N to M, and for

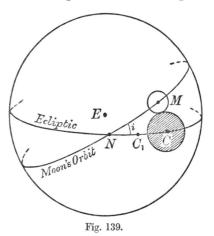

Fig. 139.

the shadow's centre to go from C_1 to C. Now the geocentric longitude of C is the sun's longitude plus $180°$. Hence, in finding the maximum value of ξ for which an eclipse of some kind is possible we are, in effect, finding the maximum angular distance of the sun from the other node.

Let θ be the rate at which the sun's longitude increases, and ϕ the angular velocity of the moon in its orbit; for simplicity, we shall assume θ and ϕ to be constant. Then

$$NM = \phi t, \quad NC = \xi + \theta t.$$

If η denotes the angular distance CM and i the inclination $M\hat{N}C$, we have, regarding the triangle MNC as plane,

$$\eta^2 = (\xi + \theta t)^2 + \phi^2 t^2 - 2\phi t \,(\xi + \theta t) \cos i,$$

or $\qquad \eta^2 = \xi^2 - 2\xi t\,(\phi \cos i - \theta) + t^2\,(\theta^2 + \phi^2 - 2\theta\phi \cos i).$

It is evident that η is a minimum when t is given by

$$\xi\,(\phi \cos i - \theta) - t\,(\theta^2 + \phi^2 - 2\theta\phi \cos i) = 0,$$

and, calling this minimum η_0, we have

$$\eta_0 = \frac{\xi\phi \sin i}{(\theta^2 + \phi^2 - 2\theta\phi \cos i)^{\frac{1}{2}}} \qquad \text{......(33),}$$

or, if q is the ratio θ/ϕ, we can write (33) in the form

$$\xi = \eta_0\,(1 - 2q \cos i + q^2)^{\frac{1}{2}} \operatorname{cosec} i \qquad \text{......(34).}$$

Now q is the ratio of the earth's orbital angular velocity to the moon's orbital angular velocity, or the ratio of the moon's sidereal period to the year; taking mean values, q is about $3/40$. Also $i = 5°\!\cdot\!2$. Hence $\qquad \xi = 10\!\cdot\!3\eta_0 \qquad \text{......(35).}$

If the moon is about to enter the umbral cone,

$$\eta_0 = s + S_1,$$

the quantity s being given by (31). For a partial eclipse to be possible it is evident that ξ must not exceed $10\!\cdot\!3\,(s + S_1)$. For a total eclipse, ξ must not exceed $10\!\cdot\!3\,(s - S_1)$. Taking the following numerical values,

$$S = 960'', \quad S_1 = 935'', \quad P = 9'', \quad P_1 = 3422'',$$

we find that, for a partial eclipse to be possible, we must have

$$\xi < 10\!\cdot\!3\,[\tfrac{51}{50}\,(P + P_1 - S) + S_1],$$

or $\qquad \xi < 9°\!\cdot\!9,$

and, for a total eclipse to be possible,

$$\xi < 10 \cdot 3 \left[\tfrac{51}{50} (P + P_1 - S) - S_1\right],$$

or $\qquad \xi < 4^\circ \cdot 6.$

These values of ξ are called the *ecliptic limits* under the circumstances stated.

The conditions for the partial or complete entrance of the moon into the penumbra may be derived in a similar manner, using the appropriate value of s' given by (32).

In the example we have just worked out it is seen that, for a partial eclipse to be possible, the distance of the sun from the node must not exceed $9^\circ \cdot 9$. But as all the quantities on which ξ depends are not constants, the ecliptic limit is a variable quantity. Its maximum value is $12^\circ \cdot 1$ (called the *superior ecliptic limit*) and its least value is $9^\circ \cdot 5$ (called the *inferior ecliptic limit*).

211. *Calculation of a lunar eclipse.*

We shall suppose that the conditions for a partial or total eclipse are satisfied. In Fig. 140 let M be the moon's position on the celestial sphere centred at E, and C the centre of the earth's

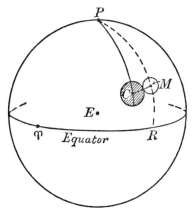

Fig. 140.

shadow. Let (α_1, δ_1) and (α_0, δ_0) be the equatorial co-ordinates of the moon's centre and of C respectively. Denote CM by η and $P\hat{C}M$ by Q. Then we have

$$\sin \eta \sin Q = \cos \delta_1 \sin (\alpha_1 - \alpha_0),$$
$$\sin \eta \cos Q = \sin \delta_1 \cos \delta_0 - \cos \delta_1 \sin \delta_0 \cos (\alpha_1 - \alpha_0),$$

or, with sufficient accuracy,

$$\eta \sin Q = (\alpha_1 - \alpha_0) \cos \delta_1 \qquad \ldots\ldots (36),$$
$$\eta \cos Q = \delta_1 - \delta_0 \qquad \ldots\ldots (37).$$

As C is diametrically opposite to the sun's centre, the R.A. α_0 of C is equal to the sun's R.A. $+ 12^h$, and the declination δ_0 of C is equal to $- \delta$ (δ being the sun's declination).

Write
$$x \equiv (\alpha_1 - \alpha_0) \cos \delta_1 = \eta \sin Q \qquad \ldots\ldots (38),$$
$$y \equiv \delta_1 - \delta_0 \qquad = \eta \cos Q \qquad \ldots\ldots (39).$$

Let x', y' denote the hourly changes in x and y. These are found by computing x and y at intervals of an hour, round about the time of the eclipse. Let T_0 be a convenient moment near opposition (say, the U.T.—to the nearest hour—of opposition in right ascension) and let x_0, y_0 be the computed values of x and y at this moment. At U.T. $(T_0 + t)$, where t is in hours, we can write
$$x = x_0 + x't, \quad y = y_0 + y't.$$

Hence, by (38) and (39),
$$\eta \sin Q = x_0 + x't,$$
$$\eta \cos Q = y_0 + y't.$$

Write
$$x_0 = m \sin M, \quad y_0 = m \cos M \qquad \ldots\ldots (40),$$
$$x' = n \sin N, \quad y' = n \cos N \qquad \ldots\ldots (41).$$

From the known numerical values of x_0, y_0, x', y' the values of m, M, n, N can be derived. We then have
$$\eta \sin Q = m \sin M + nt \sin N \qquad \ldots\ldots (42),$$
$$\eta \cos Q = m \cos M + nt \cos N \qquad \ldots\ldots (43).$$

Squaring these equations and adding, we obtain
$$\eta^2 = m^2 + 2mnt \cos (M - N) + n^2 t^2,$$
from which
$$t = -\frac{m}{n} \cos (M - N) \pm \left[\frac{\eta^2 - m^2 \sin^2 (M - N)}{n^2} \right]^{\frac{1}{2}}$$
$$\ldots\ldots (44).$$

Writing
$$m \sin (M - N) = \eta \sin \psi \qquad \ldots\ldots (45),$$
formula (44) becomes
$$t = -\frac{m}{n} \cos (M - N) \pm \frac{\eta}{n} \cos \psi \qquad \ldots\ldots (46).$$

The conditions for an eclipse can now be inserted. For the first and fourth contacts (when the moon is just entering or just leaving the umbra),

$$\eta = \tfrac{51}{50} (P + P_1 - S) + S_1.$$

Inserting this value of η in (46), we obtain the times of first and fourth contacts.

For second and third contacts (the beginning and end of totality), the value of η is given by

$$\eta = \tfrac{51}{50} (P + P_1 - S) - S_1,$$

and when this value of η is inserted in (46), the times of the beginning and the end of the total phase are derived.

The time of the middle of the eclipse, whether partial or total, is given by $T_0 + t'$, where

$$t' = - \frac{m}{n} \cos (M - N).$$

The *magnitude* of a partial eclipse is the fraction of the moon's diameter obscured and is given by

$$\frac{\eta - m \sin (M - N)}{2S_1},$$

where $m \sin (M - N)$ is taken positive for the time of mid-eclipse and η is the mean of the values used for first and last umbral contacts.

There still remains the prediction of the position angle of the point on the moon's disc, where the eclipse is just beginning or ending. Referring to Fig. 140 we see that the figure has been drawn to represent fourth contact, when the moon is just about to pass out of the umbra. The great circle joining the point of contact of the moon's disc with the umbra and the centre of the umbra makes an angle PMC with the meridian PM through the moon's centre. As position angles are measured in the sense N E S W, the position angle θ of the point of contact is given by

$$\theta = 360° - P\hat{M}C,$$

or

$$\theta = 180° + C\hat{M}R.$$

But since CM is small, $P\hat{C}M$ is approximately equal to $C\hat{M}R$. Hence

$$\theta = 180° + Q.$$

Q is determined from (42) and (43), the value of t for fourth contact being inserted in these equations. In a similar way, the position angle of the point on the moon's disc where the eclipse begins can be derived.

212. *Eclipses of the sun.*

An eclipse of the sun is due to the interposition of the moon between the sun and the earth. The moon must then be at or near conjunction with the sun, that is, it must be new moon. Also, owing to the inclination of the moon's orbit to the ecliptic, it is evident that an eclipse cannot take place unless the moon is on or near the ecliptic, that is to say, the moon must be at or near one of the nodes.

As the moon's radius is much smaller than that of the earth, the earth cannot lie wholly within the shadow cone formed by the external tangents to the solar and lunar globes; consequently, an eclipse of the sun is visible only from a limited area

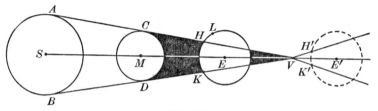

Fig. 141.

of the earth's surface. This is illustrated in Fig. 141, in which the shadow cone formed by the external tangents is shown shaded with its vertex at V; this shadow cone is called the *umbra*. As in the case of lunar eclipses, the *penumbra* is obtained by drawing the internal tangents. For points on the earth's surface between H and K, within the umbra, the moon presents a complete barrier to the light from the sun and the eclipse is then said to be a *total solar eclipse*. At a point such as L, the moon evidently conceals only part of the solar disc; the eclipse is then said to be a *partial eclipse*. The condition for a partial eclipse at L is that L must be in the penumbra.

That total eclipses of the sun are at all possible is due to the fortunate fact that the moon's angular diameter is at times greater than that of the sun; the moon then covers a greater area of the sky than the sun. As the moon's orbit has an eccentricity of about 0·055, and the mean angular diameters of the moon and sun are nearly equal, the value of the moon's diameter can fall below that of the sun (this occurs when the moon is in the neighbourhood of apogee) and consequently the moon cannot completely cover the solar disc; under these circumstances the eclipse, instead of being total, is said to be *annular*. This is illustrated in Fig. 141, where the centres of the sun, moon and earth are now supposed to be at S, M and E' respectively; for any place between H' and K', the eclipse is annular.

213. *The angle subtended at the earth's centre by the centres of the sun and moon at the beginning or end of a partial solar eclipse.*

In Fig. 142 the internal tangents between the sun and the moon are shown, forming the penumbra. Suppose that the tangent AB is also tangential to the earth's surface at C. Then

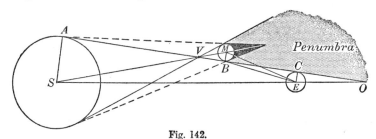

Fig. 142.

to an observer at C the sun's surface is just wholly visible, and the geometrical circumstances illustrated in the figure evidently refer to the beginning or end of the partial phase, when C is just about to enter or leave the penumbra. Let D denote the angle MES. We have

$$D = B\hat{E}S + M\hat{E}B \qquad \ldots\ldots(47).$$

But under the circumstances depicted, MB is very nearly perpendicular to EB, so that $M\hat{E}B = S_1$; hence

$$D = B\hat{E}S + S_1.$$

Also $$B\hat{E}S = O\hat{B}E + E\hat{O}B \qquad \ldots\ldots(48).$$

But $O\hat{B}E \equiv C\hat{B}E =$ horizontal parallax of B, or very nearly the horizontal parallax of M; thus $O\hat{B}E = P_1$.

Again, $$E\hat{O}B \equiv E\hat{O}A = A\hat{E}S - E\hat{A}C,$$

or $$E\hat{O}B = S - P.$$

Hence, from (47) and (48),

$$D = S + S_1 + P_1 - P \qquad \dots\dots(49).$$

Thus D can be evaluated at the time of any eclipse.

214. *Ecliptic limits.*

Consider Fig. 143, in which the moon and sun have the same geocentric longitude at M and S—corresponding to new moon—near the descending node N_1.

Let β denote the latitude SM and i the inclination $M\hat{N}_1 S$ of the moon's orbit to the ecliptic. Let M_1 and S_1 be the positions of the moon and sun a little later. Denote MM_1 by y. Regarding the triangle $MN_1 S$ as plane, we see that the moon's longitude has increased by $MM_1 \cos i$ or $y \cos i$, and the sun's longitude by SS_1. If m is the ratio of SS_1 to $y \cos i$, we have $\quad SS_1 = my \cos i$

$$\dots\dots(50).$$

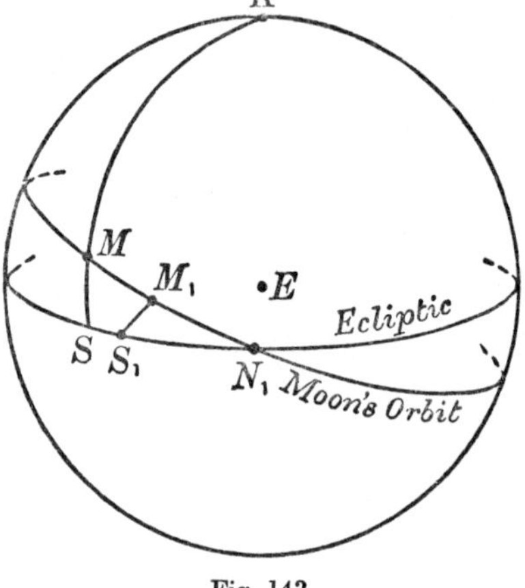

Fig. 143.

Also, $SN_1 = \beta \cot i$ and $MN_1 = \beta \operatorname{cosec} i$, so that

$$S_1 N_1 = \beta \cot i - my \cos i,$$

and $\qquad M_1 N_1 = \beta \operatorname{cosec} i - y.$

Let D denote the angular distance $M_1 S_1$. Then from the triangle $M_1 N_1 S_1$ (regarded as plane), we have

$$D^2 = (\beta \cot i - my \cos i)^2 + (\beta \operatorname{cosec} i - y)^2$$
$$- 2 (\beta \cot i - my \cos i)(\beta \operatorname{cosec} i - y) \cos i,$$

which may be written

$$D^2 = (1 - 2m \cos^2 i + m^2 \cos^2 i) \left\{ y - \frac{\beta \sin i}{1 - 2m \cos^2 i + m^2 \cos^2 i} \right\}^2$$
$$+ \frac{\beta^2 (1 - m)^2 \cos^2 i}{1 - 2m \cos^2 i + m^2 \cos^2 i}.$$

Evidently the minimum value of D (denoted by D_0) is given by

$$D_0 = \frac{\beta(1-m)\cos i}{(1-2m\cos^2 i + m^2\cos^2 i)^{\frac{1}{2}}},$$

or

$$D_0 = \frac{\beta(1-m)\cos i}{\{\sin^2 i + (1-m)^2\cos^2 i\}^{\frac{1}{2}}} \qquad \ldots\ldots(51).$$

Define an angle j by

$$\tan j = \frac{\tan i}{1-m} \qquad \ldots\ldots(52).$$

Then we have $\qquad\qquad D_0 = \beta\cos j \qquad \ldots\ldots(53).$

In (52), $\tan i$ is a small quantity (about $1/11$), since $i = 5°\cdot2$; also $m = 3/40$ approximately. Using these values in (52), we deduce that

$$\cos j = \cos i - 0\cdot0006\cos i,$$

so that, with sufficient accuracy, we can write simply

$$D_0 = \beta\cos i = \beta\left(1 - 2\sin^2\frac{i}{2}\right).$$

For an eclipse to be possible, D_0 must not exceed the value of D given by (49); hence

$$\beta\left(1 - 2\sin^2\frac{i}{2}\right) < S + S_1 + P_1 - P,$$

or, with sufficient accuracy, since $\sin^2\dfrac{i}{2}$ is a small quantity,

$$\beta < S + S_1 + P_1 - P + 2(S + S_1 + P_1 - P)\sin^2\frac{i}{2} \quad\ldots(54).$$

To evaluate the last term of (54), it is sufficient to use the mean values of the quantities concerned; the last term is thus $0'\cdot4$. We can also, with sufficient accuracy, put $P = 0'\cdot1$. Hence

$$\beta < S + S_1 + P_1 + 0'\cdot3 \qquad \ldots\ldots(55).$$

The maximum values of S, S_1 and P_1 are $16'\cdot3$, $16'\cdot8$ and $61'\cdot5$ respectively. Hence, from (55), if the moon's north or south latitude β at the time of new moon is greater than $1° 34'\cdot9$, a solar eclipse is not possible near this particular conjunction.

The *superior ecliptic limit* is defined to be the maximum distance of the sun from a node at the time of new moon if an eclipse is just possible. In Fig. 143 let x denote SN_1. Then we have

$$\sin x = \tan\beta\cot i \qquad \ldots\ldots(56).$$

x is thus a maximum when β is greatest ($1°\ 34'\!\cdot\!9$) and i is least ($4°\ 58'\!\cdot\!8$). With these values, $x = 18°\!\cdot\!5$. Hence, if the distance of the sun from a node at new moon exceeds $18°\!\cdot\!5$, an eclipse is impossible.

To calculate the *inferior ecliptic limit*, we take the minimum value of β and the maximum value of i. The minimum values of S, S_1 and P_1 are $15'\!\cdot\!8$, $14'\!\cdot\!7$ and $53'\!\cdot\!9$ respectively; hence, from (55), the minimum value of β is $1°\ 24'\!\cdot\!7$. The maximum value of i is $5°\ 18'\!\cdot\!6$. Hence, from (54), the inferior ecliptic limit is $15°\!\cdot\!4$. We thus see that if the sun is within $15°\!\cdot\!4$ of a node at new moon an eclipse must take place.

215. *The Besselian elements for a solar eclipse.*

The method used in the prediction of eclipses is analogous to that already described in connection with occultations.

Through the earth's centre E, let a line EC (Fig. 144) be drawn parallel, at a given moment, to the straight line joining

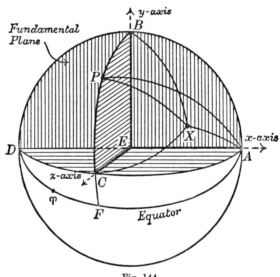

Fig. 144.

the centres of the moon and sun and meeting a sphere, centred at E, in C. EC is the z-axis and the plane DBA (shaded), to which EC is normal at E, is the *fundamental plane*. If P is the north celestial pole, the plane of the great circle through C and

P intersects the fundamental plane in EB. As in the corresponding figure for occultations, EA and EB are the x and y axes respectively.

(i) *The elements* x, y *and* d. We shall take (a, δ) to be the apparent right ascension and declination of the sun and (a_1, δ_1) the corresponding co-ordinates of the moon.

Let (a, d) be the right ascension and declination of the point C on the sphere.

Let (x, y, z) be the rectangular co-ordinates of the sun, with reference to the axes described, in terms of the earth's equatorial radius as the unit. Then if X is the sun's position on the sphere,
$$x = r \cos AX, \quad y = r \cos BX, \quad z = r \cos CX \quad ...(57),$$
r being the sun's geocentric distance.

Now A is the pole of CPB and must therefore be on the equator; we have, then, $PA = 90°$, $FA = 90°$. Hence the right ascension of A is $90° + a$ and therefore $X\hat{P}A = 90° + a - \alpha$. As $PX = 90° - \delta$, we have, from the first of (57), using the cosine-formula,
$$x = r \cos \delta \sin (a - a) \qquad ......(58).$$
In the triangle PBX, $BP = d$. Also, since $A\hat{P}B = 90°$, $X\hat{P}B = 180° + a - \alpha$. Hence
$$y = r [\sin \delta \cos d - \cos \delta \sin d \cos (a - a)] ......(59).$$
In the triangle PCX, $PC = 90° - d$, $PX = 90° - \delta$ and $X\hat{P}C = a - a$; hence
$$z = r [\sin \delta \sin d + \cos \delta \cos d \cos (a - a)] ......(60).$$
In the same way we derive the corresponding equations for the moon:
$$x_1 = r_1 \cos \delta_1 \sin (a_1 - a),$$
$$y_1 = r_1 [\sin \delta_1 \cos d - \cos \delta_1 \sin d \cos (a_1 - a)],$$
$$z_1 = r_1 [\sin \delta_1 \sin d + \cos \delta_1 \cos d \cos (a_1 - a)].$$
But since the z-axis is parallel to the line joining the centres of the moon and sun, we must have
$$x = x_1, \quad y = y_1.$$
The co-ordinates (x, y) or (x_1, y_1) are the *co-ordinates of the centre of the shadow* on the fundamental plane. Hence
$$r \cos \delta \sin (a - a) = r_1 \cos \delta_1 \sin (a_1 - a) \quad ......(61),$$
$$r [\sin \delta \cos d - \cos \delta \sin d \cos (a - a)]$$
$$= r_1 [\sin \delta_1 \cos d - \cos \delta_1 \sin d \cos (a_1 - a)] ...(62).$$

At any instant, the values of r, r_1, a, δ, α_1 and δ_1 may be presumed known; the formulae (61) and (62) thus enable us to calculate a and d. These formulae, however, can be put in a simpler form, since, at or near the time of eclipse, a and δ are little different from α_1 and δ_1 respectively. Let us write

$$\frac{r_1}{r} = b \qquad \ldots\ldots(63),$$

which can also be expressed as

$$b = \frac{\sin P}{\sin P_1} \qquad \ldots\ldots(64),$$

since, by definition, $1/r = \sin P$ and $1/r_1 = \sin P_1$. Thus b can be calculated at any time; it is a small quantity of the order of $1/400$. Writing $[\alpha_1 - a + (a - a)]$ for $(\alpha_1 - a)$ on the right of (61) and expanding, we find

$$\sin(a - a)\{1 - b \sec \delta \cos \delta_1 \cos(\alpha_1 - a)\}$$
$$= b \sec \delta \cos \delta_1 \cos(a - a)\sin(\alpha_1 - a),$$

or, with sufficient accuracy,

$$a = a - \frac{b \sec \delta \cos \delta_1}{1 - b}(\alpha_1 - a) \qquad \ldots\ldots(65).$$

In a similar way, from (62), we find that

$$d = \delta - \frac{b}{1 - b}(\delta_1 - \delta) \qquad \ldots\ldots(66).$$

The calculation of the quantities a and d is made at intervals of 10 minutes.

In addition, the variations x', y' (per minute) of the co-ordinates (x, y) of the centre of the shadow are derived from (58) and (59), and tabulated in the almanacs at intervals of one hour.

(ii) *The element μ.* For any meridian the hour angle of C (Fig. 144) is the hour angle of the vernal equinox Υ less the right ascension of C. In particular, if μ denotes the hour angle of C for the Greenwich meridian when the Greenwich sidereal time is G, then $\mu = G - a$. Hence, a being found from (65), the value of μ can be calculated at any instant. Also the values of μ' (the variation of μ per minute) can evidently be found by simple processes.

(iii) *The elements* f_1, f_2. In Fig. 145 let CD be the section of the fundamental plane with the plane of the paper. Consider first the penumbral cone with its vertex at V_1. Let f_1 denote the

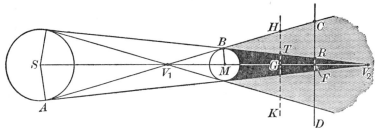

Fig. 145.

angle AV_1S or FV_1C. Let R be the linear radius of the sun and k that of the moon. Then

$$\sin f_1 = \frac{R}{SV_1} = \frac{k}{V_1M} = \frac{R+k}{SM}.$$

But during any phase of the eclipse, $SM = r - r_1$, with sufficient accuracy. Hence, using (63),

$$\sin f_1 = \frac{R+k}{r(1-b)} \qquad \ldots\ldots(67).$$

In (67), the numerator $(R + k)$ is a constant, being the sum of the linear radii of the sun and moon.

Denoting the semi-vertical angle, BV_2M, of the umbral cone by f_2, we derive, in the same way,

$$\sin f_2 = \frac{R-k}{r(1-b)} \qquad \ldots\ldots(68).$$

The angles f_1 and f_2 are easily calculated from (67) and (68).

(iv) *The elements* l_1, l_2. Referring to Fig. 145, we see that MF is the z-co-ordinate of the moon's centre, with reference to the axes already described. Hence $MF = z_1$. Also $V_1M = k \operatorname{cosec} f_1$. Denoting V_1F—the z-co-ordinate of the vertex of the penumbral cone—by c_1, we have

$$c_1 = z_1 + k \operatorname{cosec} f_1 \qquad \ldots\ldots(69),$$

and similarly, if c_2 denotes the z-co-ordinate of the vertex V_2 of the umbral cone, $c_2 = z_1 - k \operatorname{cosec} f_2$ $\ldots\ldots(70)$.

In each instance, c_1 and c_2 are algebraic quantities measured positively in the sense $\overrightarrow{FM}$ (the positive direction of the z-axis).

Let l_1 and l_2 denote the radii of the circles in which the penumbral and umbral cones intersect the fundamental plane. Then

$$l_1 \equiv FC = c_1 \tan f_1 \quad \text{and} \quad l_2 \equiv FR = c_2 \tan f_2 \quad ...(71),$$

or, using (69) and (70),

$$l_1 = z_1 \tan f_1 + k \sec f_1 \qquad ......(72),$$
$$l_2 = z_1 \tan f_2 - k \sec f_2 \qquad ......(73).$$

These formulae—(72) and (73)—enable the numerical values of l_1 and l_2 to be calculated.

In the almanacs, the quantities x, y, log sin d, log cos d, μ, l_1 and l_2 are tabulated at intervals of 10 minutes for each eclipse, and the quantities log x', log y', log μ', log $\tan f_1$ and log $\tan f_2$ at intervals of one hour. These are the *Besselian elements* and we shall now show how these elements are to be used in predicting the circumstances of a solar eclipse at any particular point on the earth's surface.

216. *Eclipse calculations for any station.*

In Fig. 145 let KH be the intersection, with the plane of the paper, of the plane through the observer parallel to the fundamental plane. Let (ξ, η, ζ) be the co-ordinates of the observer at any instant with reference to the fundamental axes. Then the plane KH is given by $z = \zeta$. Consider now the radii of the circles on the plane $z = \zeta$, given by the intersection of this plane with the penumbral and umbral cones, and let L_1 and L_2 be these radii respectively. Then in the figure $L_1 = GH$ and $L_2 = GT$. We then have, since $FG = \zeta$,

$$L_1 = (c_1 - \zeta) \tan f_1,$$
$$L_2 = (c_2 - \zeta) \tan f_2,$$

or, using (71),

$$L_1 = l_1 - \zeta \tan f_1 \qquad ......(74),$$
$$L_2 = l_2 - \zeta \tan f_2 \qquad ......(75).$$

L_1 is always positive. Since c_2 has been used in its algebraic sense, L_2 is negative when the vertex V_2 of the umbral cone is situated (as in Fig. 145) to the right of G. This is the geometrical condition that certain areas on the earth's surface can lie within the umbral cone. Hence for any particular observer, at a distance ζ from the fundamental plane, a condition for a total eclipse is that L_2, computed by means of (75), shall be a negative quantity.

Let ϕ' and ρ be respectively the geocentric latitude and distance of the observer, and λ his longitude west of Greenwich. In Fig. 144 let X now represent the observer's geocentric zenith on the celestial sphere. We then have

$$\xi = \rho \cos AX, \quad \eta = \rho \cos BX, \quad \zeta = \rho \cos CX \quad ...(76).$$

Now PX is the meridian of the observer and, since μ is the Greenwich hour angle of C, the hour angle of C for the observer —the angle XPC—is $\mu - \lambda$. Hence $X\hat{P}A = 90° - (\mu - \lambda)$. Also $PX = 90° - \phi'$. The first of (76) then becomes, applying the cosine-formula to the triangle APX,

$$\xi = \rho \cos \phi' \sin (\mu - \lambda) \qquad ......(77).$$

Similarly,

$$\eta = \rho \left[\sin \phi' \cos d - \cos \phi' \sin d \cos (\mu - \lambda)\right] \quad ...(78),$$
$$\zeta = \rho \left[\sin \phi' \sin d + \cos \phi' \cos d \cos (\mu - \lambda)\right] \quad ...(79).$$

The variations ξ', η' and ζ' per minute in the values of ξ, η and ζ are calculated. For example,

$$\xi' = \mu' \rho \cos \phi' \cos (\mu - \lambda),$$

where μ' is the variation of μ per minute.

The values of ξ, η and ζ are usually computed for the assumed time of contact; then by means of the calculated quantities ξ', η' and ζ', the values of ξ, η and ζ are obtained at 10-minute intervals. The values of L_1 and L_2 can now be derived from (74) and (75) at the appropriate instants.

We now consider the conditions for the beginning and end of a partial or total eclipse at any station. When the partial phase is just beginning or ending, the observer is situated on the boundary of the penumbral cone, and his distance from the axis of the shadow is L_1—the radius of the circle in which the plane $z = \zeta$ cuts the penumbral cone. But the centre of this circle has co-ordinates (x, y) and the observer's co-ordinates are (ξ, η)— both referred to axes in the plane $z = \zeta$ drawn parallel to the fundamental axes. Hence the condition required is

$$(x - \xi)^2 + (y - \eta)^2 = L_1{}^2 \qquad ......(80).$$

Similarly, the condition for the beginning or end of the total eclipse for the station concerned is

$$(x - \xi)^2 + (y - \eta)^2 = L_2{}^2 \qquad \ldots\ldots(81).$$

We shall now show how this last equation is to be used in calculating the beginning and end of the total phase. Let T denote a suitably chosen U.T. near the time of totality, and let $T + t$ be the true U.T. of the beginning (or end) of totality. Let x_0, y_0 be the values of the co-ordinates x, y at time T, and ξ_0, η_0 the corresponding co-ordinates of the observer. Then we shall have, at time $T + t$, t being expressed in minutes,

$$x = x_0 + x't, \quad y = y_0 + y't; \quad \xi = \xi_0 + \xi't, \quad \eta = \eta_0 + \eta't.$$

Now, by (75), $\qquad L_2 = l_2 - \zeta \tan f_2 \qquad \ldots\ldots(82),$

and, since f_2 is a small angle, the value of $\zeta \tan f_2$ at time $T + t$ will not differ appreciably from its value at time T; also l_2 varies very slowly. Hence it will be sufficient to use, in (82), the value of L_2 computed for time T. We then have, for the beginning or end of the total phase (or of the annular phase),

$$[x_0 - \xi_0 + t(x' - \xi')]^2 + [y_0 - \eta_0 + t(y' - \eta')]^2 = L_2{}^2 \ldots(83).$$

All the quantities $L_2, x_0, \ldots \eta'$ being known, (83) is a quadratic equation in t, which will give the instants at which totality begins and ends.

Let auxiliary quantities m, M, n, N be determined by means of

$$m \sin M = x_0 - \xi_0, \quad m \cos M = y_0 - \eta_0 \quad \ldots\ldots(84),$$
$$n \sin N = x' - \xi', \quad n \cos N = y' - \eta' \quad \ldots\ldots(85).$$

As $\tan M = (x_0 - \xi_0)/(y_0 - \eta_0)$, there are two values of M, differing by $180°$, which can satisfy (84). Taking m as the positive square root of $[(x_0 - \xi_0)^2 + (y_0 - \eta_0)^2]$, we shall choose that value of M which will give to $\sin M$ the same sign as $(x_0 - \xi_0)$. The procedure for n and N is similar. Geometrically, m and M are evidently the distance and position angle of the shadow axis relative to the observer; n and N, in the same way, give respectively the magnitude and direction of the motion of the centre of the shadow relative to the observer. Inserting (84), (85) in (83), we obtain

$$n^2 t^2 + 2mnt \cos(M - N) + m^2 - L_2{}^2 = 0 \quad \ldots\ldots(86).$$

The two roots of this equation give the beginning and end of totality. The solution is usually performed as follows. Let an angle ψ be defined as follows:

$$L_2 \sin \psi = m \sin (M - N) \qquad \ldots\ldots(87).$$

The formula (87) gives two values of ψ—we shall refer to these immediately. We have, from (86),

$$n^2 t^2 + 2mnt \cos (M - N) + m^2 \cos^2 (M - N)$$
$$= L_2^2 - m^2 + m^2 \cos^2 (M - N)$$
$$= L_2^2 \cos^2 \psi \quad \text{by (87).}$$

Hence

$$t = -\frac{m}{n} \cos (M - N) \pm \frac{L_2 \cos \psi}{n} \qquad \ldots\ldots(88).$$

If τ is the numerical value of $\dfrac{L_2 \cos \psi}{n}$, the beginning of totality

occurs at G.C.T. $\left(T - \dfrac{m}{n} \cos \overline{M - N} - \tau\right)$ and the end of totality

at G.C.T. $\left(T - \dfrac{m}{n} \cos \overline{M - N} + \tau\right)$. The duration of totality is 2τ. For some purposes, this result is not sufficiently accurate and we proceed as follows. Let T_1 be the approximate time of the beginning of totality. Calculate the quantities in (88) for T_1. As before, there will be two roots of (88), one referring to the beginning and the other to the end of the total phase. Let τ_1 correspond to the former, so that the beginning of totality is at G.C.T.

$$T_1 - \frac{m_1}{n_1} \cos (M_1 - N_1) - \tau_1,$$

where τ_1 is the numerical value of $\dfrac{L_2 \cos \psi}{n}$ computed for T_1, and the subscripts refer to the values of m, n, M, N at T_1. Let T_2 be the approximate time of the ending of totality. Then, similarly, the end of the total phase is at G.C.T.

$$T_2 - \frac{m_2}{n_2} \cos (M_2 - N_2) + \tau_2,$$

where the subscripts refer to the values of m, n, M, N at T_2. The difference between the computed times gives the duration of totality.

If ψ_1, ψ_2 are the values of ψ at the beginning and end of totality, the quadrants of these angles are defined as follows, taking into account the two possible values of ψ as given by (87). For the beginning of the total phase, the quadrant is such that cos ψ_1 is positive, and, for the end of the total phase, the quadrant is such that cos ψ_2 is negative.

If it is considered necessary, the computations can be repeated, choosing a more accurate value of T.

For the calculations of the beginning and end of the partial phase at any station, a similar procedure is adopted, the value of L_1 being, of course, used in the calculations.

As in the case of occultations, the position angle θ for which the partial phase is beginning or ending is given by

$$L_1 \sin \theta = x_0 - \xi_0 + t\,(x' - \xi'),$$
$$L_1 \cos \theta = y_0 - \eta_0 + t\,(y' - \eta'),$$

the appropriate value of t being that corresponding to the beginning or to the end of the partial phase, or, using the appropriate values of N and ψ, by $\theta = N + \psi$.

The corresponding position angles for the total phase are found in the same way.

In the almanacs, the eclipse phenomena, as calculated by the preceding methods, are shown on charts.

217. *The frequency of eclipses.*

The interval between new moon and new moon—the synodical month or lunation—is 29·53 mean solar days. Also we have seen that the nodes of the moon's orbit (section 83) make a complete backward revolution of the ecliptic in 6798·3 days (about 18·6 years). The interval between two consecutive passages of the sun through a node is then 346·62 days. The sun thus separates from a node at the rate of 30·67 degrees per synodical month.

In considering the frequency of eclipses, we first summarise the relevant data concerning ecliptic limits:

	Superior	Inferior
Ecliptic limits (lunar eclipse)	12°·1	9°·5
,, ,, (solar eclipse)	18°·5	15°·4

In each instance the ecliptic limit refers to the distance of the sun from a node when a partial eclipse (lunar or solar) is just possible.

Consider now Fig. 146, in which N and N' are the nodes. Let $NM_1 = NM_2 = N'M_1' = N'M_2'$ represent the ecliptic limits in the case of a lunar eclipse, and $NS_1 = NS_2 = N'S_1' = N'S_2'$ the ecliptic limits in the case of a solar eclipse.

Now the arc $S_1 S_2$ is at least $30°\cdot8$ (corresponding to the inferior limit for solar eclipses) and therefore this arc is always greater than the angle through which the sun separates from a node in a synodical month. Hence one new moon at least and, in consequence, one solar eclipse must occur when the sun is within the arc $S_1 S_2$. Similarly one solar eclipse must occur when the sun is within the arc $S_1' S_2'$.

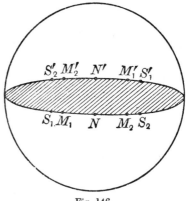

Fig. 146.

Again, the least value of $M_1 M_2$ is twice $9°\cdot5$, or $19°$, which is approximately two-thirds of the motion of the sun with reference to a node in a synodical month. It is thus possible for a full moon to occur, following a solar eclipse, near a node but outside the appropriate inferior limits; it follows that under these circumstances there will not be a lunar eclipse during the year.

We conclude that the *least* possible number of eclipses during a year is 2, and these are both solar eclipses.

Again, 6 synodical months = 177·2 days and the time required by the sun to pass from N to N' is 173·3 days. Therefore if it is full moon 2 days before the sun passes through N, it will also be full moon 2 days after the sun passes through N'. The angular distance of the sun from a node is then well within the limits for a lunar eclipse and hence, under these circumstances, there will be 2 lunar eclipses.

After $14\frac{3}{4}$ days from full moon, it will be new moon, and during this interval the sun will move through $15\frac{1}{3}°$ with reference to a node, and in 2 days will move $2°$ approximately. Hence if it is full moon 2 days before the sun reaches N, the sun will be $13\frac{1}{3}°$ beyond N at the next new moon; at the preceding new moon it is $17\frac{1}{3}°$ from N. Also when it is full moon 2 days after the sun passes through N', the sun will be $17\frac{1}{3}°$ beyond N' at the next new moon; at the preceding new moon it is $13\frac{1}{3}°$ from N'. Both $13\frac{1}{3}°$ and $17\frac{1}{3}°$ are within the superior limits for solar eclipses. Hence 2 solar eclipses are possible near each node, that is to say, 4 solar eclipses are possible within the interval of 346·6 days.

If the eclipses—when the sun is between S and S_1—occur in January, the sun will be back again between S and S_1 in the following December, that is, after 12 synodical months, and in the next half-lunation a solar and a lunar eclipse will be possible. Thus, in $12\frac{1}{2}$ synodical months, it is possible to have in all 5 solar and 3 lunar eclipses. But $12\frac{1}{2}$ lunations just exceed 365 days. Hence, in counting the possible eclipses in any year, we must omit either 1 solar eclipse or 1 lunar eclipse.

We conclude that the greatest number of eclipses possible in a year is 7, of which 4 are solar and 3 lunar or 5 solar and 2 lunar.

218. *The repetition of eclipses.*

The interval between two successive passages of the sun through a node is 346·62 days, and 19 such intervals are equivalent to 6585·8 days. The average value of the synodic month or lunation is 29·5306 days, and 223 lunations are equivalent to 6585·3 days. We thus have the approximate relation:

223 lunations = 19 revolutions of the sun with respect to a node
= 18 years 11 days, approximately.

This period of 18 years 11 days is called the *Saros*. Its significance with regard to eclipses is as follows. Suppose at new moon that the sun is within the ecliptic limits, say at S, so that a solar eclipse takes place. After 19 revolutions with respect to the node N, the sun will again be at the same distance SN from N and it will also be again new moon, for a period equal to 223 lunations has elapsed. Hence a solar eclipse will, in general, take place again.

The same argument applies to lunar eclipses. We thus see that eclipses are generally repeated at intervals of 18 years 11 days. As 19 revolutions of the sun are not, however, exactly equal to 223 lunations, it follows that the circumstances pertaining to one eclipse are not reproduced at the end of the Saros; in particular, the area on the earth's surface from which a total eclipse is visible will, in general, not be the same as that from which the next one (with reference to the Saros) is visible.

We add another interesting relation:

235 lunations = 19 years of $365\frac{1}{4}$ days.

(235 lunations = 6939·69 days and 19 years = 6939·75 days.) This is the *Metonic cycle*. Thus if a full moon occurs on a certain date, there will be a full moon 19 years later. Clearly, the Metonic cycle applies in the same way to the repetition of new moons.

EXERCISES

1. If the inclination of the moon's orbit to the ecliptic be 5° 20′ 6″, show that the moon will at some time occult every star whose latitude is numerically less than 6° 38′ 24″. *[Coll. Exam.]*

2. At midnight, on a certain date, the declination of the moon was 4° 36′ 46″·7 and its right ascension and declination were increasing at the rates of 23^s·0 and 164″ per 10 minutes, respectively. Show that a star in conjunction in right ascension with the moon at midnight could not be occulted at or near the time of this conjunction if the star's declination were less than 3° 10′·4, the sum of the moon's semi-diameter and horizontal parallax being 78′·0. *[Ball.]*

3. On the assumption of a spherical earth, show that, at a time when the moon's horizontal parallax and semi-diameter are respectively P and S, any star of which the geocentric angular distance from the centre of the moon is less than $\sin^{-1}(\sin P + \sin S)$ is occulted by the moon at some part of the earth.

Show also that the greatest possible difference in declination of a star and the centre of the moon at the instant of conjunction in R.A., which will admit of an occultation occurring somewhere on the earth, is approximately

$$\sin^{-1}\{(\sin P + \sin S)(m^2\cos^2\delta + n^2)^{\frac{1}{2}}/(m\cos\delta)\},$$

where δ is the declination of the moon at conjunction, and m and n its rates of variation in R.A. and declination respectively. *[M.T. 1919.]*

4. Show that at the time of a partial or total lunar eclipse the geocentric angular distance of the moon's centre from the axis of the earth's shadow must be less than

$$\sin^{-1}(\sin P_1 + \sin S_1) - \sin^{-1}(\sin S - \sin P),$$

where P, P_1, S, S_1 are the horizontal parallaxes and semi-diameters of the sun and moon, respectively. The earth, sun and moon are considered to be spherical. *[M.T. 1900.]*

5. Show that the maximum duration of totality of a lunar eclipse is

$$\frac{2(P + P_1 - S - S_1)}{m}\left(1 + \frac{s\cos^2 i}{m}\right)\cos i \text{ hours},$$

approximately, if the atmospheric influence be neglected, where P, P_1, S, S_1, s, m are the horizontal parallaxes, semi-diameters and hourly motions in longitude of the sun and moon, respectively, and i is the inclination of the moon's orbit to the ecliptic.

14

SA

6. Show that the interval between the middle of a lunar eclipse and the time of opposition is approximately

$$\frac{m\Delta}{m^2 + n^2 \cos \delta \cos \delta_1} \text{ hours,}$$

where m and n are the differences of hourly motion of the moon and the centre of the earth's shadow in declination and right ascension respectively, Δ is the difference in declination of the moon and the centre of the earth's shadow at the time of opposition, and δ, δ_1 are the average declinations of the shadow and the moon during the eclipse.   [*Coll. Exam.*]

7. Show that the duration of a lunar eclipse does not necessarily contain the instant of opposition in longitude if the eclipse be partial, but must do so if the eclipse be total.   [*Ball.*]

8. The horizontal parallaxes of the sun and moon being known, find the maximum inclination of the moon's orbit to the ecliptic which would ensure a solar eclipse every month.

9. Explain how a solar eclipse can be total at one place and annular at another. If at the place where the eclipse is just total at apparent noon the sun is in the zenith, what is the greatest possible breadth of the annulus at any other place?   [*Lond.* 1925.]

10. Prove that at the instant of conjunction in right ascension the ratio of the distances of the sun from the moon and from the earth is

$$\{\sin P_1 - \sin P \cos (\delta - \delta_1)\}/\sin P_1,$$

δ and δ_1 being the declinations of the sun and moon, P and P_1 their horizontal parallaxes, and the square of $\sin P$ being neglected.

Also prove that if $\dot{a}$ and $\dot{a}_1$ be the hourly changes in right ascension of the sun and moon, respectively, at the same instant, and $\dot{A}$ the hourly change in the right ascension of the line from the centre of the earth parallel to the line joining the centres of the sun and moon, then

$$\dot{A} = \dot{a} + \frac{\sin P \cos \delta_1}{\sin P_1 \cos \delta} (\dot{a} - \dot{a}_1). \qquad [Coll.\ Exam.]$$

11. In 1917, within a few days of Dec. 21, an annular eclipse of the sun took place, visible near the South Pole. According to the *Nautical Almanac* the eclipse was exactly central at *midnight* in Latitude 89° 57′ S, Longitude 142° W. According to the *Connaissance des Temps* the eclipse was central at *noon* also in Latitude 89° 57′ S, but in Longitude 38° E. Show that the difference between the two statements would be accounted for if one almanac had made its calculations for the sea-level, and the other for a plateau about 15,000 feet above sea-level.

Show also that the difference would be accounted for if the positions of the moon adopted by the two almanacs differed by about $2\frac{1}{2}''$ in declination, the moon's parallax being 57′.   [*M.T.* 1917.]

12. Obtain the following construction for the approximate determination of the point Q at which a solar eclipse is central at any particular moment.

M is the point of the earth's surface at which the moon is in the zenith, and S the point at which the sun is in the zenith. Draw the great circle through M and S, and on SM produced take Q such that

$$\sin SQ = (a_{\mathbb{C}} - a_{\odot}) \cos \delta/(P_{\mathbb{C}} - P_{\odot}) \sin \eta,$$

where

$$\tan \eta = (a_{\mathbb{C}} - a_{\odot}) \cos \delta/(\delta_{\mathbb{C}} - \delta_{\odot}).$$

The following data refer to the solar eclipse of 1922 September 20:

G.M.A.T. of conjunction ...	$16^{\text{h}}\ 47^{\text{m}}\ 18^{\text{s}}$
Equation of time	$+6^{\text{m}}\ 38^{\text{s}}$
Sun's declination	$+1°\ 1'\ 42''\cdot7$
Moon's declination ...	$+0°\ 48'\ 0''\cdot3$
Sun's horizontal parallax	$8''\cdot8$
Moon's horizontal parallax	$61'\ 24''\cdot1$

Show that the point at which the central eclipse takes place at noon is approximately in Longitude 106° 30′ E and Latitude 12° S. [*M.T.* 1922.]

APPENDIX I

The Method of Dependences

219. *Introduction.*

The method of dependences was first developed by F. Schlesinger* in 1911 in connection with the determination of stellar parallaxes and was subsequently applied to the measurement of the positions of asteroids and comets from photographs.†

In section 167 we obtained linear formulae connecting the standard co-ordinates of a star, or other object, in terms of the measured co-ordinates and the "plate constants" a, b, etc. To evaluate the plate constants, we make use of at least three comparison stars. In the astrographic problem, for example, if we have several plates giving the position of an asteroid, the plate constants have to be determined from each plate and this involves a large amount of numerical work. Consider a series of plates with the same plate-centre and in which the positions of the asteroid differ little from one plate to another. The same comparison stars can be used for each plate and instead of computing the plate constants for each plate we calculate certain quantities which depend on the comparison stars selected and on one position of the asteroid; these quantities, called *dependences*, are thus independent of the particular plate under investigation. The position of the asteroid is then expressed in terms of the dependences and certain measured quantities. A similar procedure is adopted for the measurement of stellar parallaxes.

220. *The astrographic problem* (3 *comparison stars*).

In this section we shall suppose that we employ three comparison stars. For a given plate, the measured and standard co-ordinates of the comparison stars are given by (57) and (58) on p. 297. We consider the measures in x only—the procedure

---

* *Astrophysical Journal*, vol. xxxiii, p. 161 (1911).

† *Astronomical Journal*, vol. xxxvii, p. 77 (1926). See also H. C. Plummer, *Monthly Notices*, vol. xcii, p. 892 (1932).

for the measures in y is similar. We then have for the three comparison stars

$$\xi_1 - x_1 = a\xi_1 + b\eta_1 + c \qquad \ldots\ldots(1),$$
$$\xi_2 - x_2 = a\xi_2 + b\eta_2 + c \qquad \ldots\ldots(2),$$
$$\xi_3 - x_3 = a\xi_3 + b\eta_3 + c \qquad \ldots\ldots(3),$$

and for the asteroid

$$\xi - x = a\xi + b\eta + c \qquad \ldots\ldots(4).$$

If (ξ_0, η_0) are the standard co-ordinates of the asteroid for one of the plates—we shall refer to it as the "selected plate"—we can write (4) as

$$\xi - x = a\xi_0 + b\eta_0 + c + a\,(\xi - \xi_0) + b\,(\eta - \eta_0)$$
$$\ldots\ldots(5).$$

We are assuming for all the plates concerned that $(\xi - \xi_0)$ and $(\eta - \eta_0)$ are small quantities. Also, the constants a and b, which involve respectively the scale-correction and the orientation of the plate, are to be regarded as small quantities; accordingly, we neglect $a\,(\xi - \xi_0)$ and $b\,(\eta - \eta_0)$ in (5) which then becomes

$$\xi - x = a\xi_0 + b\eta_0 + c \qquad \ldots\ldots(6).$$

It is to be remembered that, in (1), (2), (3) and (6), x_1, x_2, x_3 and x are measured quantities obtained with all necessary accuracy. The standard co-ordinates (ξ_1, η_1) etc. of the comparison stars are supposed to be known. We can then, if we please, solve (1), (2) and (3) in order to obtain a, b and c and then substitute their values in (6). Suppose for the moment that we know the values of ξ_0 and η_0. We then derive from (6) the value of ξ for the plate concerned.

But this procedure is equivalent to the elimination of a, b and c between the four equations (1), (2), (3) and (6), and we can effect this elimination as follows.

Multiply (1), (2), (3) and (6) by D_1, D_2, D_3 and -1 respectively and add. We obtain

$$D_1\,(\xi_1 - x_1) + D_2\,(\xi_2 - x_2) + D_3\,(\xi_3 - x_3) - (\xi - x)$$
$$= a\,\{D_1\xi_1 + D_2\xi_2 + D_3\xi_3 - \xi_0\}$$
$$+ b\,\{D_1\eta_1 + D_2\eta_2 + D_3\eta_3 - \eta_0\}$$
$$+ c\,\{D_1 + D_2 + D_3 - 1\} \qquad \ldots\ldots(7).$$

The elimination of a, b and c is effected if

$$D_1\xi_1 + D_2\xi_2 + D_3\xi_3 = \xi_0 \qquad \ldots\ldots(8),$$

$$D_1\eta_1 + D_2\eta_2 + D_3\eta_3 = \eta_0 \qquad \ldots\ldots(9),$$

$$D_1 + D_2 + D_3 = 1 \qquad \ldots\ldots(10),$$

and these are three equations from which D_1, D_2 and D_3 can be obtained.

The factors D_1, D_2 and D_3 are called the *dependences*. We then have from (7)

$$\xi - x = D_1(\xi_1 - x_1) + D_2(\xi_2 - x_2) + D_3(\xi_3 - x_3)$$
$$\ldots\ldots(11),$$

from which ξ can be determined, all the other quantities being now supposed known.

Now $(\xi_1 - x_1)$, $(\xi_2 - x_2)$ and $(\xi_3 - x_3)$ are all small quantities and it will thus be sufficiently accurate to determine D_1, D_2 and D_3 if we substitute in (8) and (9) the *measured* co-ordinates of the comparison stars and the asteroid for the selected plate. Denoting these by (X_1, Y_1), (X_2, Y_2), (X_3, Y_3) and (X_0, Y_0) respectively, the equations to determine D_1, D_2 and D_3 become

$$D_1 X_1 + D_2 X_2 + D_3 X_3 = X_0 \qquad \ldots\ldots(12),$$

$$D_1 Y_1 + D_2 Y_2 + D_3 Y_3 = Y_0 \qquad \ldots\ldots(13),$$

$$D_1 + D_2 + D_3 = 1 \qquad \ldots\ldots(14),$$

from which, solving in determinant form, we have for D_1

$$\frac{D_1}{\begin{vmatrix} X_0, & X_2, & X_3 \\ Y_0, & Y_2, & Y_3 \\ 1, & 1, & 1 \end{vmatrix}} = \frac{1}{\begin{vmatrix} X_1, & X_2, & X_3 \\ Y_1, & Y_2, & Y_3 \\ 1, & 1, & 1 \end{vmatrix}} \qquad \ldots\ldots(15).$$

In Fig. 147, let S_1, S_2 and S_3 be the positions on the selected plate of the images of the comparison stars, and A_0 the position of the image of the asteroid. Then the determinant under D_1 in (15) is simply twice the algebraic measure of the area of the triangle $A_0 S_2 S_3$ and the second determinant in (15) is twice the area of the triangle $S_1 S_2 S_3$. We obtain similar results for D_2 and for D_3 and we then have, as the solutions of (12), (13) and (14),

$$\frac{D_1}{A_0 S_2 S_3} = \frac{D_2}{A_0 S_3 S_1} = \frac{D_3}{A_0 S_1 S_2} = \frac{1}{S_1 S_2 S_3} \qquad \ldots\ldots(16).$$

Let the straight lines through A_0 and the vertices of the triangle cut the sides in P_1, P_2 and P_3.

Then
$$\frac{A_0 S_2 S_3}{S_1 S_2 S_3} = \frac{A_0 P_1}{S_1 P_1}.$$

Hence
$$D_1 = \frac{A_0 P_1}{S_1 P_1}, \quad D_2 = \frac{A_0 P_2}{S_2 P_2} \text{ and } D_3 = 1 - D_1 - D_2 \quad ...(17).$$

The values of D_1, D_2 and D_3 can be readily obtained, with sufficient accuracy, by plotting the positions of the comparison stars and asteroid on squared paper, from the measured co-ordinates of the selected plate. It can be shown that the results of applying the method are most accurate when A_0 in Fig. 147 coincides with the centroid of the triangle $S_1 S_2 S_3$. In most instances it is possible to choose the comparison stars to satisfy this condition approximately. In any event, we choose the comparison stars so that the asteroid is within the triangle which they form.

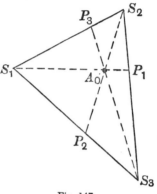

Fig. 147.

We rewrite (11) as
$$\xi = \sum_{i=1}^{3} D_i \xi_i + x - \sum_{i=1}^{3} D_i x_i \qquad ......(18).$$

Similarly
$$\eta = \sum_{i=1}^{3} D_i \eta_i + y - \sum_{i=1}^{3} D_i y_i \qquad ......(19).$$

In (18) the expression $(x - \sum_{1}^{3} D_i x_i)$ is obtained, let us say, in millimetres and its actual value will depend on the scale of the plate taken with a particular telescope whose focal length we denote by f_1. The standard co-ordinates, on the other hand, are generally expressed in terms of the scale of the "astrographic telescope" the focal length of which we denote by f_2 ($f_2 = 3 \cdot 44$ metres so that 1 mm. on the plate corresponds to $60''$). Consequently, we require to multiply $(x - \Sigma D_i x_i)$ by f_2/f_1 so that ξ may be given by (18) in astrographic units. A similar action is to be taken with respect to (19).

It is to be remembered that (ξ_i, η_i) in (18) and (19) denote the standard co-ordinates of a comparison star with respect to the centre of the plate measured. If we make use of an Astrographic Catalogue we then obtain the standard co-ordinates of the comparison stars with respect to the "astrographic centre", that is, the centre of the astrographic plate from which the published standard co-ordinates are derived. If ξ_1', ξ_2', ξ_3' and ξ' denote the standard co-ordinates of the three comparison stars and the asteroid relative to the astrographic centre, we have, with sufficient accuracy,

$$\xi_1' = \xi_1 + k, ..., \xi' = \xi + k,$$

where k is the standard co-ordinate of the centre of the measured plate with respect to the astrographic centre. Thus we obtain from (18), since $D_1 + D_2 + D_3 = 1$,

$$\xi' = \sum_{i=1}^{3} D_i \xi_i' + x - \sum_{i=1}^{3} D_i x_i \qquad ......(20),$$

with a similar equation for η'.

Example.

The following example has been given by H. E. Wood,* relating to Comet 1911 *b*. The values of ξ_1', ξ_2' and ξ_3' are taken from the *Oxford Astrographic Catalogue*. We give the data in tabular form for deriving the standard co-ordinate ξ' of the comet.

Star	ξ'	x	y	D	$D\xi'$	Dx
1	7·692	5·6067	11·6605	0·233	1·7922	1·3064
2	17·650	15·8160	22·0935	0·273	4·8184	4·3178
3	23·547	21·6011	9·3951	0·494	11·6322	10·6709
				1·000	18·2428	16·2951

For the comet,

$$x = 16·3437, \quad y = 13·3462 \qquad ......(21).$$

The dependences, D, in column 5 are found by means of a figure using the measured co-ordinates in columns 3 and 4 and the values of x and y for the comet as given in (21).

From (21) and the last column, we find that

$$x - \Sigma D_i x_i \equiv 16·3437 - 16·2951 = + 0·0486.$$

* *Journal of the Brit. Ast. Association*, vol. xxxix, p. 201 (1929).

We can obtain the value of the factor f_2/f_1 from the above data with sufficient accuracy, as follows. We have

$$\xi_2' - \xi_1' = 9\cdot958 \quad \text{and} \quad x_2 - x_1 = 10\cdot2093.$$

Hence in astrographic co-ordinates

$$x - \Sigma D_i x_i = + 0\cdot0486 \times \frac{9\cdot958}{10\cdot209} = + 0\cdot0478.$$

Accordingly, from (20),

$$\xi' = 18\cdot2428 + 0\cdot0478$$
$$= 18\cdot2906.$$

This is the ξ standard co-ordinate of the comet referred to the astrographic centre.

The η co-ordinate is obtained in a similar way. If desired, the right ascension and declination of the comet can be computed by the formulae (22) and (23) of p. 285—or by equivalent formulae.

221. *The astrographic problem* (n *comparison stars*).

We suppose now that the number, n, of comparison stars is greater than 3. We now have n equations of the form (1) to give the values of the plate constants a, b and c. They are

$$\left.\begin{array}{l} a\xi_1 + b\eta_1 + c = \xi_1 - x_1 \\ a\xi_2 + b\eta_2 + c = \xi_2 - x_2 \\ \dots\dots\dots\dots\dots\dots\dots\dots \\ a\xi_n + b\eta_n + c = \xi_n - x_n \end{array}\right\} \quad \dots\dots(22).$$

The values of a, b and c are to be obtained by the method of least squares. Multiply each by the corresponding coefficient of a and, adding, we obtain

$$a\{\xi_1^2 + \xi_2^2 + \dots + \xi_n^2\} + b\{\xi_1\eta_1 + \xi_2\eta_2 + \dots + \xi_n\eta_n\}$$
$$+ c\{\xi_1 + \dots + \xi_n\} = \xi_1(\xi_1 - x_1) + \xi_2(\xi_2 - x_2).$$

This is usually written in the notation

$$a[\xi_i^2] + b[\xi_i\eta_i] + c[\xi_i] = [\xi_i(\xi_i - x_i)] \quad \dots\dots(23),$$

the square brackets denoting summations for $i = 1, 2, \dots, n$.

Similarly, by multiplying each equation of (22) by the corresponding coefficient of b, and adding, we obtain

$$a[\xi_i\eta_i] + b[\eta_i^2] + c[\eta_i] = [\eta_i(\xi_i - x_i)] \quad \dots\dots(24).$$

Also, adding the n equations of (22), we have

$$a\,[\xi_i] + b\,[\eta_i] + nc = [\xi_i - x_i] \qquad \ldots\ldots(25).$$

As in (6), the equation for the asteroid is

$$a\xi_0 + b\eta_0 + c = \xi - x \qquad \ldots\ldots(26),$$

in which, as before, ξ_0 and η_0 denote the standard co-ordinates of the asteroid for a selected plate.

To eliminate a, b and c from the four equations (23)–(26) multiply these in order by P, Q, R and -1 and add. Then if P, Q and R are given by

$$P\,[\xi_i^2] + Q\,[\xi_i\eta_i] + R\,[\xi_i] = \xi_0 \qquad \ldots\ldots(27),$$
$$P\,[\xi_i\eta_i] + Q\,[\eta_i^2] + R\,[\eta_i] = \eta_0 \qquad \ldots\ldots(28),$$
$$P\,[\xi_i] + Q\,[\eta_i] + Rn = 1 \qquad \ldots\ldots(29),$$

we have

$$\xi - x = P\,[\xi_i\,(\xi_i - x_i)] + Q\,[\eta_i\,(\xi_i - x_i)] + R\,[\xi_i - x_i]$$

which can be written

$$\xi - x = \sum_{i=1}^{n} (P\xi_i + Q\eta_i + R)\,(\xi_i - x_i).$$

Set

$$D_i \equiv P\xi_i + Q\eta_i + R \qquad \ldots\ldots(30).$$

Then

$$\xi - x = \sum_{i=1}^{n} D_i\,(\xi_i - x_i) \qquad \ldots\ldots(31).$$

In this last equation, D_i is a function of the standard co-ordinates of the n comparison stars and of ξ_0 and η_0. Since $(\xi_i - x_i)$ is a small quantity for any one of the stars it will be sufficient to regard D_i as a function of the measured co-ordinates $x_1, \ldots, y_n, x_0, y_0$ for the selected plate; as before, we denote these co-ordinates by X_i, Y_i, X_0 and Y_0. Thus P, Q and R are now to be determined from (27), (28) and (29) in which we replace ξ_i etc. by X_i etc. Hence

$$P\,[X_i^2] + Q\,[X_iY_i] + R\,[X_i] = X_0 \qquad \ldots\ldots(32),$$
$$P\,[X_iY_i] + Q\,[Y_i^2] + R\,[Y_i] = Y_0 \qquad \ldots\ldots(33),$$
$$P\,[X_i] + Q\,[Y_i] + Rn = 1 \qquad \ldots\ldots(34).$$

We can simplify the calculation of P, Q and R by supposing that for the selected plate the values of X_i, Y_i, X_0 and Y_0 are measured from the centroid of the n comparison stars. We then have

$$\sum_{i=1}^{n} X_i = \sum_{i=1}^{n} Y_i = 0$$

and P, Q and R are now to be determined from

$$P[X_i^2] + Q[X_iY_i] = X_0 \qquad \ldots\ldots(35),$$
$$P[X_iY_i] + Q[Y_i^2] = Y_0 \qquad \ldots\ldots(36),$$
$$Rn = 1 \qquad \ldots\ldots(37).$$

Then $\qquad\qquad D_i = PX_i + QY_i + R \qquad \ldots\ldots(38).$

The quantities D_i are the *dependences*.

The practical procedure is first to calculate P, Q and R by means of (35), (36) and (37) and then to form the dependence D_i for each star by means of (38).

As in the previous section,

$$\sum_{i=1}^{n} D_i = 1 \qquad \ldots\ldots(39),$$

as we can see from (38) and (37), remembering that

$$\Sigma X_i = \Sigma Y_i = 0.$$

When the dependences have been calculated the value of ξ for the asteroid can be then found from (31) for any number of plates.

It is to be remarked that the application of (31) requires that the values of ξ_i for the n comparison stars should be accurately known and that the values of x and of x_i for the n comparison stars should be accurately measured.

222. *Application of the method of dependences to the determination of stellar parallax.*

The method of dependences is now used extensively in the reduction of parallax measures.

Let ξ_0 denote the heliocentric standard co-ordinate of the parallax star at time t_0. As in section 175 we denote the parallax factor for an observation made at time t by F, and the component of the annual proper motion by μ_x ($\equiv \mu_a \cos \delta$). The geocentric standard co-ordinate ξ at time t is then given by

$$\xi - \xi_0 + F\Pi + T\mu_x \qquad \ldots\ldots(40),$$

where Π is the parallax of the star and $T \equiv t - t_0$ in years. But from the measurement of a plate obtained at time t, we have the value of ξ given by (31) as

$$\xi = x + \sum_{i=1}^{n} D_i (\xi_i - x_i) \qquad \ldots\ldots(41).$$

Hence from (40) and (41)

$$F\Pi + T\mu_x + \xi_0 - \sum_{i=1}^{n} D_i \xi_i = x - \sum_{i=1}^{n} D_i x_i \quad \ldots\ldots(42).$$

Let
$$m = x - \sum_{i=1}^{n} D_i x_i \qquad \ldots\ldots(43),$$

$$C = \xi_0 - \sum_{i=1}^{n} D_i \xi_i \qquad \ldots\ldots(44).$$

Then
$$F\Pi + T\mu_x + C = m \qquad \ldots\ldots(45).$$

In (43), m depends on the measured quantities x_i for the comparison stars and on the measured quantity x for the parallax star. In (44), C is a constant for each plate if we suppose the comparison stars to have negligible parallaxes and proper motions (actually, Π and μ_x in (45) are the relative parallax and proper motion, that is, relative to the mean of the comparison stars). Thus each exposure provides an equation of the form of (45) and the values of Π, μ_x and C (this is of no further interest) are determined from a least squares solution of all such equations. As the number of plates measured is generally about a score, the convenience of the method and its economy in calculation can be readily appreciated.

Example.

The data below (in terms of a particular scale unit) are taken from the measures of a star, as given by Schlesinger.* Four comparison stars were used.

Star	X	Y	x	D	Dx
1	− 378	+ 62	198·766	0·25	49·6915
2	− 8	+ 200	569·351	0·48	273·2885
3	+ 84	− 130	660·934	0·11	72·7027
4	+ 302	− 132	879·485	0·16	140·7176
Parallax star	− 40·3	+ 75·5	535·133	—	$\Sigma D_i x_i = 536 \cdot 400$

In columns 2 and 3 the co-ordinates of the four comparison stars and of the parallax star are with reference to the centroid

* *Astrophysical Journal*, vol. xxxiii, p. 163 (1911).

of the four comparison stars. From the values for the comparison stars we find

$$\sum_{i=1}^{4} X_i^2 = 241,208; \quad \sum_{i=1}^{4} Y_i^2 = 78,168; \quad \sum_{i=1}^{4} X_i Y_i = -75,820.$$

Inserting these values in (32), (33) and (34) and putting $X_0 = -40.3$, $Y_0 = +75.5$ (from the last line of the table), we find that

$$P = 0.000196, \quad Q = 0.00116, \quad R = 0.250.$$

The values of D_i are then calculated from (38)—they are shown in the fifth column. The products $D_i x_i$ are then formed for each star and the sum $\Sigma D_i x_i$ is shown at the bottom of the last column. Thus for this particular exposure the value of m, given by (43), is

$$535.133 - 536.400 \equiv -1.267.$$

Also, for this particular exposure $F = +0.880$ and $T = +0.723$, so that we have from (45),

$$0.880\Pi + 0.723\mu_x + C = -1.267 \quad \ldots\ldots(46).$$

From the least square solution of all the equations of condition similar to (46), the values of Π and μ_x in terms of the scale unit were found to be 0.1055 and -0.1307 respectively, yielding the final results

$$\Pi = 0''.281 \pm 0''.004; \quad \mu_x = -0''.348.$$

APPENDIX II

Stellar Magnitudes

223. *Apparent magnitude.*

The first classification of the stars, visible to the naked eye, according to their apparent brightness was made by Hipparchus about two thousand years ago. The fifteen brightest stars were designated "stars of the first magnitude" and stars just visible to the naked eye "stars of the sixth magnitude", stars of intermediate brightness being assigned to intermediate magnitude classes. With accurate instruments at present available for measuring the relative brightness of the stars, it is essential to have an accurate classification according to brightness and "magnitude" has now come to mean a number, on a certain scale, associated with the brightness of a star. If m_1 and m_2 denote the magnitudes of two stars on this scale, and l_1 and l_2 their apparent brightness or luminosity, the difference of magnitude, $m_2 - m_1$, is defined by the formula

$$\frac{l_1}{l_2} = 10^{-0\cdot4(m_1-m_2)} \qquad \qquad \ldots\ldots(1).$$

A difference of 5 magnitudes thus corresponds to a ratio of 100 : 1 in brightness and a difference of one magnitude to a ratio of 2·512 : 1. The zero of the magnitude scale is chosen arbitrarily. On the visual magnitude scale as adopted in practice the magnitude 1·0 corresponds to the mean brightness of the two nearly equally bright stars Altair and Aldebaran. The visual magnitude scale corresponds to that part of the spectrum to which the eye is most sensitive.

The photographic magnitude scale corresponds to that part of the spectrum to which the ordinary photographic plate is most sensitive; the ratio $l_1 : l_2$ for two stars can be obtained photographically by measuring the relative diameters or the relative density of the images on plates of standard quality. If m_v and m_p denote the visual and photographic magnitudes of a star, the difference $m_p - m_v$ is called the *colour-index* of the star.

224. *Absolute magnitude.*

Although one star may appear to be brighter than another, it does not follow that the first star is *intrinsically* more luminous than the second. The brightness of a star varies inversely as the square of its distance from us; if l denotes its apparent brightness corresponding to its distance d and L denotes what its apparent brightness would be if it were at a standard distance D from us, we have

$$\frac{l}{L} = \frac{D^2}{d^2} \qquad \ldots\ldots(2).$$

Let M denote what the star's apparent magnitude would be if it were at the standard distance D. Then, by (1) and (2),

$$10^{-0\cdot4\,(m-M)} = \frac{D^2}{d^2}$$

from which $\qquad M - m = 5 \log_{10}(D/d) \qquad \ldots\ldots(3).$

But the star's parallax Π (in seconds of arc) is defined by

$$\Pi = \frac{a}{d} \operatorname{cosec} 1'' \qquad \ldots\ldots(4),$$

where a is the radius of the earth's orbit. Similarly, if Π_0 is the parallax corresponding to the distance D,

$$\Pi_0 = \frac{a}{D} \operatorname{cosec} 1'' \qquad \ldots\ldots(5).$$

Hence, from (4) and (5),

$$\frac{D}{d} = \frac{\Pi}{\Pi_0}$$

and, from (3),

$$M = m + 5 \log_{10} \Pi - 5 \log_{10} \Pi_0 \qquad \ldots\ldots(6).$$

The standard distance, D, is by convention ten parsecs, so that $\Pi_0 = 0''\cdot1$. Hence we have

$$M = m + 5 + 5 \log_{10} \Pi \qquad \ldots\ldots(7),$$

in which M is called the *absolute magnitude*.

It is found that the absolute magnitudes of the stars range from about -5 to $+15$, corresponding to a range in intrinsic luminosity of $100^4 : 1$ or $100,000,000 : 1$.

The Coelostat

225. *General principles.*

For eclipse and certain astrophysical observations it is convenient to use a telescope which is fixed in position. Consequently, it is necessary to have subsidiary apparatus whose
function is to reflect the light from the heavenly body under
observation into the fixed telescope. Such an auxiliary instrument is called a *coelostat*. Its principal feature is a plane mirror

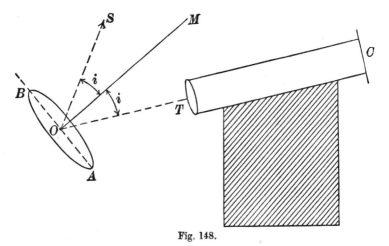

Fig. 148.

which can be made to rotate about a diameter at a uniform rate,
this diameter being parallel to the earth's axis.

In Fig. 148, the mirror (which is generally circular) is shown
with its axial diameter AB—the remainder of the instrument is
not shown—and OM is the normal to the mirror. The telescope
is mounted as illustrated in the figure, T being the object glass
and C the focal plane in which a photographic plate can be
placed. The mirror is initially adjusted with its normal OM
in such a position that the beam of rays from a heavenly body S
is reflected along OT, which is collinear with the optical axis

of the telescope. As we shall see later, the mirror is made to rotate at half the angular rate of the diurnal motion of S and this is a necessary condition that the rays from S shall continue to fall normally on the object glass at T after reflection at the mirror. We denote by i the angle between the normal of the mirror and the direction OS or OT.

Let O, the centre of the mirror, be the centre of the celestial sphere in Fig. 149. Then S gives the direction of the heavenly

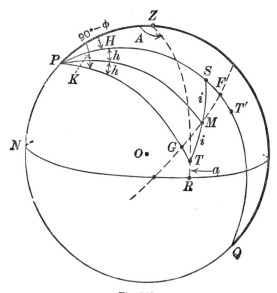

Fig. 149.

body, M the direction of the normal of the mirror and T the direction of the rays after reflection. Since the axis AB of the mirror is parallel to the earth's axis of rotation, M in Fig. 149 is on the celestial equator FG. If δ is the declination of S, the arc FS, measured along the meridian PS, is δ. Let the meridian PT intersect the equator in G. Then by the laws of reflection OS, OM and OT are coplanar and therefore S, M and T lie on a great circle; also the arcs SM and MT are equal and each is equal to i. It follows that GT is equal to SF, so that $PT = 90° + \delta$. Again, $GM = MF$ and consequently $T\hat{P}M = S\hat{P}M$; we denote these angles by h.

Since the telescope is fixed in position, T is a fixed point on the celestial sphere. Let the altitude and azimuth of T be a and A respectively; then a and A are constants for a given position of the telescope.

Let H be the hour angle of S, H' the hour angle of M and K (a constant) the hour angle of T. We have
$$H' = H + h \quad \text{and} \quad H + 2h = K$$
from which
$$2H' = H + K.$$
Hence
$$\frac{dH'}{dt} = \frac{1}{2}\frac{dH}{dt} \qquad \ldots\ldots(1).$$

This formula gives the rate at which the mirror must be rotated about its axis AB. For the sun, the rate may be expressed as $360°$ in 48 hours or $7\frac{1}{2}°$ per mean solar hour, very nearly. For a star, the rate is $7\frac{1}{2}°$ per sidereal hour.

226. *Formulae for the position of the telescope.*

We investigate now the various ways in which the telescope may be set up in a given latitude ϕ.

(i) From the triangle PZT, in which
$$PZ = 90° - \phi, \ ZT = 90° - a, \ PT = 90° + \delta \ \text{ and } \ P\hat{Z}T = A,$$
we have by formula A,
$$- \sin \delta = \sin \phi \sin a + \cos \phi \cos a \cos A \qquad \ldots\ldots(2),$$
which gives the necessary relation between a and A for rays to be reflected into the telescope. Thus, if it is decided to have the telescope inclined to the horizontal at a certain angle, the azimuth A is easily calculated from (2).

(ii) In particular, it may be found convenient to mount the telescope horizontally; in this case $a = 0°$ and A is given by
$$\cos A = - \sin \delta \sec \phi \qquad \ldots\ldots(3),$$
with the condition for the numerical values of δ and ϕ that
$$\delta + \phi < 90°.$$

(iii) Again, it may be found convenient to mount the telescope in the meridian, in which event, $A = 180°$ and (2) becomes
$$\sin \delta = \cos (\phi + a)$$
whence
$$a = 90° - \phi - \delta \qquad \ldots\ldots(4).$$

(iv) A further practical consideration concerns the magnitude of the angle i which, as a general rule, ought to be kept fairly small.

The cross section of the reflected beam is an ellipse with principal axes in the ratio $\cos i : 1$. It is always desirable to "fill" the object glass of the telescope and the condition for this is that $\cos i > \dfrac{d}{D}$, where d and D are respectively the diameters of the object glass and the mirror.

From the triangle PSM, in which $PM = 90°$, we have, by formula **A**,

$$\cos i = \cos \delta \cos h \qquad \ldots\ldots(5).$$

It is seen from (5) that i is a minimum when $h = 0$. Thus the minimum value of i is δ. It follows that, in this case, M and T lie on the meridian PS, with M at F and T at T' where $FT' = FS = \delta$. The altitude, a', of T' is then easily found from the triangle PZT' in which $PT' = 90° + \delta$: we have

$$\sin a' = - \sin \delta \sin \phi + \cos \delta \cos \phi \cos H \quad \ldots\ldots(6).$$

For example, with regard to a total eclipse of the sun, the hour angle H for mid-totality can be found with sufficient accuracy and hence a' can be calculated from (6). The azimuth, A', of T' can then be found from the formula

$$- \sin \delta = \sin \phi \sin a' + \cos \phi \cos a' \cos A' \quad \ldots\ldots(7).$$

APPENDIX A

ASTRONOMICAL CONSTANTS

Dimensions of the earth:

Equatorial radius, a = 6378·39 kms. or 3963·35 miles

Polar radius, b = 6356·91 kms. or 3950·01 miles

Compression, $c \equiv \dfrac{a-b}{a} = \dfrac{1}{297}$

Length of the day:

Sidereal day = 23^h 56^m 4^s·091 mean solar time

Mean solar day = 24 3 56 ·555 sidereal time

Length of the month:

Synodical = 29^d·530 588 = 29^d 12^h 44^m 2^s·8
Tropical = 27 ·321 582 = 27 7 43 4 ·7
Sidereal = 27 ·321 661 = 27 7 43 11 ·5
Anomalistic = 27 ·554 550 = 27 13 18 33 ·1
Nodical = 27 ·212 220 = 27 5 5 35 ·8

Length of the year:

Tropical = 365^d·2422
Sidereal = 365 ·2564
Anomalistic = 365 ·2596
Eclipse = 346 ·6200

Solar parallax = 8''·80
Moon's equatorial horizontal parallax = 57' 2''·70
Constant of aberration = 20''·47
Constant of nutation = 9''·21
General precession = 50''·2564 + 0''·0222T*
Precession in R.A., m = 46''·0850 + 0''·0279T
= 3^s·07234 + 0^s·00186T
Precession in Dec., n = 20''·0468 − 0''·0085T
Obliquity of ecliptic, ϵ = 23° 27' 8''·26 − 46''·84T
Mean geocentric distance of moon = 384,400 kms. = 238,900 miles
Mean geocentric distance of sun = 149,674,000 kms. = 93,003,000 miles
(1 astronomical unit)
1 Parsec = 30·87 × 10^{12} kms. = 19·18 × 10^{12} miles
1 Light-year = 9·46 × 10^{12} kms. = 5·88 × 10^{12} miles
Velocity of light = 299,791 kms. or 186,279 miles per second

Light travels 1 astronomical unit in 498·4 seconds
Constant of gravitation, G = 6·670 × 10^{-8} C.G.S. units
Solar apex (*Boss*), R.A. = 18^h or 270°, Dec. = +34°
Pole of galactic plane (1900), R.A. = 12^h 40^m or 190°, Dec. = +28°

* T is measured in Julian centuries from 1900·0.

APPENDIX B

DIMENSIONS OF THE SUN, MOON AND PLANETS

	Symbol	Semi-diameter in		Mass (fraction of Sun's mass) $1 \div$	Density (water $=1$)	Mean orbital velocity (kms./ sec.)
		Kms.	Miles			
Sun	☉	696,000	432,000	—	1·41	—
Moon	☽	1,738	1,080	27,158,000	3·34	1·02
Mercury	☿	2,495	1,550	6,000,000	5·13	47·8
Venus	♀	6,200	3,850	408,000	4·97	35·0
Earth	⊕	{6,378 (Eql.)	3,963	332,000	5·52	29·8
		{6,357 (Polar)	3,950			
Mars	♂	3,400	2,100	3,093,500	3·94	24·2
Jupiter	♃	{71,350 (Eql.)	44,350	1,047·35	1·33	13·1
		{66,600 (Polar)	41,400			
Saturn	♄	{60,400 (Eql.)	37,550	3,501·6	0·69	9·7
		{54,050 (Polar)	33,600			
Uranus	♅	24,850	15,450	22,870	1·56	6·8
Neptune	♆	26,500	16,500	19,315	2·27	5·4
Pluto	PL	7,000?	4,350?	3,300,000?	5·6?	4·7

APPENDIX C

ELEMENTS OF THE PLANETARY ORBITS FOR THE EPOCH 1960 JANUARY 1, G.M.NOON

Name	Semi-major axis a (in astronomical units)	Eccentricity e	Sideral period in tropical years	Sideral mean daily motion n	Synodic period (in days)
				°	
Mercury	0·387 099	0·205 627	0·240 85	4·092 339	115·88
Venus	0·723 332	0·006 793	0·615 21	1·602 131	583·92
Earth	1·000 000	0·016 726	1·000 04	0·985 609	—
Mars	1·523 691	0·093 368	1·880 89	0·524 033	779·94
Jupiter	5·202 803	0·048 435	11·862 23	0·083 091	398·88
Saturn	9·538 843	0·055 682	29·457 72	0·033 460	378·09
Uranus	19·181 951	0·047 209	84·013 31	0·011 732	369·66
Neptune	30·057 779	0·008 575	164·793 45	0·005 981	367·48
Pluto	39·438 71	0·250 236	247·686	0·003 979	366·72

Name	Inclination i	Of the ascending node θ	Mean longitude Of perihelion ϖ	Mean longitude At the epoch ϵ
	°	°	°	°
Mercury	7·003 99	47·857 14	76·833 09	222·621 65
Venus	3·394 23	76·319 72	131·008 31	174·294 31
Earth	0·0	0·0	102·252 53	100·158 15
Mars	1·849 91	49·249 03	335·322 69	258·767 29
Jupiter	1·305 36	100·044 44	13·678 23	259·831 12
Saturn	2·489 91	113·307 47	92·264 47	280·671 35
Uranus	0·773 06	73·796 30	170·010 83	141·304 96
Neptune	1·773 75	131·339 80	44·273 95	216·940 90
Pluto	17·169 9	109·885 62	224·160 24	181·646 32

APPENDIX D

ELEMENTS AND DIMENSIONS OF THE SATELLITES

Planet		Satellite	Mean distance from planet (unit $= 10^3$ kms.)	Sidereal period (in days)	Eccentricity of mean orbit	Diameter (in kms.)
Earth		Moon	384·4	27·321 66	0·054 90	3476
Mars	I	Phobos	9·4	0·318 91	0·0210	16?
	II	Deimos	23·5	1·262 44	0·0028	8?
Jupiter	I	Io	422	1·769 14	0	3200
	II	Europa	671	3·551 18	0	2800
	III	Ganymede	1 071	7·154 55	0	4800
	IV	Callisto	1 884	16·689 02	0	4480
	V	Unnamed	181	0·498 18	0·003	160
	VI	„	11 480	250·57	0·157 98	100
	VII	„	11 740	259·65	0·207 19	40
	VIII*	„	23 500	738·9	0·378	40
	IX*	„	23 700	758	0·275	20
	X	„	11 860	263·55	0·130 29	20
	XI*	„	22 600	692·5	0·206 78	25
	XII	„	21 200	631·1	0·168 70	20
Saturn	I	Mimas	186	0·942 42	0·0201	480
	II	Enceladus	238	1·370 22	0·004 44	640
	III	Tethys	295	1·887 80	0	950
	IV	Dione	378	2·736 92	0·002 21	950
	V	Rhea	527	4·517 50	0·000 98	1350
	VI	Titan	1 222	15·945 45	0·0289	4800
	VII	Hyperion	1 481	21·276 66	0·104	400
	VIII	Iapetus	3 562	79·330 82	0·028 28	1200
	IX*	Phoebe	12 960	550·45	0·163 26	240
Uranus	I	Ariel	192	2·520 38	0·0028	640
	II	Umbriel	267	4·144 18	0·0035	480
	III	Titania	438	8·705 88	0·0024	950
	IV	Oberon	587	13·463 26	0·0007	800
	V	Miranda	124	1·414	<0·01	150?
Neptune	I*	Triton	354	5·876 83	0	4000
	II	Nereid	5 570	359·4	0·76	320?

* Motion retrograde.

Visual magnitude	Galactic latitude									
	0°	10°	20°	30°	40°	50°	60°	70°	80°	90°
	″	″	″	″	″	″	″	″	″	″
6·0	0·045	0·045	0·047	0·050	0·054	0·058	0·062	0·066	0·069	0·070
7·0	0·033	0·033	0·034	0·037	0·039	0·042	0·046	0·048	0·050	0·051
8·0	0·024	0·024	0·025	0·027	0·029	0·031	0·033	0·035	0·037	0·037
9·0	0·017	0·018	0·018	0·019	0·021	0·022	0·024	0·026	0·027	0·027
10·0	0·013	0·013	0·013	0·014	0·015	0·016	0·018	0·019	0·020	0·020
11·0	0·009	0·009	0·010	0·010	0·011	0·012	0·013	0·014	0·014	0·014
12·0	0·007	0·007	0·007	0·007	0·008	0·009	0·009	0·010	0·010	0·010
13·0	0·005	0·005	0·005	0·006	0·006	0·006	0·007	0·007	0·007	0·008

[Annual parallax = 0·243 × secular parallax.]

APPENDIX F

EPHEMERIS TIME (E.T.)

The rotation of the earth is the basis for time-measurement and as regards G.M.T., G.C.T. and U.T. the rate of rotation is implied to be uniform. Recently, crystal and atomic clocks—accurate in performance to 1 or 2 parts in 10^{10}—have shown that the rotation is at times irregular, the deviations from uniformity being minute—of the order of 1 or 2 milliseconds per day—and unpredictable. In the gravitational theories of the bodies of the solar system, the passage of time is postulated to be *uniform*; this time is defined as *Ephemeris Time* (E.T.) and it is in terms of E.T. that astronomical quantities are now tabulated in the almanacs. The epoch from which E.T. is measured is

1900 January 0·5 [E.T.],

more elaborately defined in 1958 as "the instant near the beginning of the calendar year A.D. 1900 when the mean longitude of the sun was 279° 41′ 48″·04, at which instant the measure of E.T. was 1900 January 0, 12 h. *precisely*". The epoch for U.T. is 1900 January 0, 12 h. [U.T.]. Although the epochs for E.T. are apparently denoted by the same expression, they do not correspond to the same *instant* of time, the epoch of E.T. being 4 s. *later* than that of U.T.

It may be added that the fundamental unit of time is 1 second (E.T.) defined as 1/31556925·975 of the length of the tropical year for 1900·0.

INDEX

The numbers refer to the pages